IEE CONTROL ENGINEERING SERIES 67

Series Editors: Professor D. Atherton
Professor S. Spurgeon

Motion Vision: Design of Compact Motion Sensing Solutions for Autonomous Systems Navigation

Other volumes in print in this series:

Volume 2	**Elevator traffic analysis, design and control** G. C. Barney and S. M. dos Santos
Volume 8	**A history of control engineering, 1800–1930** S. Bennett
Volume 14	**Optimal relay and saturating control system synthesis** E. P. Ryan
Volume 15	**Self-tuning and adaptive control: theory and application** C. J. Harris and S. A. Billings (Editors)
Volume 16	**Systems modelling and optimisation** P. Nash
Volume 18	**Applied control theory** J. R. Leigh
Volume 20	**Design of modern control systems** D. J. Bell, P. A. Cook and N. Munro (Editors)
Volume 21	**Computer control of industrial processes** S. Bennett and D. A. Linkens (Editors)
Volume 23	**Robotic technology** A. Pugh (Editor)
Volume 26	**Measurement and instrumentation for control** M. G. Mylroi and G. Calvert (Editors)
Volume 27	**Process dynamics estimation and control** A. Johnson
Volume 28	**Robots and automated manufacture** J. Billingsley (Editor)
Volume 30	**Electromagnetic suspension – dynamics and control** P. K. Sinha
Volume 31	**Modelling and control of fermentation processes** J. R. Leigh (Editor)
Volume 32	**Multivariable control for industrial applications** J. O'Reilly (Editor)
Volume 34	**Singular perturbation methodology in control systems** D. S. Naidu
Volume 35	**Implementation of self-tuning controllers** K. Warwick (Editor)
Volume 36	**Robot control** K. Warwick and A. Pugh (Editors)
Volume 37	**Industrial digital control systems (revised edition)** K. Warwick and D. Rees (Editors)
Volume 38	**Parallel processing in control** P. J. Fleming (Editor)
Volume 39	**Continuous time controller design** R. Balasubramanian
Volume 40	**Deterministic control of uncertain systems** A. S. I. Zinober (Editor)
Volume 41	**Computer control of real-time processes** S. Bennett and G. S. Virk (Editors)
Volume 42	**Digital signal processing: principles, devices and applications** N. B. Jones and J. D. McK. Watson (Editors)
Volume 43	**Trends in information technology** D. A. Linkens and R. I. Nicolson (Editors)
Volume 44	**Knowledge-based systems for industrial control** J. McGhee, M. J. Grimble and A. Mowforth (Editors)
Volume 46	**Neural networks for control and systems** K. Warwick, G. W. Irwin and K. J. Hunt (Editors)
Volume 47	**A history of control engineering, 1930–1956** S. Bennett
Volume 48	**MATLAB toolboxes and applications for control** A. J. Chipperfield and P. J. Fleming (Editors)
Volume 49	**Polynomial methods in optimal control and filtering** K. J. Hunt (Editor)
Volume 50	**Programming industrial control systems using IEC 1131-3** R. W. Lewis
Volume 51	**Advanced robotics and intelligent machines** J. O. Gray and D. G. Caldwell (Editors)
Volume 52	**Adaptive prediction and predictive control** P. P. Kanjilal
Volume 53	**Neural network applications in control** G. W. Irwin, K. Warwick and K. J. Hunt (Editors)
Volume 54	**Control engineering solutions: a practical approach** P. Albertos, R. Strietzel and N. Mort (Editors)
Volume 55	**Genetic algorithms in engineering systems** A. M. S. Zalzala and P. J. Fleming (Editors)
Volume 56	**Symbolic methods in control system analysis and design** N. Munro (Editor)
Volume 57	**Flight control systems** R. W. Pratt (Editor)
Volume 58	**Power-plant control and instrumentation** D. Lindsley
Volume 59	**Modelling control systems using IEC 61499** R. Lewis
Volume 60	**People in control: human factors in control room design** J. Noyes and M. Bransby (Editors)
Volume 61	**Nonlinear predictive control: theory and practice** B. Kouvaritakis and M. Cannon (Editors)
Volume 62	**Active sound and vibration control** M. O. Tokhi and S. M. Veres
Volume 63	**Stepping motors: a guide to theory and practice** P. P. Acarnley
Volume 64	**Control theory, 2nd edition** J. R. Leigh
Volume 65	**Modelling and parameter estimation of dynamic systems** J. R. Raol, G. Girija and J. Singh
Volume 66	**Variable structure systems: from principles to implementation** A. Sabanovic, L. Fridman and S. Spurgeon

Motion Vision: Design of Compact Motion Sensing Solutions for Autonomous Systems Navigation

by Julian Kolodko and
Ljubo Vlacic

The Institution of Electrical Engineers

Published by: The Institution of Electrical Engineers, London,
United Kingdom

©2005: The Institution of Electrical Engineers

This publication is copyright under the Berne Convention and the Universal
Copyright Convention. All rights reserved. Apart from any fair dealing for the
purposes of research or private study, or criticism or review, as permitted under the
Copyright, Designs and Patents Act, 1988, this publication may be reproduced,
stored or transmitted, in any forms or by any means, only with the prior permission
in writing of the publishers, or in the case of reprographic reproduction in
accordance with the terms of licences issued by the Copyright Licensing Agency.
Inquiries concerning reproduction outside those terms should be sent to the
publishers at the undermentioned address:

The Institution of Electrical Engineers,
Michael Faraday House,
Six Hills Way, Stevenage,
Herts., SG1 2AY, United Kingdom

www.iee.org

While the authors and the publishers believe that the information and
guidance given in this work are correct, all parties must rely upon their own
skill and judgment when making use of them. Neither the authors nor the
publishers assume any liability to anyone for any loss or damage caused by
any error or omission in the work, whether such error or omission is the result
of negligence or any other cause. Any and all such liability is disclaimed.

The moral rights of the authors to be identified as authors of this work have been
asserted by them in accordance with the Copyright, Designs and Patents Act 1988.

British Library Cataloguing in Publication Data

Kolodko, Julian
   Motion vision : design of compact motion sensing solutions for autonomous
systems navigation
   1. Motion – Measurement 2. Detectors – Design and construction 3. Motion
control devices – Design and construction
   I. Title II. Vlacic, Ljubo III. Institution of Electrical Engineers
   629.8'315

ISBN 0 86341 453 2

Typeset in India by Newgen Imaging Systems (P) Ltd., Chennai, India
Printed in the UK by MPG Books Ltd., Bodmin, Cornwall

To Kath,
…love she holds for him as she ran away to… [J.K.]

To Mirjana, Bojan and Jelena [L.V.]

Contents

Preface	xiii
List of abbreviations	xix
Symbols	xxi
Typographical conventions	xxv
Acknowledgements	xxvii

1 Introduction — 1
   1.1 The intelligent motion measuring sensor — 2
      1.1.1 Inputs and outputs — 2
      1.1.2 Real-time motion estimation — 3
      1.1.3 The motion estimation algorithm — 5
      1.1.4 The prototype sensor — 7

PART 1 – BACKGROUND — 9

2 Mathematical preliminaries — 11
   2.1 Basic concepts in probability — 11
      2.1.1 Experiments and trials — 12
      2.1.2 Sample space and outcome — 12
      2.1.3 Event — 12
      2.1.4 Computation of probability — 13
      2.1.5 Conditional probability — 13
      2.1.6 Total probability — 14
      2.1.7 Complement — 14
      2.1.8 OR — 14
      2.1.9 AND — 15
      2.1.10 Independent events — 16

	2.1.11	Bayes theorem	16
	2.1.12	Order statistics	16
	2.1.13	Random variable	16
	2.1.14	Probability Density Function (PDF)	17
	2.1.15	Cumulative Distribution Function (CDF)	17
	2.1.16	Joint distribution functions	18
	2.1.17	Marginal distribution function	19
	2.1.18	Independent, identically distributed (iid)	19
	2.1.19	Gaussian distribution and the central limit theorem	19
	2.1.20	Random or stochastic processes	19
	2.1.21	Stationary processes	20
	2.1.22	Average	21
	2.1.23	Variance	21
	2.1.24	Expectation	22
	2.1.25	Likelihood	22
2.2	Simple estimation problems		23
	2.2.1	Linear regression	23
	2.2.2	Solving linear regression problems	26
	2.2.3	The Hough transform	27
	2.2.4	Solving Hough transform problems	28
	2.2.5	Multiple linear regression and regularisation	29
	2.2.6	Solving the membrane model	33
	2.2.7	Location estimates	39
	2.2.8	Solving location estimation problems	40
	2.2.9	Properties of simple estimators	42
2.3	Robust estimation		43
	2.3.1	Outliers and leverage points	43
	2.3.2	Properties of robust estimators	47
	2.3.3	Some robust estimators	50

3 Motion estimation — 63

3.1	The motion estimation problem		64
3.2	Visual motion estimation		65
	3.2.1	Brightness constancy	66
	3.2.2	Background subtraction and surveillance	69
	3.2.3	Gradient based motion estimation	69
	3.2.4	Displaced frame difference	76
	3.2.5	Variations of the OFCE	79
	3.2.6	Token based motion estimation	81
	3.2.7	Frequency domain motion estimation	85
	3.2.8	Multiple motions	87
3.3	Temporal integration		95
3.4	Alternate motion estimation techniques		96
3.5	Motion estimation hardware		98
3.6	Proposed motion sensor		100

PART 2 – ALGORITHM DEVELOPMENT — 101

4 Real-time motion processing — 103
 4.1 Frequency domain analysis of image motion — 103
 4.2 Rigid body motion and the pinhole camera model — 106
 4.3 Linking temporal aliasing to the safety margin — 109
 4.4 Scale space — 111
 4.5 Dynamic scale space — 113
 4.6 Issues surrounding a dynamic scale space — 114

5 Motion estimation for autonomous navigation — 117
 5.1 Assumptions, requirements and principles — 117
  5.1.1 Application — 118
  5.1.2 Data sources — 118
  5.1.3 Motion — 121
  5.1.4 Environment — 123
 5.2 The motion estimation algorithm — 124
  5.2.1 Inputs and outputs — 124
  5.2.2 Constraint equation — 125
  5.2.3 Derivative estimation – practicalities — 126
  5.2.4 Effect of illumination change — 131
  5.2.5 Robust average — 132
  5.2.6 Comparing our robust average to other techniques — 137
  5.2.7 Monte Carlo study of the LTSV estimator — 146
  5.2.8 Computational complexity — 153
  5.2.9 Dynamic scale space implementation — 153
  5.2.10 Temporal integration implementation — 154
  5.2.11 The motion estimation algorithm — 156
  5.2.12 Simulation results — 156
 5.3 Navigation using the motion estimate — 164

PART 3 – HARDWARE — 171

6 Digital design — 173
 6.1 What is an FPGA? — 174
 6.2 How do I specify what my FPGA does? — 174
 6.3 The FPGA design process in a nutshell — 175
 6.4 Time — 177
 6.5 Our design approach — 178
 6.6 Introducing VHDL — 178
  6.6.1 VHDL entities and architectures — 179
  6.6.2 VHDL types and libraries — 182
  6.6.3 Concurrent and sequential statements — 190
  6.6.4 Inference — 199
 6.7 Timing constraints — 201

x   Contents

	6.8	General design tips	204
		6.8.1 Synchronisation and metastability	204
		6.8.2 Limit nesting of `if` statements	206
		6.8.3 Tristate buffers for large multiplexers	206
		6.8.4 Tristate buffers	206
		6.8.5 Don't gate clocks	207
		6.8.6 Register outputs for all blocks	208
		6.8.7 Counters	208
		6.8.8 Special features	208
		6.8.9 Sequential pipelining	208
		6.8.10 Use of hierarchy	208
		6.8.11 Parentheses	210
		6.8.12 Bit width	210
		6.8.13 Initialisation	211
		6.8.14 Propagation delay	211
	6.9	Graphical design entry	211
		6.9.1 State machines	212
		6.9.2 A more complex design	221
	6.10	Applying our design method	230
7	**Sensor implementation**		**231**
	7.1	Components	232
		7.1.1 Image sensor	232
		7.1.2 Range sensor	234
		7.1.3 Processing platform	236
		7.1.4 PC	236
	7.2	FPGA system design	237
		7.2.1 Boot process	238
		7.2.2 Order of operations	239
		7.2.3 Memory management	240
		7.2.4 `RAMIC`	244
		7.2.5 Buffers	250
		7.2.6 Data paths	253
	7.3	Experimental results	270
		7.3.1 Experimental setup	270
		7.3.2 Aligning the camera and range sensors	270
		7.3.3 Stationary camera	273
		7.3.4 Moving camera – effect of barrel distortion	273
		7.3.5 Moving camera – elimination of barrel distortion	276
		7.3.6 Moving camera – image noise	276
		7.3.7 Moving camera – noise motion and high velocities	277
	7.4	Implementation statistics	277
	7.5	Where to from here?	278
		7.5.1 Dynamic scale space	278
		7.5.2 Extending the LTSV estimator	280

	7.5.3	Temporal integration	280
	7.5.4	Trimming versus Winsorising	281
	7.5.5	Rough ground	281
	7.5.6	Extending the hardware	281

PART 4 – APPENDICES — 283

A System timing — 285
- A.1 Timing for a 512×32 pixel image — 286
- A.2 Control flow for a 512×32 pixel image — 287
- A.3 Timing for a 32×32 pixel image — 288
- A.4 Control flow for a 32×32 pixel image — 289
- A.5 Legend for timing diagrams — 290
    - A.5.1 Note 1: image data clobbering — 294
    - A.5.2 Note 2: the use of n in the timing diagram — 295
    - A.5.3 Note 3: scale space change over process — 295
    - A.5.4 Note 4: the first frame — 295

B SDRAM timing — 297
- B.1 Powerup sequence — 297
- B.2 Read cycle — 298
- B.3 Write cycle — 298

C FPGA design — 301
- C.1 Summary of design components — 301
- C.2 Top level schematic — 306
- C.3 RAMIC — 307
- C.4 Buffers — 318
    - C.4.1 Camera data path — 333
- C.5 Laser and PC data paths — 340
- C.6 Processing — 344
- C.7 Miscellaneous components — 360
- C.8 User constraints file — 368

D Simulation of range data — 377

Bibliography — 395

Index — 417

Preface

Understanding motion is a principal requirement for a machine or an organism to interact meaningfully with its environment.

Murray and Buxton [210]

This research monograph focuses on the design, development and implementation of a motion measuring sensor. We have opted to present our work in more of a project style, beginning with background information before moving onto the design and test of our algorithm and the implementation of that algorithm in hardware. Our project-oriented style makes this book a valuable source of information for all those eager to learn about how to proceed with the development of a motion sensor from the idea and initial concept to its prototype. Extensive references are given for those readers wishing to further research a particular topic.

We begin the book with an introduction to estimation theory – the fundamental concept behind our work. Based on this we give a broad introduction to the idea of motion estimation. Although most attention is given to the gradient based techniques used in our sensor, we present all main motion estimation techniques. From here we develop and analyse our algorithm and its components in detail using simulation tools.

The inspiration for this book stems from our work[1] in developing intelligent vehicles and in particular decision and control systems to enable them to autonomously (without human assistance) drive and cooperate with each other [167, 292].

While implementation of sensing systems for static obstacle avoidance was relatively straightforward, these systems did not directly allow a very simple form of interaction: intelligent avoidance of moving objects (i.e. other vehicles). How were the vehicles to make intelligent navigation decisions in a dynamic environment without a sensor providing environmental motion data?

---

[1] Undertaken jointly with our colleagues from Griffith University's Intelligent Control Systems Laboratory [140].

In response to this need, we developed a prototype intelligent motion sensor. This sensor was implemented as a 'system on a chip', using Field Programmable Gate Array (FPGA) technology and it uses a combination of visual information (from a camera) and range information (from a scanning laser range finder) to give an estimate of the motion in the environment. The process of building this sensor led us through a range of fields from robust statistics and motion estimation to digital design. As we progressed, we developed a number of novel techniques that seemed to have value to the broader machine vision community and the motion estimation community in particular. It also became apparent that our design process would serve as a useful case study not only for those with an interest in machine vision systems design but also those with an interest in digital systems design since the book provides direct illustration of practical design issues.

As such, this book is primarily intended for those interested in real-time motion measuring sensor design where the sensor does not simply *estimate* motion in pixels per frame, but gives a *measurement* of motion in (centi)metres per second. Also, it is aimed at assisting undergraduate/postgraduate students and graduate engineers who have a grasp of programming and basic digital logic and are interested in learning about programmable logic devices. This book considers FPGAs in particular and shows how an FPGA can be used to successfully implement computational algorithms under real-time constraints. We present the design of our digital motion measurement sensor in a way that will make the leap from programming PCs to programming FPGAs easier for those who already have a grasp of standard programming techniques.

We believe our project-oriented approach in composing this book gives a more solid grounding since the reader immediately sees the relevance of design issues – these issues do not remain theoretical as they so often do in textbooks.

Our review of estimation in general and motion estimation in particular, also makes this book a valuable introduction to the concepts of motion processing for those working in the fields of computer vision and image processing motion. Finally, our presentation of a number of novel techniques (the LTSV estimation, dynamic scale space, use of range to segment motion) makes this book valuable for researchers in the fields of estimation theory and motion estimation techniques.

This book is structured to lead you through our design process. Introductory material on the topics of estimation theory (Chapter 2), motion estimation (Chapter 3) and digital design (Chapter 6) ensure the remainder of the material is accessible to most readers; however, a basic knowledge of calculus, image/signal processing and digital design concepts would be helpful. Those seeking a deeper understanding of any particular topic can research further using the extensive references provided. The discussion of our particular algorithm's design begins with a detailed analysis of our application and then proceeds to build our motion estimation algorithm in stages. At each stage we analyse the performance of the components being used to ensure that the final algorithm fulfils the design goals. Given the algorithm we then show how this algorithm can be implemented as an intelligent sensor using FPGA technology. The resulting design is one of our first prototypes and leaves room for eager readers to further improve the design. Some hints for improving the design and further research are included in the final chapter.

To solve the motion estimation problem, we take a lesson from human motion perception. The process of motion perception in humans is often considered to have two parts: a short-range process and a long-range process [66][2]. The short-range process appears to operate directly and uniformly on visual information to give an estimate of small motions based on a small temporal support. The long-range process tracks higher-level image structures such as points and edges over longer periods and distances.

Our sensor emulates the short-range process. Motion is considered to be a fundamental quantity (like brightness) and the aim is to estimate it quickly, simply and consistently across the visual field without explicit reference to particular image features. In the context of autonomous vehicles and their operation under real-time constraints, a relatively rough motion estimate calculated quickly is more useful than an accurate motion estimate that is updated slowly. An additional long-range estimation process could be built on top of our algorithm if further refinement of the motion estimate is required; however, we show that a short-range style of motion estimation can generate information suitable for autonomous navigation applications.

This book is divided into three parts. The first part, *Background*, gives the reader an introduction to fields of robust estimation and motion estimation. Chapter 2 provides a straightforward introduction to estimation and robust estimation theory, beginning from basic probability theory, moving onto linear regression and location estimation before considering some of the more common robust estimators.

Chapter 3 gives an insight into the range of motion estimation techniques available. The discussion begins with a consideration of the brightness constancy assumption – a fundamental assumption made in the majority of motion estimation techniques. Based on this, a range of elementary motion estimation techniques are introduced before discussion moves onto the concept of robustness in motion estimation and the use of temporal integration to improve motion estimates. While the majority of this chapter focuses on offline motion estimation using a single camera, the final sections broaden the discussion to include real-time, custom hardware based motion estimation and a number of alternate motion estimation techniques.

The second part of this book, *Algorithm Development*, leads the reader through the algorithm development process. Since Chapter 3 illustrated the available techniques, the purpose of this section is to adapt these techniques to our design problem. Chapter 4 is devoted to consideration of the issue of real-time motion estimation in the context of autonomous navigation. The need for scale space to avoid temporal aliasing is illustrated, then the dynamics of the navigation problem (summarised by the idea of a safety margin) and the availability of range data are used to simplify the concept of scale space to 'dynamic scale space' where motion estimation is focused on the nearest object.

In Chapter 5 we consider our proposed motion estimation algorithm in detail. The chapter begins by considering the assumptions and design constraints for the

---

[2] While this model does have its critics, e.g. Reference 244, it gives us a convenient means of describing the intent of our work. We do not imply that our work is a model of biological vision.

system, then each component of the algorithm is developed and analysed in turn. Initially, a constraint equation is developed giving a means of calculating motion using a combination of visual and range data. This constraint equation requires the calculation of image derivatives, thus derivative calculation is analysed in detail next. To obtain useful motion estimates, results must be integrated over space, however, this must be done carefully to ensure that the presence of multiple motions within an image region does not skew results. To achieve this we propose a novel[3], simple robust location estimator based on iterative averaging. We show the uniqueness of this algorithm by comparing it with a range of other robust estimators, and we examine our estimator's performance in a Monte Carlo study. Finally, the performance of the complete algorithm is assessed using a specially developed simulation system and a brief discussion regarding the use of the resulting motion estimates in an autonomous navigation system is presented.

While the focus of the earlier chapters on the functions our sensor will perform is at a conceptual level, Part 3, *Hardware*, considers the implementation of these functions using FPGA technology. In Chapter 6 we introduce the VHDL hardware description language, which is commonly used in FPGA design. In addition to the basic syntax of VHDL, this chapter introduces a design approach that will make the leap from programming with inherently sequential languages (like C, C++) to programming with the explicitly parallel VHDL language easier for the reader. Our design approach is based on a combination of state machine and schematic design entry. State machines allow the designer to create 'sequential' components that are similar to sequential programs and schematic design entry allows the user to combine these components so that each component can operate in parallel.

Our design and implementation work revolved around a range of tools from Xilinx[4] so this book is skewed toward these products. Of course, technology moves quickly and sometime in the future these products may well be obsolete. Therefore, we introduce FPGA design in a general way that will be applicable no matter what the current technology.

Chapter 7 leaves the details of VHDL behind and considers design of our sensor and the specific issues that arose in the development of our sensor. Because these issues are presented in the context of a specific design, the reader can more clearly see how these issues relate to real designs. Our complete design is included in the appendices so that the reader can see, in detail, how we approach each issue. We show how FPGA technology can be used to allow the implementation of both the algorithm and all interfacing logic in a single chip resulting in a potentially very compact motion sensor. Instead of being connected to a range sensor, the current motion sensor prototype uses range data simulated by a ray tracing algorithm. The result is a sensor that can operate 16 Hz (limited by the rate at which range information can be gathered) with a latency of 3 frames – a rate that can be considered 'real-time'

---

[3] To the best of our knowledge, this location estimator has not been proposed by other researchers in the field.
[4] www.xilinx.com

in our test environment. The FPGA component of the design can operate at up to approximately 23 Hz assuming a 40 MHz clock.

The final section of Chapter 7 is intended for the reader interested in extending on the basic design given in this book. It lists potential improvements to the design and algorithm, and gives some ideas for future research.

Julian Kolodko and Ljubo Vlacic
Brisbane, August 2004

List of abbreviations

Abbreviation	Description
BCA	Brightness Constancy Assumption
CDF	Cumulative Distribution Function
CHP	Convex Hull Peeling
DBD	Displaced Block Difference
DFD	Displaced Frame Difference
DSM	Downhill Simplex Method
FDM	Finite Difference Method
FIFO	First-In, First-Out
FOBD	First-Order Backward-Difference estimate of the derivative
FOCD	First-Order Central-Difference estimate of the derivative
FOFD	First-Order Forward-Difference estimate of the derivative
FOV	Field of View
FPGA	Field Programmable Gate Array
FPS	Frames per Second
GNC	Graduated Non-Convexity
ICM	Iterated Conditional Modes
IRLS	Iteratively Reweighted Least Squares
LAR	Least Absolute Residuals
LMedS	Least Median of Squares
LS	Least Squares
LTS	Least Trimmed Squares
LTSV	Least Trimmed Square Variant – our estimation algorithm
MAD	Median Absolute Deviation from the Median
MAP	Maximum a posteriori
MIMD	Multiple Instruction stream, Multiple Data stream
MLE	Maximum Likelihood Estimate
MMSE	Minimum Mean Squared Error
MPEG	Motion Picture Experts Group
MRF	Markov Random Field

Abbreviation	Description
MSM	Mean Shift Method
MTL	Maximum Trimmed Likelihood
MVE	Minimum Volume Ellipsoid
OFCE	Optical Flow Constraint Equation
PDF	Probability Distribution Function
PE	Processing Element
RAMIC	RAM (Random Access Memory) Interface and Controller
RANSAC	Random Sample Consensus
SAD	Sum of Absolute Differences
SDRAM	Synchronous Dynamic RAM (Random Access Memory)
SIMD	Single Instruction (stream) – Single Data (stream)
SNR	Signal to Noise Ratio
SOR	Successive Over-Relaxation
SSD	Sum of Squared Differences
TLS	Total Least Squares
VHDL	Very High Speed Integrated Circuit Hardware Description Language
VLSI	Very Large Scale Integration

Symbols

Camera

Symbol	Description
ζ	Camera pixel pitch (metres).
ζ_{eff}	Effective pixel pitch taking dynamic scale into account (metres).
(x_r, y_r)	Image resolution – 512 × 32 pixels.
f	Camera focal length (metres).
F	Frame interval (seconds). Equal to 1/FPS.

Image formation

Symbol	Description
$(\omega_x, \omega_y, \omega_t)$	Spatiotemporal frequency of $I(x, y, t)$.
$C(x, y, t)$	Additive component of spatiotemporal brightness offset field.
$I(x, y, t)$	True spatiotemporal intensity function on the camera sensor surface.
$I(x, y, t)$	Spatiotemporal intensity function measured by the camera sensor.
I_x, I_y, I_t	Measured spatiotemporal partial derivatives of the intensity function.
Level	Level in the image pyramid, with the original image labelled level 0.
$M(x, y, t)$	Multiplicative component of spatiotemporal brightness offset field.

Coordinate systems and positions

Symbol	Description
(x, y)	Image sensor coordinate system, measured in pixels.
(x, y)	Image sensor coordinate system. Measured in metres with origin at the intersection of optical axis and image plane.
(X, Y, Z)	Camera centred world coordinate system. Defined with origin at the focal point, the XY plane parallel to the xy plane and the Z axis coinciding with the optical axis (metres).
Z_X, Z_Y, Z_Z	Depth derivatives.
$\boldsymbol{p} = (p_x, p_y)$	A point in the camera sensor coordinate system (metres).
$\mathbf{p} = (p_x, p_y)$	A point in the camera sensor coordinate system (pixels).
$\mathbf{P} = (P_x, P_y, P_z)$	A point in the camera centred world coordinate system (metres).

Operational parameters

Symbol	Description
D_{min}, D_{max}	Minimum and maximum operational ranges for sensor.
S	Safety margin defined as D_{min}/V_{max}. Equal to the minimum available stopping time.
V_{max}	Maximum allowable velocities for sensor.

Motion

Symbol	Description
$\tilde{U}_X$	Motion estimate calculated by our sensor.
ψ	Conversion factor from apparent motion (u, v) to optical flow (u, v). Equal to F/ζ.
$\boldsymbol{\Omega} = (\Omega_X, \Omega_Y, \Omega_Z)$	Rotational velocity of a point P in the camera centred world coordinate system about the camera axes.
$(\delta x, \delta y)$	Displacement – pixels.
(u, v)	Horizontal and vertical components of apparent velocity (metres/second). Projection of T, $\boldsymbol{\Omega}$ onto the image sensor.

Continued

Symbol	Description
(u, v)	Horizontal and vertical components of optical flow (velocity – pixels/frame).
(U, V, W)	Three-dimensional range flow.
$\mathbf{T} = (T_X, T_Y, T_Z)$	Relative translational velocity of a point P in the camera centred world coordinate system.
U_X	Motion projected onto the X axis (metres/s).

Probability and estimation

Symbol	Description
$p_\mathbf{R}()$	Probability density function for random variable $\mathbf{R}$.
$P_\mathbf{R}()$	Cumulative distribution function of random variable $\mathbf{R}$.
ξ	An outcome of an experiment – the result of a trial.
Φ	Sample space.
η	A random variable representing a noise.
α	Line field energy.
λ	Regularisation parameter in membrane function.
σ	Standard deviation.
$\rho()$	Estimator function.
$\psi()$	Influence function corresponding to estimator function $\rho()$.
$\boldsymbol{\theta}, \theta$	Vector of parameters to be estimated, a single parameter.
$\mathbf{A}, \mathbf{A}^{-1}$	An event, the complement of an event.
b	Bias is an estimate.
$\mathbf{d}, d$	A dataset, a data element.
E[ ]	Expectation operator.
E_d	Data term of membrane energy function.
E_s	Smoothness term of membrane energy function.
$f(\boldsymbol{\theta})$	An arbitrary function of the parameters $\boldsymbol{\theta}$.
h	Fraction of residuals used in an LTS estimator.
l	Line process.
$l()$	Likelihood function.
m, c	Parameters in a simple two-dimensional linear regression problem.
N	Number of elements in a dataset.
p	Number of parameters in vector $\boldsymbol{\theta}$.
Pr{ }	Probability of an outcome or event.
R	Region of support in regression/regularisation.
r	A residual.

Continued

Symbol	Description
R( )	A random variable.
R_s	Set of candidate translations in block based motion estimation.
t	Time variable.
W( )	Window function.
x, x	A collection of measurements of a random variable, a measurement of a random variable(s).

Markov random fields

Symbol	Description
Φ	Solution (sample) space for an MRF.
R	Set of random variables in an MRF.
φ	Realisation of a random variable drawn from set of labels L.
L	Set of labels that a realisation of a random variable can take.
$p_\mathbf{R}$( )	Probability distribution function.
R	Local neighbourhood in MRFs.
S	Set of sites on which an MRF is defined.
T	Temperature in a Gibbs distribution.
U( )	Energy function in a Gibbs distribution.
V_c( )	Clique potential function in a Gibbs distribution.
Z( )	Partition function of a Gibbs distribution.

Algorithmic and hardware parameters

Symbol	Description
$\tilde{U}_X$	Motion estimate calculated by our sensor.
`changeScale`	A flag indicating whether the current frame of data indicates the need for a change of scale.
`ImgWidth`	Current image width = 512/`SS`.
I_xs, I_ts	Smoothed spatial and temporal derivatives of the intensity function used in our algorithm.
`SD`	$2^{\mathtt{SD}} = \mathtt{SS}$.
`SS`	Subsampling rate.
SU_1	Scaling factor for computation of optical flow in FPGA.
SU_2	Scaling factor for computation of velocity in FPGA.
TC	Temporal integration coefficient. Weight given to previous motion estimate.

Typographical conventions

Example	Font description	Details
`Component`	Lucida Console	Used when referring to specific system components
`Signal`	Courier	Used when referring to specific system signals
n	Times – Bold	Used to denote vector quantities

- In equations:
    - Subscripts can indicate derivative with respect to the subscript variable, a component of a vector, or may simply be descriptions (e.g. max, worst case, etc.). The usage is clear from the context.
    - Superscripts indicate iteration number (this can be a time instant if iteration occurs over time).
    - The ˆ symbol over a variable indicates the Fourier transform of that variable, an arrow over a symbol indicates an ordered dataset and a bar indicates average.
    - Positional variables are usually dropped for clarity, correct use of equations is clear from the context.

Acknowledgements

Kudos to Dr Julian Kolodko, one of the book's co-authors. His affiliation with Griffith University's Intelligent Control Systems Laboratory (ICSL) produced a fertile breeding ground for numerous enlightening discussions regarding intelligent artefacts. His affiliation with the ICSL also assisted greatly in developing our understanding of how intelligent motion sensors should operate to assist intelligent vehicles in making robust and safe navigation decisions in dynamic, city road environments.

We would like to express our appreciation to the reviewers of the initial book manuscript for their critiques, comments and valuable suggestions. We trust that the current content and presentation of the book reflect their insightful remarks. We accept full responsibility, however, for the views expressed in the book and the blame for any shortcomings.

When all is said and done, the book could not have been produced without the able assistance and professional support of the IEE's editorial staff. Their trust in our work, patience and expert guidance were instrumental in bringing the book to the market in a timely fashion. Specifically, we are sincerely appreciative to the IEE Control Engineering Series Editors, Professor Derek Atherton and Professor Sarah Spurgeon, Commissioning Editor Ms Sarah Kramer and the Editorial Assistant Ms Wendy Hiles.

Also, we would like to thank Xilinx for permitting us to base some of the FPGA and VHDL information in this book on their resources and for their willingness to promote the book among their clients, the FPGA based real-time embedded system designers. Thanks also to GMD.AiS (now known as Fraunhofer AiS) in Germany for their generous donation of the Lyr SignalMaster and GatesMaster prototype platforms and the Fuga camera used in this work. Finally, we are profoundly indebted to our families for their endless encouragement and invaluable moral support throughout the development of this book.

Chapter 1
Introduction

> Vehicle navigation in dynamic environments is an important challenge, especially when the motion of the objects populating the environment is unknown.
>
> Large et al. [186]

For some time we have been dealing with the design and development of cooperative autonomous vehicles. That is, we build vehicles that are not only intelligent in themselves (e.g. able to follow a lane or to avoid a collision with road obstacles such as other vehicles, bikes, pedestrians, etc.), but that can also explicitly cooperate with other autonomous vehicles so that both the individual and shared goals of each vehicle can be expediently achieved. In the road environment this includes the ability to perform driving manoeuvres such as traversal of unsignalised intersections, cooperative overtaking, cooperative formation driving, distance control, stop and go cruise control, intelligent speed adaptation, etc. [292]. Figure 1.1 illustrates some of our experiments in cooperative overtaking, distance control and unsignalised intersection traversal.

While developing the sensing and communication structures required to achieve our aims, it became clear that it would be relatively easy to avoid static obstacles; however, intelligent avoidance of moving obstacles would be a greater challenge. Any attempt to avoid moving obstacles using sensors and algorithms initially developed for static obstacles is not sustainable because they do not allow the prediction of the obstacles' future positions. This can lead to poor or unstable path plans at best, and catastrophic collisions at worst. What we needed was a sensor able to describe the vehicle's surrounding environment in terms of motion, rather than in terms of distance (e.g. radar), brightness (e.g. camera) or 'objects' (e.g. the result of processing range or camera data). An appropriate navigation control system can use this information not only to avoid static obstacles (which can be thought of as moving obstacles with a velocity of zero) but also to determine the best course of action to avoid a collision with moving obstacles before continuing towards the goal location.

2 *Motion vision: design of compact motion sensing solutions*

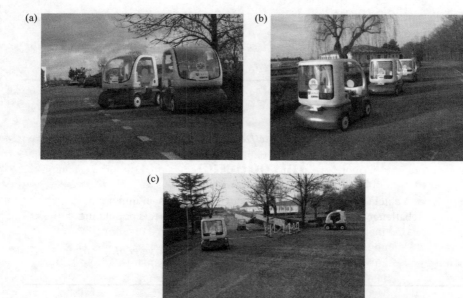

Figure 1.1 *Testing of the ICSL's cooperative behaviours using INRIA's Cycab Vehicles. (a) Cooperative overtaking, (b) distance control, (c) unsignalised intersection traversal*

Therefore this book provides know-how on the designing of a sensor that allows static and dynamic obstacle avoidance to be neatly merged.

1.1 The intelligent motion measuring sensor

Responding to the need for a source of motion information, we set out to develop a prototype intelligent sensor. This sensor was to provide *timely* information about the motion in the environment assuming that all motion occurs on a *smooth ground plane*. In order to have a compact sensing solution, we decided to implement our sensor using Field Programmable Gate Array (FPGA) technology. This, in principle, allowed us to incorporate all processing and interface logic in a single chip – the only external devices are memory, power supplies and sensors.

1.1.1 Inputs and outputs

Perhaps because vision is our dominant sense and visual information can be captured passively at a high rate, motion estimation has traditionally been performed using visual information. In principle, however, other time varying data sources (such as range information) can be used in place of visual information to determine how things are moving around.

Based on this idea we decided to see if a combination of range and visual data could simplify or improve the motion estimation process. This is a logical combination since motion is essentially a geometric property (in that it relates to changes in position) of a three-dimensional world. A single camera only gives *directly* a two-dimensional sense of the world, thus a range sensor could be used to complement the visual data. Of course, it is well known that range information can be inferred from a pair of images, however, this processes is complex in itself and can be error prone. For simplicity, we decided to use a single, greyscale camera and a scanning laser radar sensor as data sources. Furthermore, we assume that motion occurs on a smooth ground plane to avoid the complicating issue of sensor stabilisation. This is a relatively complex topic in itself and is not considered in detail here.

In designing any sensor, it is also important to understand the expected output. In this case, the motion information given by the sensor must be directly useful for navigation planning and this defines three characteristics for the output.

First, results in terms of *pixels per frame*, as provided by many motion estimation algorithms, are inappropriate since things move in a world coordinate system, not the camera's coordinate system. To overcome this we use the combination of range and visual data to calculate *relative* motion in *(centi)metres per second*. Second, because motion is limited to a ground plane, the sensor must give data that express the *horizontal* and *longitudinal* (relative to the camera) motion in the environment. Again, the combination of visual and range information assists in this task. Longitudinal motion is most easily measured using range data (especially if certain conditions related to the processing rate and object motion are met) while a combination of range and visual data can easily determine horizontal motion.

The fact that only longitudinal and horizontal motion are important for ground plane motion allows us to define the third characteristic of the output data. Obtaining a motion estimate for each pixel in the image is unnecessary because, in general[5], the horizontal and longitudinal motion of a pixel will be the same as the horizontal and longitudinal motion of the pixels above and below. Therefore, we use a narrow input image and a column-wise processing approach to compute a single motion estimate for each column in the image.

1.1.2 Real-time motion estimation

The primary consideration in our sensor's design was the need to measure motion in real-time, but what is real-time motion estimation? The idea of real-time processing has two aspects: *processing time* is the time required to calculate a result assuming all required data are available and *latency* is the time from when sampling of input data for time instant n begins to when the corresponding measurement (or output)

---

[5] Of course, this is not always true, and strategies for dealing with cases where this assumption fails are explained in the book.

is made available. In the context of a motion sensor, both terms have more specific meanings.

The idea of processing time raises three interrelated issues – environmental dynamics, algorithmic considerations and implementation considerations. In Chapter 4 we show that many motion estimation techniques place a limit on the largest measurable motion, and we show how this limit, combined with the camera's parameters and environmental dynamics can be used to calculate a maximum allowable processing time[6]. Put simply, the processing time must be quick enough to ensure that the dynamics of the environment can be reliably captured. To ensure successful measurement under a wide range of dynamic conditions we also introduce the idea of safety margin and a novel technique we call *dynamic scale space*.

Now that we know the maximum permissible processing time, we require an algorithm that can achieve this rate. However, the algorithm cannot be considered in isolation – we must consider it in the context of its final implementation. In principle, a modern PC could meet our processing time constraint, however, a device as large as a PC could hardly be considered as a compact sensing solution. More compact embedded PC/DSP hardware (e.g. PC104) is only a partial solution since they often do not have sufficient processing and I/O flexibility for such applications. Therefore, we have chosen to implement the algorithm using Field Programmable Gate Array (FPGA) technology where we can build our sensor as a 'system on a chip'. That is, all processing and interface functions are incorporated into a single chip leading to an extremely compact intelligent sensing solution. FPGAs have ample I/O resources and they have the added benefit of allowing us to directly and extensively utilise parallelism while standard microprocessors do not. Given that the algorithm will be implemented in FPGA, we seek an algorithm with the following properties so that we can more easily meet our processing time constraint:

- Algorithms that can be implemented in a parallel fashion are desirable since they provide a relatively simple means of decreasing processing times.
- If an iterative algorithm must be used, then each iteration should be as simple as possible and the smallest practical number of iterations should be used.
- The algorithm should consist of a small number of simple arithmetic operations (such as addition, multiplication and division) since these can be implemented simply and quickly. More complex functions such as exponentiation are generally built from a combination of these simpler blocks and therefore will take a relatively long time to compute.
- Deterministic algorithms are preferred over stochastic algorithms since processes such as random sampling are difficult to implement efficiently using digital logic. Deterministic algorithms often have the added advantage of a known run time.
- If possible, sorting should be avoided as it can add significant overheads.

---

[6] More specifically, we determine a minimum *processing rate*.

- Algorithms that can be implemented using integer or fixed point arithmetic are preferred over algorithms requiring floating point calculations since floating point implementations require significant additional resources.

Having determined the processing time constraint and the broad characteristics of an algorithm that can meet that constraint, we now consider the other aspect of real-time motion estimation – latency. Latency is a function of both the algorithm itself and its implementation and it depends on whether the implementation of an algorithm is *causal* (i.e. requires only present and/or past data) or *non-causal* (i.e. requires past, present, *and future* data). For example, gradient based motion estimation can be implemented in both causal and non-causal ways, depending upon how derivates are calculated. Non-causal algorithms are not as counter-intuitive as they may seem. To implement a non-causal algorithm all we need to do is buffer data so that the appropriate elements of data are available for processing.

If an implementation is causal then its latency is determined entirely by the processing time and the time it takes to move data and results in and out of the processing device[7] (Figure 1.2a). For non-causal implementations, latency also includes the time it takes to buffer the required data (Figure 1.2b). Naturally we must look for an algorithm with minimal latency so that navigation decisions can be made using the most up-to-date data possible.

1.1.3 The motion estimation algorithm

In the previous sections, we listed the input/output properties our motion estimation algorithm must have and some properties that will simplify real-time implementation of motion estimation in FPGA. Before describing the algorithm that we developed to fulfil these requirements, we describe our philosophical views towards motion estimation so that we can describe the algorithm within its full context.

First, we seek algorithms that do not have data dependent parameters[8]. Tuning such an algorithm to work in one application environment (e.g. a low noise environment) does not imply that it will work correctly in other environments (e.g. a high noise environment). Second, we believe that motion should be treated (as far as possible) as a fundamental quantity much like brightness or range. This ideal is not completely achievable because motion cannot be directly measured, it can only be inferred from spatiotemporal changes in other data sources (such as brightness or range[9]) so in practice, we require that motion be calculated as quickly, consistently and directly from the underlying data sources as possible. A rough, quickly calculated motion estimate is arguably more useful for navigation than a more accurate,

---

[7] For simplicity we define latency as the time from the moment we begin sampling input data to the time we finish transmitting results to the destination (e.g. a buffer in the host).

[8] Environmental parameters regarding such factors as anticipated maximum object velocity are unavoidable. We seek to avoid parameters whose value cannot be specified in advance.

[9] One can argue that range is not directly measurable either, but must be inferred from some other parameter such as time of flight or stereo disparity.

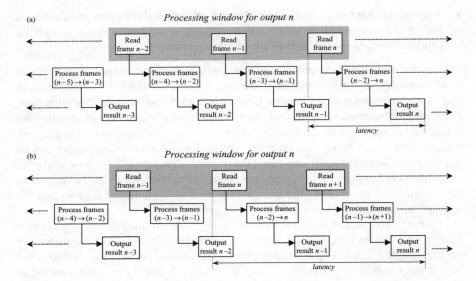

Figure 1.2 Latency for (a) causal and (b) non-causal algorithms that use three data elements to calculate a result. Note that, depending on how the algorithm is implemented, it may be possible to read in new data while existing data are processed or existing results are output

but slowly calculated estimate. Furthermore, colour, shape, size, and other specific features are irrelevant when trying to avoid a moving object so they should not directly enter the motion estimation process. This is essentially the same as the 'short-range' view of biological motion estimation [66][10].

Based on this ideal, we chose to use gradient based motion estimation techniques. Gradient based techniques use the spatiotemporal image derivatives to estimate motion which is far more direct than the feature based techniques that rely on extraction of image features and far simpler than frequency domain techniques that require frequency domain analysis of an image sequence. Gradient based techniques have a number of further advantages including:

- Relative ease with which range and visual information can be combined.
- Easy to implement.
- Easy to extend to operate over a wide range of dynamic conditions. Estimation of higher velocities can actually reduce computation.

---

[10] It is generally believed that biological motion processing has two distinct components – the short and the long-range processes. The short-range process gives a rough motion estimate as directly and consistently as possible using data from a relatively small temporal window. The long-range motion process seems to track points and features of the image over longer periods and distances. While it is possible to augment our algorithm with a long-range process, we show that a short-range process can generate data that is useful for navigation tasks.

- Easy to formulate the estimation problem in a robust framework so that noise and failure of the assumptions underlying the algorithm can be dealt with gracefully.

Based on this combination of input/output, implementation and philosophical considerations, we have developed an algorithm that combines range and visual data using a gradient based motion estimation technique together with the ground plan rigid body motion model. While this particular combination is not entirely unique from an algorithmic standpoint, our approach does have a number of distinctive features.

- The first among them is the ability to directly combine range and visual information for motion estimation. This combination of range and visual information has allowed us to develop the concept of *dynamic scale space* where processing is focused on the nearest object – the object that is most likely to cause collision in the short term. Dynamic scale space reduces computational complexity, therefore making it easier to meet our real-time constraints.
- The availability of range information also allows us to segment the environment based on range. Traditionally, motion segmentation has been performed in conjunction with, or as a post-processing step to, motion estimation. Instead, we perform segmentation as a pre-processing step – we divide the world up into *blobs* based on range discontinuities and process motion in a blob-wise fashion. A blob may, or may not, correspond to a single 'object' in the environment, however, this is irrelevant in the context of autonomous navigation where the goal is to avoid all blobs, not to identify or characterise objects.
- The final distinctive feature of our approach is our use of the novel *Least Trimmed Squares Variant* (LTSV) estimator. This is a location estimator that has many properties making it especially suitable for implementation in digital hardware. The LTSV is entirely deterministic, it requires only addition, multiplication and division, it requires very few iterations to converge, it does not have any data-dependent parameters that require tuning, and it does not require reordering of data. The trade off for the LTSV's high computational efficiency is a relatively low statistical efficiency (i.e. high variance); however, we show that the performance of the estimator is sufficient in our context. We also detail some ways the variance of the LTSV estimator can be improved.

Although our approach has a number of novel components (which we can compare to the work of others), we do not claim that the overall algorithm is superior to other motion estimation techniques because our approach has been specifically crafted to solve a very particular problem. Based on a series of assumptions, it takes in both visual and range data and produces a one-dimensional motion estimate given in *(centi)metres per second*. To the best of our knowledge, at the current time there is no other sensor (or algorithm) that performs a similar function, therefore it is impossible to directly compare the performance of our approach with that of others.

1.1.4 The prototype sensor

Our final prototype intelligent sensor was implemented using a commercial FPGA prototype platform so it does not reach the ideal for compactness, however, this ideal is clearly attainable if a fully custom implementation was built. The sensor processes images of up to 512 × 32 pixels resolution in real-time – at 16 Hz[11] with a latency of three frames. For the purpose of the experiments described in the book we simulated the laser radar device using a digital signal processor (DSP); however, we took care to ensure that the nature of the simulated range data matched the nature of the true laser radar device. Therefore, we are confident that the results described here will match those that would be obtained if an appropriate laser radar device were used. Our experimental results show that while our sensor operates very well at low object speeds, it tends to underestimate the speed of fast moving objects; however, we are able to show that this problem can easily be corrected (Section 7.5.4).

---

[11] This is limited by the rate at which range data can be obtained. Our core motion-processing algorithm (implemented in FPGA) can operate at 23 Hz assuming a 40 MHz clock.

Part 1
Background

Chapter 2
Mathematical preliminaries

> More recently, some people began to realize that real data do not completely satisfy the classical assumptions.
>
> Rousseeuw and Leroy [239]

Problems requiring a best guess[12] for the value of some parameters based on noisy input data are estimation problems. Alternatively, you might view estimation problems as problems where you try to characterise a dataset using a mathematical construct such as a point or a line. This chapter is intended as a self-contained introduction to the basic concepts in the field of *linear* estimation. As such, it will give you an understanding of the analysis in the following chapters where we focus specifically on motion estimation. The discussion here is quite informal: our aim is to introduce the broad ideas of estimation – more detail is available from a vast range of sources [64, 122, 148, 201, 239, 267, 303]. This chapter begins with a brief introduction to probability theory and some key terms before moving onto the basics of estimation, which then allows us to spring into the field of robust estimation. It is assumed you are familiar with basic calculus.

2.1 Basic concepts in probability

This section defines a range of basic concepts from probability theory that appear in the estimation literature. Concepts are presented in order with earlier concepts building on later concepts, thus it may help if this section is read in order. Readers familiar with basic probability theory can skip this section, which is based on the presentation in Reference 226.

---

[12] We will come to the meaning of 'best guess' later.

2.1.1 Experiments and trials

We consider an *experiment* to be an activity whose result cannot be predicted in advance and a *trial* to be a single execution of such an experiment. In such an experiment we can only talk about the relative likelihood of a particular result occurring. Tossing a coin fits into this category. An experiment can also be an activity whose outcome cannot be measured with complete certainty. The measurement is said to contain some *error*, or *noise*. Measuring the length of a stick is an example of such an experiment.

2.1.2 Sample space and outcome

Sample space (denoted by Φ) is the exhaustive list for all possible results of an experiment. An *outcome* (denoted by ξ) is one of the possible results of a trial. If the sample space is *continuous* (e.g. your location on the Earth's surface) outcomes correspond to points in the sample space, while for a *discrete sample space* (e.g. the numbers 1 to 6 that result from the throw of a dice) outcomes correspond to elements of the sample space (Figure 2.1). If all possible outcomes have equal probability then the sample space is said to be *uniform*, otherwise it is *non-uniform*.

$$\Pr\{\Phi\} = \sum_{\xi \in \Phi} \Pr\{\xi\} = 1 \qquad (2.1)$$

2.1.3 Event

An *event* (denoted by **A**) is a subset of the sample space, or a group of outcomes (Figure 2.2). An event can correspond to a single outcome, and can be an empty set (i.e. the event where anything happens). Each time we repeat an experiment and detect a new event we say that we have another *sample* or *observation* or *measurement*.

$$\forall \mathbf{A} \in \Phi | \Pr\{\mathbf{A}\} \geq 0 \qquad (2.2)$$

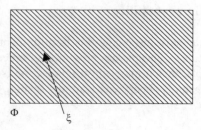

Figure 2.1   An outcome is an element of sample space

Mathematical preliminaries 13

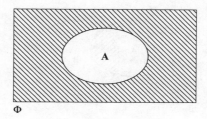

Figure 2.2   An event is a group of outcomes

2.1.4   Computation of probability

We represent the probability of event **A** by Pr{**A**}. In a uniform sample space the probability of an event occurring can be computed as

$$\Pr\{\mathbf{A}\} = \frac{\text{number of outcomes in } \mathbf{A}}{\text{total number of outcomes}} \tag{2.3}$$

If the sample space is non-uniform, the probability of **A** can be computed by adding together the probabilities of all points in the sample space corresponding to event **A**.

$$\Pr\{\mathbf{A}\} = \sum_{\xi \in \mathbf{A}} \Pr\{\xi\} \tag{2.4}$$

In either case, we can see that the probability of event **A** can be computed in advance (before any trials are carried out) based on knowledge of the experiment. This sort of probability is called the *a priori* or the *prior* probability of event **A**.

Sometimes it is not possible to determine the probability in advance. In this case we can determine the probability of event **A** by conducting a number of trials and determining the relative frequency of occurrence of event **A**.

$$\Pr\{\mathbf{A}\} \approx \frac{\text{number of trials giving an outcome from event } \mathbf{A}}{\text{total number of trials}} \tag{2.5}$$

Of course this will only be an approximation to the true probability. As the number of trials approaches ∞, this estimate will converge to the true result.

2.1.5   Conditional probability

Conditional probability Pr{**A**|**B**} is the probability of event **A** occurring given the condition that event **B** has occurred. Referring to Figure 2.3, we can imagine that the sample space **S** has been reduced to the region occupied by event **B** (i.e. **S** now equals **B**). Pr{**A**|**B**} is the fraction of **B** covered by **A**.

Conditional probability is also called *posterior* or *a posteriori* probability in the situation where we are trying to find the probability of event **A** and the occurrence of event **B** forces us to review this probability (that is, we review the probability of event **A** after the occurrence of event **B**).

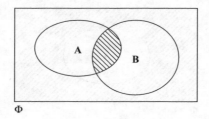

Figure 2.3   Conditional probability

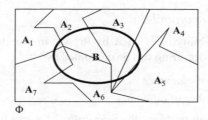

Figure 2.4   Total probability

2.1.6   Total probability

The total probability is the probability of **B** given an independent set of events $\mathbf{A}_i$ (Figure 2.4).

$$\Pr\{\mathbf{B}\} = \sum_{i=1}^{N} \Pr\{\mathbf{A}_i|\mathbf{B}\} \Pr\{\mathbf{B}\} \tag{2.6}$$

2.1.7   Complement

The *complement* of event **A** is represented by **A**′. It is the event where event **A** does not occur (Figure 2.5).

$$\begin{aligned}\Pr\{\Phi\} &= \Pr\{\mathbf{A}\} + \Pr\{\mathbf{A}'\} = 1 \\ \Pr\{\mathbf{A}'\} &= 1 - \Pr\{\mathbf{A}\}\end{aligned} \tag{2.7}$$

2.1.8   OR

OR is the probability of event **A** or **B** or both occurring (Figure 2.6).

$$\Pr\{\mathbf{A} \cup \mathbf{B}\} = \Pr\{\mathbf{A} + \mathbf{B}\} = \Pr\{\mathbf{A}\} + \Pr\{\mathbf{B}\} - \Pr\{\mathbf{AB}\} \tag{2.8}$$

If events **A** and **B** are independent, they will not overlap, therefore

$$\Pr\{\mathbf{A} \cup \mathbf{B}\} = \Pr\{\mathbf{A}\} + \Pr\{\mathbf{B}\} \tag{2.9}$$

Mathematical preliminaries 15

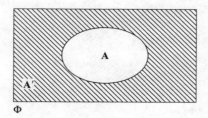

Figure 2.5   Complement of event **A**

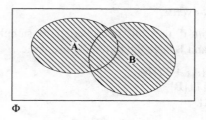

Figure 2.6   Probability **A** *OR* **B** *assuming non-independent events*

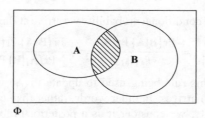

Figure 2.7   Probability **A** *AND* **B** *assuming non-independent events*

2.1.9 AND

AND is the probability of **A** and **B** occurring together (Figure 2.7).

$$\Pr\{\mathbf{A} \cap \mathbf{B}\} = \Pr\{\mathbf{AB}\} = \Pr\{\mathbf{A}|\mathbf{B}\}\Pr\{\mathbf{B}\} = \Pr\{\mathbf{B}|\mathbf{A}\}\Pr\{\mathbf{A}\} \tag{2.10}$$

Intuitively, $\Pr\{\mathbf{A}|\mathbf{B}\}$ is the fraction of **B** covered by **A** (see posterior probability Section 2.1.5) and $\Pr\{\mathbf{B}\}$ is the fraction of Φ covered by **B**. The multiple of these two values will give the fraction of Φ covered by $\mathbf{A} \cap \mathbf{B}$. If **A** and **B** are independent then

$$\Pr\{\mathbf{AB}\} = \Pr\{\mathbf{A}\}\Pr\{\mathbf{B}\} \tag{2.11}$$

16   *Motion vision: design of compact motion sensing solutions*

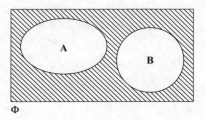

Figure 2.8   Independent events

2.1.10   Independent events

Events are *independent* if they do not overlap in sample space (Figure 2.8). Independent events **A** and **B** have the following properties.

$$\begin{aligned}&\Pr\{\mathbf{A}+\mathbf{B}\} = \Pr\{\mathbf{A}\} + \Pr\{\mathbf{B}\} \\ &\Pr\{\mathbf{AB}\} = \Pr\{\mathbf{A}\}\Pr\{\mathbf{B}\} \\ &\Pr\{\mathbf{A}|\mathbf{B}\} = \Pr\{\mathbf{A}\} \\ &\Pr\{\mathbf{B}|\mathbf{A}\} = \Pr\{\mathbf{B}\}\end{aligned} \quad (2.12)$$

2.1.11   Bayes theorem

Using the AND rule and conditional probability we arrive at Bayes theorem as follows:

$$\Pr\{\mathbf{A}|\mathbf{B}\} = \frac{\Pr\{\mathbf{BA}\}}{\Pr\{\mathbf{B}\}} = \frac{\Pr\{\mathbf{B}|\mathbf{A}\}\Pr\{\mathbf{A}\}}{\Pr\{\mathbf{B}\}} = \frac{\Pr\{\mathbf{B}|\mathbf{A}\}\Pr\{\mathbf{A}\}}{\sum_i \Pr\{\mathbf{A}_i|\mathbf{B}\}\Pr\{\mathbf{B}\}} \quad (2.13)$$

You can think of this rule being used to update Pr(**A**) when new information (occurrence of event **B**) arrives. In other words, rather than considering Pr{**A**|**B**} as a conditional probability, we consider it as a posterior probability. While we can sometimes use a direct method to compute posterior probability this is not always possible. In such cases it is often easier to compute Pr{**B**|**A**} – prior probabilities Pr{**A**} and Pr{**B**} are usually known in advance.

2.1.12   Order statistics

The *n*th *order statistic* is defined as the *n*th smallest element in a set. Thus the minimum is the first order statistic and the median is the $\frac{1}{2}n$th order statistic.

2.1.13   Random variable

A *random variable* $\mathbf{R}(\xi)$ is simply a function that maps each possible outcome of an experiment (ξ) to a real number x which is called the *measurement* (Figure 2.9). Often this mapping is direct ($\mathbf{R}(\xi) = \xi$) as in the case measuring temperature or people's height, however, this is not always the case. For example, if we throw two dice, the mapping may be the sum of the outcomes for each die. The measurement

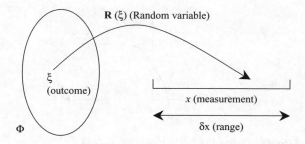

Figure 2.9   A random variable (adapted from Reference 226)

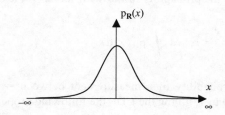

Figure 2.10   Example probability density function

can be discrete (i.e. map onto a defined set of values) as in the dice experiment, or continuous (i.e. map onto a continuous range of values) as in the height experiment. The concept of random variables allows us to define functions that describe how likely a particular measurement is.

2.1.14   Probability Density Function (PDF)

The *probability density function* (PDF) $p_\mathbf{R}(x)$ represents the probability of obtaining measurement x from random variable $\mathbf{R}$ (Figure 2.10). For example, if we are throwing two dice, the PDF can describe the probability of throwing a total value of x. The PDF is sometimes called the *probability distribution* or simply the *distribution*.

$$p_\mathbf{R}(x) = \Pr\{\mathbf{R}(\mathbf{A}) = x\} = \frac{dP_\mathbf{R}(x)}{dx} \tag{2.14}$$

In a PDF the probability of all measurements is always greater than or equal to zero (i.e. $p_\mathbf{R}(x) \geq 0$) and the area under the PDF (i.e. $\int_{-\infty}^{\infty} p_\mathbf{R}(\lambda)d\lambda$) is always equal to one indicating that we are guaranteed to get some measurement each time we perform a trial.

2.1.15   Cumulative Distribution Function (CDF)

The *cumulative distribution function* (CDF) represents the probability of a measurement less than x occurring (Figure 2.11). For example, if we are throwing two dice,

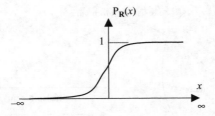

Figure 2.11   Example cumulative distribution function

the CDF can describe the probability of throwing a total value of less than x.

$$P_{\mathbf{R}}(x) = \Pr\{\mathbf{R}(\mathbf{A}) \leq x\} = \int_{-\infty}^{x} p_{\mathbf{R}}(\lambda) d\lambda \qquad (2.15)$$

In a CDF, the probability of a measurement between $-\infty$ and ∞ increases monotonically from 0 to 1, giving the CDF the following properties:

1. $P_{\mathbf{R}}(-\infty) = 0$
2. $P_{\mathbf{R}}(\infty) = 1$
3. $0 \leq P_{\mathbf{R}}(x) \leq 1$
4. If $x_1 < x_2$ then $P_{\mathbf{R}}(x_1) \leq P_{\mathbf{R}}(x_2)$
5. $\Pr\{x_1 < x < x_2\} = P_{\mathbf{R}}(x_2) - P_{\mathbf{R}}(x_1)$

2.1.16 Joint distribution functions

A *joint distribution function* is a PDF that combines the PDFs with a number of variables $\mathbf{R}_1, \mathbf{R}_2, \ldots, \mathbf{R}_n$. Joint distribution functions are random variables themselves and are composed using the AND rule (2.10).

$$\begin{aligned} \text{JCDF} \rightarrow P_{\mathbf{R}_1,\mathbf{R}_2,\ldots,\mathbf{R}_n}(x_1, x_2, \ldots, x_n) &= P_{\mathbf{R}}(\mathbf{x}) \\ &= \Pr\{\mathbf{R}_1 \leq x_1 \text{ AND } \mathbf{R}_2 \leq x_2 \text{ AND} \ldots \text{AND } \mathbf{R}_n \leq x_n\} \end{aligned} \qquad (2.16)$$

$$\text{JPDF} \rightarrow p_{\mathbf{R}_1,\mathbf{R}_2,\ldots,\mathbf{R}_n}(x_1, x_2, \ldots, x_n) = \frac{\partial^n P_{\mathbf{R}}(\mathbf{x})}{\partial x_1, \partial x_2, \ldots, \partial x_n} \qquad (2.17)$$

If random variables $\mathbf{R}_1, \mathbf{R}_2, \ldots, \mathbf{R}_n$ are independent then the joint probability will be

$$\text{JPDF} \rightarrow p_{\mathbf{R}_1,\mathbf{R}_2,\ldots,\mathbf{R}_n}(x_1, x_2, \ldots, x_n) = \prod_i p_{\mathbf{R}_i}(x_i) \qquad (2.18)$$

This follows directly from (2.11). The pi symbol represents multiplication of each element, much as the sigma symbol represents addition of elements.

2.1.17 Marginal distribution function

Occasionally, we might wish to convert a joint distribution function to a distribution over a single (or a few) variables. For example,

$$p_{\mathbf{R}_1,\mathbf{R}_2}(x_1, x_2) \rightarrow p_{\mathbf{R}_1}(x_1) \tag{2.19}$$

We can achieve this by integrating over the random variable that we wish to eliminate. This gives us what is a known as a marginal distribution function.

$$p_{\mathbf{R}_1}(x_1) = \int_{-\infty}^{\infty} p_{\mathbf{R}_1,\mathbf{R}_2}(x_1, x_2) dx_2 \tag{2.20}$$

2.1.18 Independent, identically distributed (iid)

Random variables are *independent, identically distributed* (*iid*) if they are independent (2.12) and have the same PDF (2.14).

2.1.19 Gaussian distribution and the central limit theorem

The *Gaussian distribution* is the most commonly used distribution and often goes by the name of the *normal distribution*. The Gaussian distribution is defined as:

$$p_{\mathbf{R}}(x) = \frac{1}{\sigma\sqrt{2\pi}} e^{-((x-\mu)^2/2\sigma^2)} \tag{2.21}$$

A Gaussian distribution is completely defined by its mean (μ) and variance (σ^2). The frequent appearance of the Gaussian distribution can be attributed to the *central limit theorem* which says that if an observed variable is the sum of many random processes, the variable will tend to have a Gaussian distribution even if the individual random processes do not.

2.1.20 Random or stochastic processes

The idea of a *process* extends the idea of a random variable $\mathbf{R}(\xi)$ over time, changing $\mathbf{R}(\xi)$ to $\mathbf{R}(\xi(t))$ where t is the time variable. A process also brings with it an underlying connotation of some underlying physical process or cause and we may wish to model this cause in order to better understand the phenomenon we are measuring. The term *random* (or *stochastic*) process is used if the outcome (measurement) cannot be known in advance. Processes whose outcome can be predicted are known as *deterministic* processes.

As an example, let us consider your voice. Speech is generated by the movement of air through your vocal tract and leads to the vibration of air molecules. This vibration can be measured at distinct moments (i.e. it can be sampled) using a microphone and computer giving a sequence of random variables $\mathbf{R}_1(\xi), \mathbf{R}_2(\xi), \ldots, \mathbf{R}_N(\xi)$. We can write this more succinctly by making the outcome a function of time $\mathbf{R}(\xi(t))$ – the resulting function is a function of time is called a random process. In our speech

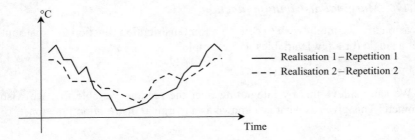

Figure 2.12   Imagine a person repeating the word 'MONEY'. Each repetition of this word is a realisation and the collection of realisations is called an ensemble. The waveforms here might represent portions of two realisations

example, the random process maps the measured air pressure to a discrete numerical value that is a function of time. Sometimes we might be able to capture a number of repetitions of the random process (for example, you might repeat the same word a number of times). Each such repetition of the random process $\mathbf{R}(\xi_n(t))$ is called a *realisation* of that random process. The collection of realisations $\mathbf{R}(\xi_{1...N}(t))$ is called an *ensemble* (Figure 2.12).

We can also extend the idea of a random variable over space or a combination of time and space $\mathbf{R}(\xi(t, l))$ – here l is a location variable. For example, temperature (whose underlying cause is the weather) can be thought of as a function of both time and space. More importantly in the context of this book, a video sequence also forms a random process that varies over time and space. This video sequence arises from light reflected from a collection of objects that may or may not be in motion. The problem we aim to solve in this book is the measurement (modelling) of the motion of the objects based on the measurements separately made by a camera and a laser radar.

2.1.21   Stationary processes

A process is said to be *stationary* if its statistical properties do not change over time. For example, a process with the following PDF

$$p_{\mathbf{R}(t)}(x) = p_{\mathbf{R}(t+\tau)}(x) \tag{2.22}$$

is said to be first order stationary. Its mean and variance (which are referred to as the *first order moments* of the distribution) will not change over time. Note that we have replaced the somewhat complex notation $\mathbf{R}(\xi_n(t))$ with the simpler $\mathbf{R}(t)$. We do this because often we only have a single realisation of a process available (e.g. we might only have a single image, a single series of temperature measurements, etc.) hence there is no need to use a notation able to represent multiple realisations. This idea of stationarity can be extended to higher orders, and if a process is stationary for all orders it is considered to be *strictly stationary*. In practice, we use the idea of a *wide sense stationary* process. The mean of a wide sense stationary process does not change over time, and its autocorrelation is a function only of time difference [127]. Most importantly, a wide sense stationary with a Gaussian distribution is also strictly stationary.

2.1.22 Average

Average is often used as a general term meaning 'central tendency'. It can refer to:

1. The *expectation* (see Section 2.1.24).
2. The *arithmetic mean*. For a realisation of some discrete random variable, the arithmetic mean can be approximated using (2.23) where N is the total number of measurements:

$$\mu_\mathbf{R} = \frac{1}{N} \sum_{i=1}^{N} \mathbf{R}(i) \qquad (2.23)$$

3. The *sample mean*. The arithmetic and sample means are different ways of stating the same thing, however, they are formulated differently: with the sample mean we simply group together equal outcomes while in the arithmetic mean all outcomes are directly added together. For discrete random variables the sample mean is calculated as follows. Assume there are η different outcomes and outcome k appears n_k times. If the total number of trials is N then the sample mean is

$$\mu_\mathbf{R} = \frac{1}{N} \sum_{k=1}^{\eta} k \times n_k \qquad (2.24)$$

4. The *median* (middle outcome in an ordered set of outcomes).
5. The *mode* (most frequently occurring outcome). The mode will correspond to the peak of the PDF.

For a large realisation with a Gaussian distribution, the arithmetic/sample mean, the median and the mode will coincide.

2.1.23 Variance

Variance describes how far the outcomes of an experiment spread about the mean. *Standard deviation* is the square root of the variance, that is, σ. For a discrete set of data (i.e. a realisation of some discrete random variable), variance can be approximated using the following equation

$$\sigma_\mathbf{R}^2 = \frac{1}{N} \sum_{i=1}^{N} (\mathbf{R}(i) - \mu_\mathbf{R})^2 \qquad (2.25)$$

2.1.24 Expectation

The *expectation* is essentially a more formal way of stating the mean – you might consider the expectation as being the 'true' (or theoretical) mean since it is defined using the PDF of the random variable. The arithmetic/sample mean is an approximation of the expectation. Given some function of a random variable g(**R**) and the PDF of **R**, the expected value of the function g(**R**) (i.e. the expectation) is given by:

$$E[g(\mathbf{R})] \equiv \int_{-\infty}^{\infty} g(\alpha) p_{\mathbf{R}}(\alpha) d\alpha \qquad (2.26)$$

Here α is a dummy variable of integration. For example, if g(**R**) = **R**, then we obtain the mean of the random variable **R**:

$$\mu_{\mathbf{R}} = E[\mathbf{R}] = \int_{-\infty}^{\infty} \alpha p_{\mathbf{R}}(\alpha) d\alpha = \sum_{k} x_k \Pr\{\mathbf{R} = x_k\} \qquad (2.27)$$

This is equivalent to the sample mean. As another example, consider the expectation of g(**R**) = **R**^{2}

$$E[\mathbf{R}^2] = \int_{-\infty}^{\infty} \alpha^2 p_{\mathbf{R}}(\alpha) d\alpha \qquad (2.28)$$

This is an important value called the *mean squared value*. We often want to make estimators (see later) that minimise this value. Just as the idea of the mean can be formalised as expectation, we can formalise the idea of variance as follows:

$$\sigma_{\mathbf{R}}^2 = E[(\mathbf{R} - E[\mathbf{R}])^2] = E[(\mathbf{R} - \mu_{\mathbf{R}})^2] = E[\mathbf{R}^2] - \mu_{\mathbf{R}}^2 \qquad (2.29)$$

2.1.25 Likelihood

Imagine we have a dataset that we want to model using a line-of-best-fit. This line-of-best-fit might have two parameters – the slope m and the intercept c. The *likelihood* of the parameters m, c being correct given an individual data element x is defined by the likelihood function

$$l(m, c|x) = p(x|m, c) \qquad (2.30)$$

In other words, the likelihood of m, c being the correct parameters is equal to the probability of the data point x arising given the condition that m, c are the parameters. You should not think of the likelihood function as a probability density – this does not make any sense since the parameters m, c are definite values, not random variables [250]. Note that the *log likelihood* (i.e. the natural logarithm of the likelihood) is often used in place of the likelihood to simplify likelihood functions that include exponential functions (such as the Gaussian distribution).

2.2 Simple estimation problems

At its simplest, an estimation problem is composed of:

1. Some parameter or set of parameters ($\boldsymbol{\theta}$) for which we seek a value. There may be one or multiple parameters and the value of each parameter may be multidimensional (e.g. a position in N-dimensional space).
2. A dataset (**d**) corrupted by some noise. Usually we assume the noise to be additive such that $\mathbf{d} = \mathbf{d}^t + \eta$. Here $\mathbf{d}^t$ is the 'true' data, unperturbed by noise and η is the noise process whose distribution is generally unknown, though it is often assumed to be Gaussian. This data might be measured or come from some other estimation process and it is assumed that it is stationary.
3. A relationship between the data **d** to the parameters $\boldsymbol{\theta}$, which we will call the *data constraint*.

Many estimation problems also have additional constraints that define the relationships between the parameters. This usually occurs when we have some extra knowledge about how the data or parameters behave, or if the data constraint does not specify a unique solution.

Estimation problems are usually formulated as *optimisation problems*, that is, problems where a set of parameters must be found that minimises (or maximises) some function. It is sometimes possible to find a closed form solution for such optimisation problems by simply taking derivatives and setting them to zero, however, this depends on the form of the problem. Furthermore, the larger the number of parameters, the more cumbersome it becomes to solve using a closed form. In cases where the closed form is unavailable or impractical, iterative techniques are used.

2.2.1 Linear regression

While the data constraint can take on any form we only consider linear constraints here. One of the more common linear constraints is the linear regression constraint that is used to find a *line-of-best-fit* for a dataset.

$$d_0 = \theta_1 d_1 + \theta_2 d_2 + \cdots + \theta_p d_p + r \qquad (2.31)$$

In this equation there are p parameters[13] and r is an error term (the use of this value will become clear shortly). We have singled out one of the data points, (d_0), as the 'output' variable[14], which you might think of as the response of a system described by the constraint (2.31) when its 'input' variables[15] are set to values d_1, d_2, etc. In other words, you might think of (2.31) as an input-output model for a system. Both the input and output variables may be corrupted by noise. This noise could come from

---

[13] The linear regression problem is said to be *p*-dimensional.
[14] Also called the response variable, or the dependent variable.
[15] Also called explanatory variables, or independent variables.

measurement error (inputs to a system generally have to be measured or estimated if we are to quantify them!) or, if the data comes from another estimation process, it will have some estimation error.

To better understand the idea of linear regression, let us consider a simple problem with only two parameters. This leads to the familiar equation for a line.

$$y = mx + c + r \tag{2.32}$$

Here we have replaced d_0 with y and d_1 with x. In this case, $d_2 = 1$. The parameters θ_1 and θ_2 have been replaced by m (the slope of the line) and c (the intercept) respectively. One can now image that x is some sort of input variable for a system (e.g. voltage, temperature) while y is an output variable (e.g. current, pressure). In a linear regression problem we will have multiple *realisations* of the data (e.g. the *ensemble* of data $\mathbf{x}$ and $\mathbf{y}$) as in Figure 2.13 and our task is to find a line that best fits this data.

Clearly, just choosing an arbitrary pair of points will not give us a good line-of-best-fit. In estimation we use all the available data as evidence as to what the line-of-best-fit might be. This is generally done by minimising some sort of *error* or *energy function* that defines how each data point contributes to the error/energy. One possible error measure arises if we rearrange (2.32) so that the *residual* (i.e. the error) r is the vertical distance between the measured output y and the line-of-best-fit $mx + c$ (see Figure 2.14a).

$$r_i = y_i - mx_i - c \tag{2.33}$$

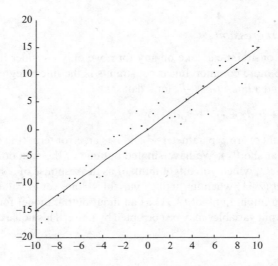

Figure 2.13   Linear regression. Dots are example data points while the line is a line-of-best-fit for the data

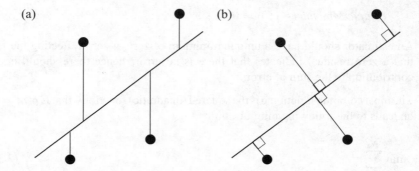

Figure 2.14   *(a) Least Squares minimises the total vertical distance to the line-of-best-fit, while (b) Total Least Squares minimises the shortest (normal) distance to the line-of-best-fit*

Residuals of this form can lead to what is called the *Least Squares* (LS) estimate[16]. Alternatively, the residuals can be defined as the *shortest* distance (i.e. the normal distance) to the line-of-best-fit as shown in Figure 2.14b. The resulting estimate is called the *Total Least Squares* (TLS) estimate [70][17]. We do not consider this type of estimate in detail here (see Reference 70 for more information) other than to say that it has some advantage over LS estimates because it takes into account errors both in the input *and* output variables. LS estimates only takes into account error in the output variable, however, the LS estimate is more common because it is easier to formulate and solve.

Given the residuals, it would seem logical to require that the line-of-best-fit minimises the sum of the residuals, however, this will not work because the positive residuals may balance the negative residuals giving a small sum when the total error is large. Therefore, we minimise the sum of *some function* of the residuals. This function is called the *estimator function* and we can think of the resulting sum as a sort of total error (i.e. a total distance from the line-of-best-fit).

$$\min_{\theta} \sum_{i=1}^{N} \rho(r_i) \tag{2.34}$$

The *estimator function* $\rho(r)$ should have the following properties:

Positive Definite – $\rho(r) \geq 0$

This ensures that positive and negative residuals do not cancel.

Even – $\rho(r) = \rho(-r)$

Ensures that positive and negative residuals have the same influence on the final result.

---

[16] Provided the quadratic estimator function is used. This is explained shortly.
[17] Again, this name arises if the quadratic estimator function is used.

Unique Minimum – $\rho(0) = 0$

An estimator should have a unique minimum of zero at zero reflecting the fact that a zero residual indicates that there is no error; hence there should be no contribution to the sum of errors.

The most common function ρ is the squared (quadratic) function – that is $\rho(r) = r^2$ which leads to the following minimisation:

$$\min_{\theta} \sum_{i=1}^{N} r_i^2 \qquad (2.35)$$

The line-of-best-fit given by the quadratic estimator function minimises the total *squared* error (or the sum of squared residuals). This is why residuals of the form in (2.33) lead to what is called a *Least Squares* (LS) estimate[18]. In our specific two-dimensional example, (2.35) becomes:

$$\min_{m,c} \sum_{i=1}^{N} (y_i - mx_i - c)^2 \qquad (2.36)$$

As we shall see later, there are other options for $\rho(r)$, however, the squared function is the easiest to solve and gives optimal results if the residuals have a Gaussian distribution [239]. In fact, minimising the sum of the squared residuals corresponds to finding the mean of the residuals (see Section 2.2.8).

The idea of the LS estimator can be formalised to the *Minimum Mean Squared Error* (MMSE) estimator (if the distribution of the residuals is available) by calculating the minimum expectation for the residuals [127]; however, we do not consider this further here. Furthermore, if the residuals are iid, zero mean Gaussian then the least squares estimate will give the value for the parameters that has the maximum likelihood (i.e. the LS estimate will be equivalent to the *maximum likelihood estimate* (MLE) [250]). This is because the maximum likelihood estimate gives the mode of the residuals (i.e. the most frequently occurring residual), which is equal to the mean (i.e. to the LS estimate) if the residuals have a Gaussian distribution.

2.2.2 Solving linear regression problems

Given a linear data constraint and a quadratic estimator function $\rho(r) = r^2$, it is possible to find a closed form solution for least squares linear regression problems. This is done by taking partial derivatives with respect to each parameter then solving

---

[18] Perhaps a more complete name would be Least *Sum of* Squares. Summation is not the only possible means of combining the residuals as we shall see in Section 2.3.

the resulting equations simultaneously[19]. For example, a closed form solution for the two-dimensional linear regression problem of (2.36) can be found as follows:

$$\min_{m,c} \sum_{i=1}^{N} (y_i - mx_i - c)^2$$

$$\frac{\partial}{\partial m} = \sum_{i=1}^{N} x_i y_i - m \sum_{i=1}^{N} x_i^2 - c \sum_{i=1}^{N} x_i = 0 \qquad (2.37)$$

$$\frac{\partial}{\partial c} = \sum_{i=1}^{N} y_i - m \sum_{i=1}^{N} x_i - Nc = 0$$

Solving for m and c simultaneously yields

$$m = \frac{N \sum_{i=1}^{N} x_i y_i - \left(\sum_{i=1}^{N} x_i\right)\left(\sum_{i=1}^{N} y_i\right)}{N \sum_{i=1}^{N} x_i^2 - \left(\sum_{i=1}^{N} x_i\right)^2}$$

$$c = -\frac{\left(\sum_{i=1}^{N} x_i\right)\left(\sum_{i=1}^{N} x_i y_i\right) - \left(\sum_{i=1}^{N} x_i^2\right)\left(\sum_{i=1}^{N} y_i\right)}{N \sum_{i=1}^{N} x_i^2 - \left(\sum_{i=1}^{N} x_i\right)^2}$$

(2.38)

A similar technique can be used to find the maximum likelihood estimate [250], however, the details are beyond the scope of this book. Obviously as the number of parameters grows it becomes more difficult (though it is possible) to find the relevant derivatives and to solve the resulting equations simultaneously. Furthermore, the form of the data constraint and of the estimator function $\rho(r)$ could make it difficult or impossible to find the derivatives. In these types of situations, one of the minimisation algorithms described in Section 2.2.6 can be used to find a minimum.

2.2.3 The Hough transform

An alternative view of linear regression arises if we consider each data point to contribute one equation to a set of over-determined equations as in (2.39).

$$\begin{aligned} y_1 &= mx_1 + c \\ y_2 &= mx_2 + c \\ &\vdots \\ y_N &= mx_N + c \end{aligned} \qquad (2.39)$$

Here we have N equations in $p = 2$ unknowns and each equation defines infinitely many lines passing through the point (x_i, y_i) in *data space*. In an approach called the *Hough transform* we rewrite these equations so that each equation defines a single

---

[19] A solution can also be found using matrix algebra [127].

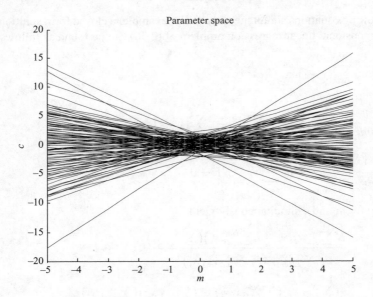

Figure 2.15   *An example of the lines in the parameter space of a linear regression problem*

constraint line in *parameter space* (e.g. m, c space) as illustrated in Figure 2.15.

$$\begin{aligned} c &= -x_1 m + y_1 \\ c &= -x_2 m + y_2 \\ &\vdots \\ c &= -x_N m + y_N \end{aligned} \quad (2.40)$$

The constraint lines will tend to intersect at a point corresponding to the parameters (e.g. m – slope and c – intercept) that define the line-of-best-fit for the data. Noise in the data means that there will not be a single clear crossing point, thus attempting to solve some subset of the equations simultaneously will not work. Linear regression finds the *location* in parameter space that is the 'best' intersection for the lines, however, it does so by minimising an error function (e.g. (2.35)). The Hough transform suggests that the least squares problem can also be solved by searching for the best intersection in parameter space.

2.2.4 Solving Hough transform problems

While we could exhaustively search the parameter space to find the best intersection point, computational complexity can be reduced by quantising the parameter space. That is, we divide the parameter space into blocks called *accumulator cell* as in Figure 2.16. We do not quantise the entire parameter space, just the region where we might expect a solution to be found ($m_{min} \ldots m_{max}, c_{min} \ldots c_{max}$).

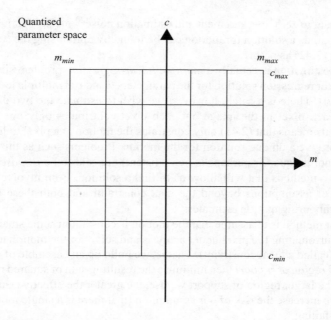

Figure 2.16 *Quantised parameter space*

Each of the accumulator cells is initialised to zero. Then, for each constraint line we find the (quantised) value of c that corresponds to each (quantised) value of m. For each such pair of m and c we add one to the corresponding accumulator cell. When this process has been applied to each line, the accumulator cell with the highest value will be the best estimate of the intersection point of the constraint lines. One potential problem in using equations of the form of (2.40) is that m can become infinite. For this reason it is common to use the polar equivalent of (2.40) rather than (2.40) directly.

2.2.5 Multiple linear regression and regularisation

So far, the residuals we have considered arose from explicit application of a linear relationship between the data and parameters (2.31), however, the residual function can also arise from first principles. For example, consider the problem of estimating motion from a sequence of grey-scale images. As we will see in Section 3.2.3, the relationship between the brightness patterns of the images and two-dimensional motion at pixel (i, j) is

$$E_{d(i,j)} = u_{(i,j)} I_{x(i,j)} + v_{(i,j)} I_{y(i,j)} + I_{t(i,j)} = 0 \qquad (2.41)$$

where (u, v) is the (horizontal and vertical) motion at a given pixel, and I_x, I_y, I_t are the spatial and temporal derivatives of the images. This equation actually has the form of linear regression (2.31) – this can be seen if we choose $-I_t = d_0$. That is, we choose $-I_t$ as the output variable and I_x and I_y as the input variables. This makes (2.41) the residual of the linear regression problem. While this residual is formally equal to zero, measurement and estimation noise[20] means that it will not be exactly zero, thus a solution for motion can be found by minimising the sum squared of residuals (2.42) as before.

Motion estimation is actually a *multiple linear regression* problem since we aim to find a linear regression solution for the parameters u and v at multiple locations (i.e. at each pixel). There is a catch, however: we wish to estimate the two dimensional motion at each pixel in the image but each pixel contributes only one data point I_x, I_y, I_t. This means that (2.41) only constrains the motion at a pixel to lie on a line – it does not give a unique solution for the motion. Problems such as this where the solution is not unique (or perhaps is non-existent) are said to be *ill-posed* [31] and there are two methods that will allow us to find a solution. Both involve additional constraints or assumptions beyond the data constraint and both these constraints couple neighbouring motion estimates.

The first method is to assume that the motion is constant in some small region of the image surrounding the pixel under analysis and solve for the motion using linear regression (called a local constraint). That is, we build up an ensemble of data points from a local *region of support* then minimise the resulting sum of squared residuals as in (2.42). The larger region of support we use, the greater the effective smoothing[21], however, we increase the risk of our assumption (that there is a single motion within the region) failing.

$$\min_{u,v} \sum_{(i,j) \in region} (u I_{x(i,j)} + v I_{y(i,j)} + I_{t(i,j)})^2 \qquad (2.42)$$

This is essentially the same as the normal regression problem – the only difference is in how the data are gathered. In a normal regression problem we take a number of measurements for the same system and attempt to find a linear model to describe that system. In multiple linear regression we essentially have a number of separate 'systems' and we can only make one measurement for each of the systems. In order to be able to find a solution for this multiple regression problem we make the assumption that all the measurements in a local region of support have the same underlying model (i.e. the same parameter values – for example, motion). This allows us to create a collection of measurements and to solve for the parameters as in normal linear regression.

The alternative to the multiple linear regression approach is to impose global constraint on the parameters (e.g. the motion). This approach is called *regularisation* and it exploits *a priori* (often heuristic) knowledge of a problem in order to overcome its ill-posedness. In motion estimation it is often assumed that neighbouring pixels will move with more or less the same velocity. That is, rather than implicitly assuming that motion is constant within a given region of the image, we can explicitly assume the motion at the neighbouring pixel is similar (i.e. motion should vary smoothly).

---

[20] Only the brightness is measured directly using a camera – the derivates are estimated from the brightness patterns in the images.
[21] The least squares solution is a sort of average hence the larger the dataset, the greater the smoothing.

Mathematical preliminaries

Mathematically, the deviation from smoothness can be measured using the gradient of the motion estimates as in the following *smoothness constraint*[22].

$$E_s = \nabla^2 u + \nabla^2 v$$
$$E_s^2 = (\nabla^2 u)^2 + (\nabla^2 v)^2 \tag{2.43}$$

Note that the square of the gradient is determined by squaring each component of the gradient. This smoothness constraint allows us to form an estimation problem where we simultaneously minimise both (a) the total error as described by E_d in (2.41) and (b) the deviation from smoothness as described by E_s in (2.43) across the entire image. This results in the following minimisation:

$$\min_{\mathbf{u},\mathbf{v}} \sum_{(i,j) \in \text{image}} (u_{(i,j)} I_{x(i,j)} + v_{(i,j)} I_{y(i,j)} + I_{t(i,j)})^2 + \lambda^2 ((\nabla^2 u)^2 + (\nabla^2 v)^2)$$

$$\tag{2.44}$$

This sort of estimation model is often referred to as a *membrane model* since we can think of the smoothness constraint as an elastic link between the parameters (i.e. the motion estimates) creating a parameter membrane. The data constraint then deforms the membrane of parameters to fit the data, as illustrated in Figure 3.4. In effect, the membrane model smooths the parameter estimates. The summation term in (2.44) is often called an *energy function* since it describes how far the membrane is from the zero energy (undeformed) state. Remember that in (2.42) we assume motion (u, v) is constant in a region of support around the pixel under analysis and perform the summation over that region of support. On the other hand, (2.44) involves the summation over the entire image and motion is not assumed to be constant – it is allowed to vary slightly from pixel to pixel. Equation (2.44) can be written more simply as:

$$\min_{\mathbf{u},\mathbf{v}} \sum_{(i,j) \in \text{image}} \left(E_{d(i,j)}^2 + \lambda^2 E_{s(i,j)}^2 \right) \tag{2.45}$$

The strength of the elastic link between parameters is controlled by the *regularisation parameter* λ. Higher λ leads to a smoother result because the smoothness constraint is being given a higher weight in the summation forcing a reduction in the difference between neighbouring estimates. Conversely, using a lower λ gives the data constraint more weight allowing the membrane to stretch more and better conform to the underlying data (Figure 2.17).

Just as the assumption of local constancy could fail in the linear regression solution, the assumption of global smoothness in the regularisation solution can also fail if there are multiple objects moving about the image. We consider solutions to this problem in Section 2.3.

---

[22] Other constraints will introduce a different conceptual framework so for simplicity we only consider the two-dimensional smoothness constraint here. Extension to higher dimensions is generally straightforward.

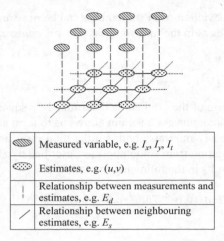

Figure 2.17   Schematic view of a global constraint (adapted from Reference 96)

Note that the membrane model can be used as a smoothing mechanism in its own right. For example, to smooth an image we could use the following data and smoothness constraints:

$$E_{d(i,j)} = I_{(i,j)} - \theta_{(i,j)}$$
$$E_{s(i,j)} = \nabla \theta_{(i,j)}$$
(2.46)

Here $I_{(i,j)}$ corresponds to the brightness of the pixel (i,j), $\theta_{(i,j)}$ is the smoothed value of pixel (i,j) and the x and y subscripts refer to horizontal and vertical partial derivatives. We can see that, as before, the data constraint will deform the membrane to approximate the input data (the intensity) while the smoothness constraint will attempt to minimise the deviation from smoothness.

While it is beyond the scope of this book, it is interesting to consider that the membrane model is related to a modelling technique called the *Finite Difference Method* (FDM) [182]. FDM is used to solve (partial) differential equations over some user-defined space with particular geometric and physical properties (for example, we might be solving for the electric field around a conductor encased in some material). The method begins with the creation of a finite difference approximation to the differential equation under consideration then the user-defined space is discretised and the finite difference approximation applied to each node in the solution space. The resulting system of equations is then solved showing how the parameter (e.g. the electric field) behaves within the user-defined space. There are two key differences between FDM problems and the membrane problem.

1. In FDM there is no data constraint (indeed there is no underlying measured data). Instead *boundary conditions* are used to define the parameters at particular locations. For example, we might know that the potential on a conductor is 10 V. We shall see later that boundary conditions can also play a role in the membrane

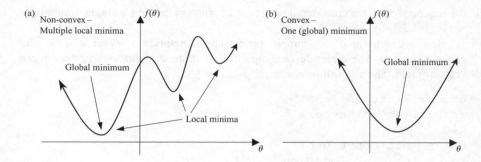

Figure 2.18   (a) Local minima versus (b) global minimum

model. Interestingly, the smoothness term in (2.46) is in fact the finite difference approximation of a partial differential equation known as the Poisson equation (or the Laplace equation if the left hand side is zero) that is used for modelling static electric fields in FDM.
2. In a membrane model, the discretisation is often done for us (e.g. we solve for motion at each *pixel* of the image). In FDM, we are modelling a physical (continuous) structure, hence the discretisation must be performed explicitly.

2.2.6   Solving the membrane model

An important difference between regression and membrane problems is that neighbouring parameters have no explicit relationship in regression problems. This allows a closed form, single step solution. In regularisation problems, neighbouring parameters are explicitly constrained to be similar and this in turn leads to an implied relationship between parameters that are not direct neighbours. For example, if neighbouring parameters θ_1 and θ_2 are constrained to be similar and neighbouring parameters θ_2 and θ_3 are also constrained to be similar, then it stands to reason that parameters θ_1 and θ_3 should also be similar[23]. If we try to solve a regularisation problem in a single pass we will have only enforced the smoothness constraint on directly neighbouring parameters. An iterative solution method is required to allow the smoothness constraint to propagate further than just the direct neighbours.

2.2.6.1   Descent algorithms

There are a number of iterative optimisation techniques (e.g. see Reference 231) that search parameter space for a local minimum of a function $f(\boldsymbol{\theta})$ of the parameters. In general $f(\boldsymbol{\theta})$ may have many local minima as illustrated in Figure 2.18a and optimisation techniques are only guaranteed to find one of these – they will not usually find the global minimum. This is not a problem for us at the moment since

---

[23] Though they will be similar to a lesser degree than θ_1 & θ_2 & θ_2 & θ_3 since we are now considering a second order relationship.

the functions we have considered so far are all *convex* (i.e. have a single minimum – see Figure 2.18b).

Here we introduce two simple optimisation methods – *direct descent* and *steepest descent*. The direct descent algorithm is quite straightforward, though not very efficient. The algorithm runs as follows.

```
choose a step size δ
initialise parameters θ
k=0
while (not Converged)
      for each location (i, j)
         for P = 1 to p           //For each parameter
             //Does this update lead downhill?
```
$$\text{if } f\bigl(\theta^k_{1(i,j)}, \ldots, \theta^k_{P(i,j)} + \delta, \ldots, \theta^k_{p(i,j)}\bigr) < f\bigl(\theta^k_{(i,j)}\bigr) \text{ then}$$
```
             //If yes then update θᵢ
```
$$\theta^{k+1}_{P(i,j)} = \theta^k_{P(i,j)} + \delta$$
```
         else
             //else update θᵢ in opposite direction
```
$$\theta^{k+1}_{P(i,j)} = \theta^k_{P(i,j)} - \delta$$
```
         end if
       end for
     end for
     k = k + 1;
end while
```

In this code fragment, superscripts indicate the iteration number. The algorithm starts with some initial guess for the solution θ^0. If we have no information telling us what a good guess may be, we can arbitrarily set the parameters to zero. Of course, if we have a rough idea of what the solution may be, using this as a starting point means the minimum will be found more quickly. The algorithm then modifies each parameter in turn. If the change in this one parameter leads to a lower value for $f(\theta)$ then the change is kept. If the change has not led to a lower value we assume that a change in the opposite direction will lead to a lower value (since we have assumed $f(\theta)$ convex) and the parameter is updated appropriately. This process is illustrated in Figure 2.19 for the two-dimensional case.

The parameters are repeatedly updated until the algorithm *converges* – that is, until the minimum of $f(\theta)$ is found. In practice, we may not be able to find the minimum exactly and will not have an infinite amount of time to search for the minimum, hence the algorithm will iterate until the change in $f(\theta)$ between iterations is below some threshold or until a certain number of iterations have been completed.

The choice of the step size δ is crucial. Larger step sizes may lead to quicker convergence, however, they also decrease the accuracy with which we can locate the minimum. Conversely, smaller step sizes allow us to locate the minimum more accurately but it will take longer for the algorithm to converge. These problems

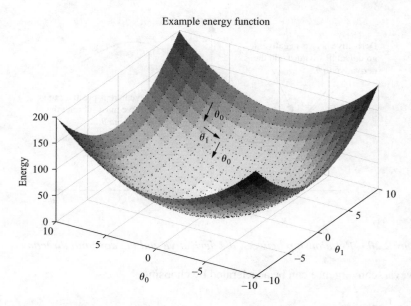

Figure 2.19 Direct descent for a two-dimensional energy function $f(\boldsymbol{\theta})$

can be overcome using the steepest descent algorithm. Rather than taking arbitrary size steps, the steepest descent algorithm takes steps of a size related to the derivative of $f(\boldsymbol{\theta})$. In this way, the algorithm will take larger steps where the derivative is larger (far from the minimum) and the step size will reduce as we approach the minimum so that it can accurately be located. The steepest descent method can be implemented as follows.

```
initialise parameters θ
k=0
while (not Converged)
        for each location (i, j)
                for P = 1 to p         // For each parameter
```

$$\theta^{k+1}_{P(i,\ j)} = \theta^{k}_{P(i,\ j)} - \omega^{T} \frac{\partial f(\boldsymbol{\theta}^{k}_{(i,\ j)})}{\partial \theta_{P(i,\ j)}} \quad (2.47)$$

```
                end for
        end for
        k=k+1
end while
```

Equation (2.47) is the called an *update equation* since it updates the value of the parameter at each iteration. In this equation we subtract the derivative since the derivative always points 'up-hill' as illustrated in Figure 2.20. The variable ω^{T} allows us to control the step size. When $\omega^{T} = 1$ the standard steepest descent algorithm arises;

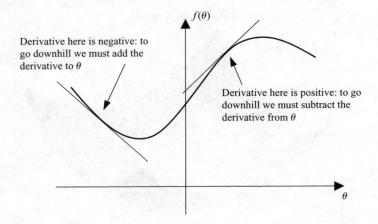

Figure 2.20   *Relationship between the derivative and the downhill direction*

however, convergence can be accelerated by choosing

$$\omega^T = \omega \frac{1}{T(\theta_i)} \quad (2.48)$$

Here $T(\theta_i)$ is a constant equal to the upperbound on the second derivative $\partial f(\boldsymbol{\theta})/\partial \theta_i$ and ω is a constant between 0 and 2 [35]. Maximum acceleration of convergence tends to occur when ω is approximately 1.9, though this changes depending on $f(\boldsymbol{\theta})$ [35]. When we choose ω^T to be of the form of (2.48) the minimisation algorithm is called *successive overrelaxation* (SOR)[24] rather than steepest descent.

While the steepest descent algorithm and SOR can converge more quickly than direct descent, all these algorithms have one particular drawback. Because they move in a zigzag pattern down the slope of $f(\boldsymbol{\theta})$ the direct and steepest descent methods can be very slow in situations where there are long narrow valleys in $f(\boldsymbol{\theta})$. In this case, the algorithms will zigzag along the valley[25] rather than reaching the valley floor and moving directly towards the minimum. In such cases, algorithms such as the *conjugate gradients method* are more appropriate [231]; however, we do not consider those algorithms here.

Let us apply the idea of steepest decent to our example membrane problem (3.9)[26]. First, we require our function $f(\boldsymbol{\theta})$. For motion estimation this is

$$f(u_{(i,j)}, v_{(i,j)}) = (u_{(i,j)} I_{x(i,j)} + v_{(i,j)} I_{y(i,j)} + I_{t(i,j)})^2 + \lambda^2 ((\nabla^2 u)^2 + (\nabla^2 v)^2) \quad (2.49)$$

---

[24] SOR can also be used to solve sets of simultaneous equations (matrix equations). In this case the form of the solution is slightly different. See [182].

[25] Unless the valley happens to be aligned with one of the parameters.

[26] Note that there is another way of solving this problem (see Section 3.2.3.2.2); however, it is useful to see an example of the steepest descent method.

Notice that we have not included the summation term in this equation because we wish to solve the parameters at each pixel individually – we are not trying to calculate the sum over all pixels. The gradient of the parameters u and v are approximated using the following *first order central difference approximation* (see Section 3.2.3.3 or Reference 109 for more detail):

$$\begin{aligned}\nabla^2 u_{(i,j)} &= (u_{(i-1,j)} + u_{(i+1,j)} + u_{(i,j-1)} + u_{(i,j+1)} - 4u_{(i,j)}) \\ \nabla^2 v_{(i,j)} &= (v_{(i-1,j)} + v_{(i+1,j)} + v_{(i,j-1)} + v_{(i,j+1)} - 4v_{(i,j)})\end{aligned} \quad (2.50)$$

Thus the update equations become

$$\begin{aligned}u_{(i,j)}^{k+1} &= u_{(i,j)}^k - \omega^{\mathrm{T}} \\ &\quad \times \left[2I_{x(i,j)}\left(u_{(i,j)}^k I_{x(i,j)} + v_{(i,j)}^k I_{y(i,j)} + I_{t(i,j)}\right) - 8u_{(i,j)}\lambda^2\left(\nabla^2 u_{(i,j)}^k\right)\right] \\ v_{(i,j)}^{k+1} &= v_{(i,j)}^k - \omega^{\mathrm{T}} \\ &\quad \times \left[2I_{y(i,j)}\left(u_{(i,j)}^k I_{x(i,j)} + v_{(i,j)}^k I_{y(i,j)} + I_{t(i,j)}\right) - 8v_{(i,j)}\lambda^2\left(\nabla^2 v_{(i,j)}^k\right)\right]\end{aligned}$$
$$(2.51)$$

These update equations can be applied iteratively as per the steepest descent algorithm to find a motion estimate.

2.2.6.2 Boundary conditions

Special care must be taken at the boundaries (and corners) of the image since our approximations to the gradient of the motion estimate will change at these locations. For example, at the left boundary of the image ($i = 0$) there is no estimate available at $i - 1$ thus we need to make some sort of modification to the gradient estimate.

Generally we can make use of further information (or assumptions) regarding what occurs at the boundaries. Such information is called a *boundary condition*, which can take one of two forms. *Dirichlet boundary conditions* assume that value of the unavailable parameter is some constant value. For example, we might assume that the motion is zero at the left boundary:

$$\nabla^2 \theta_{(i,j)} = (0 + \theta_{(i+1,j)} + \theta_{(i,j-1)} + \theta_{(i,j+1)} - 4\theta_{(i,j)}) \quad (2.52)$$

An alternative to the Dirichlet boundary condition is the *Neumann boundary condition*, which specifies the derivative of the parameter rather than its value. One simple, yet common, Neumann boundary condition is to assume a zero derivative across the boundary. A zero derivative implies that the unavailable parameter is equal to the parameter at the boundary – for example, at the left boundary a zero derivative implies that $\theta_{(i-1,j)} = \theta_{(i,j)}$. One can easily see that such a boundary condition leads to the following equation for the left boundary.

$$\nabla^2 \theta_{(i,j)} = (\theta_{(i+1,j)} + \theta_{(i,j-1)} + \theta_{(i,j+1)} - 3\theta_{(i,j)}) \quad (2.53)$$

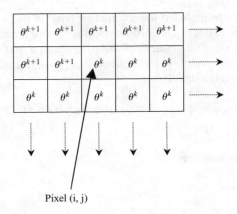

Figure 2.21   *Illustrating the update scheme for multiple linear regression*

The boundary conditions we use depend entirely on the problem at hand. For example, in motion estimation the zero-derivative Neumann condition is the most logical since we have assumed that neighbouring motion estimates are similar. Remember that once you choose how to handle boundaries, you must make sure to modify the update equations appropriately at the boundaries.

2.2.6.3   Iteration techniques

In implementing the steepest descent algorithm we have to consider how we use the data that we have. If we update the parameters for each location (e.g. pixel) in turn parameters will be available as illustrated in Figure 2.21.

Given these parameters, the most natural way of calculating the parameter gradient is using

$$\nabla^2 \theta_{(i,j)}^k = \left(\theta_{(i-1,j)}^{k+1} + \theta_{(i+1,j)}^k + \theta_{(i,j-1)}^{k+1} + \theta_{(i,j+1)}^k - 4\theta_{(i,j)}^k\right) \tag{2.54}$$

This type of update, where the most recently calculated values for the parameters are used, is called *Gauss–Siedel iteration*. Rather than using the most recent values, you could also use only the parameter values from the same iteration.

$$\nabla^2 \theta_{(i,j)}^k = \left(\theta_{(i-1,j)}^k + \theta_{(i+1,j)}^k + \theta_{(i,j-1)}^k + \theta_{(i,j+1)}^k - 4\theta_{(i,j)}^k\right) \tag{2.55}$$

This type of update is called *Jacobi iteration*. Jacobi iteration is not often used in multiple linear regression problems since it requires extra memory to store the previous values of the parameters and can be slower to converge than Gauss–Siedel iteration [101].

It is interesting to note that while the steepest descent method can be implemented sequentially (as we have assumed thus far), it is inherently parallel. If we assume that there is a processor associated with each pixel, all that is required to update the parameter estimate at a given pixel are the parameters of the neighbouring pixels. In a parallel implementation Gauss–Siedel iteration is inefficient since it assumes sequential update of the parameters. Instead, a *chequerboard update* is more

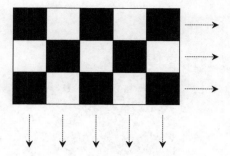

Figure 2.22 Chequerboard update scheme for a parallel solution of a membrane model using the steepest descent algorithm

appropriate. In this update scheme each iteration has two passes. On the first pass, the 'white' squares of the chequerboard (Figure 2.22) are held constant while the 'black' squares are updated. Then, on the second pass, the 'black' squares are held constant and the 'white' squares updated [35].

2.2.7 Location estimates

Another common form for the data constraint is the *location estimate*. Rather than trying to find a line that best models a dataset (as in linear regression), a location estimate seeks to model the data using a point. The constraint equation for a location estimate is simply

$$\mathbf{d} = \mathbf{\theta} + r \tag{2.56}$$

Note that our data points and parameters are vectors since they can be multidimensional. Multidimensional location estimation problems are often called *multivariate* problems. For example, we might try to find a location estimate for a dataset scattered in two-dimensional space as illustrated in Figure 2.23.

Importantly, even though the data points and parameters may be multidimensional, the residuals remain as scalars. The residuals for location estimation problems are given by (2.57)

$$r = \|\mathbf{d} - \mathbf{\theta}\| \tag{2.57}$$

where the residual r is the *norm* (magnitude) of the vector $\mathbf{d} - \mathbf{\theta}$. In other words, the residual is a measure of the distance between the data point and location estimate $\mathbf{\theta}$. In one dimension, the norm is a simple subtraction; however, in higher dimensions other distance measures must be used. Perhaps the most familiar distance estimate is the *Euclidian distance* (also called the L_2 *norm* or the *sum of squared differences* (SSD)):

$$r_i = \sqrt{\sum_{n=1}^{N} (d_{i,\dim(n)} - \theta_{\dim(n)})^2} \tag{2.58}$$

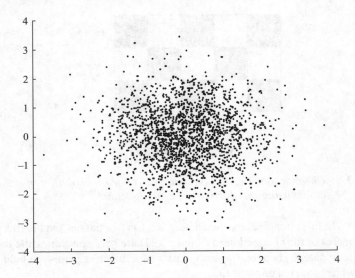

Figure 2.23 Example data for a two-dimensional location estimate

Here the summation is over each dimension of the location (i.e. in the two-dimensional case, summation is over the X and Y dimensions). Other distance measures such as the *sum of absolute differences* (SAD) (also called L_1 *norm*) and the *Mahalanobis distance* [2, 4] can also be used, depending on the application. Given the residuals, we can use (2.35) to obtain an LS estimate of the position parameter.

Be careful to understand the difference between linear regression and location estimates. In linear regression we search for a point in p-dimensional parameter space that defines the line-of-best-fit for the data. In location estimation we search for a point in p-dimensional data space that best represents the dataset. Put another way, linear regression seeks to find the relationship between the inputs and the outputs of a system. A location estimate simply tries to find a point that best represents a set of data – there is no concept of input and output variables for a location estimate. Location estimation and linear regression coincide only in the simple case where $p = 1$ and $d_1 = 1$.

2.2.8 Solving location estimation problems

In the one-dimensional case, the least squares location estimate can be written as:

$$\min_{\boldsymbol{\theta}} \sum_{i=1}^{N} (d_i - \theta)^2 \tag{2.59}$$

We can find a value for θ by finding where the derivative of this equation is zero as follows:

$$\frac{d}{d\theta} = -2\sum_{i=1}^{N}(d_i - \theta) = 0$$
$$\sum_{i=1}^{N} d_i = \sum_{i=1}^{N} \theta \qquad (2.60)$$
$$\frac{1}{N}\sum_{i=1}^{N} d_i = \theta$$

Thus, the one-dimensional least squares location estimate is simply the arithmetic mean of the data. In an p-dimensional location estimate with an L_2 norm, the above analysis becomes

$$\begin{aligned}\min_{\boldsymbol{\theta}} &\sum_{i=1}^{N}\left((d_{i,\dim(1)} - \theta_{\dim(1)})^2 \right.\\ &\left. + (d_{i,\dim(2)} - \theta_{\dim(2)})^2 + \cdots + (d_{i,\dim(p)} - \theta_{\dim(p)})^2\right)\\ \min_{\boldsymbol{\theta}} &\left[\sum_{i=1}^{N}(d_{i,\dim(1)} - \theta_{\dim(1)})^2 \right.\\ &\left. + \sum_{i=1}^{N}(d_{i,\dim(2)} - \theta_{\dim(2)})^2 + \cdots + \sum_{i=1}^{N}(d_{i,\dim(p)} - \theta_{\dim(p)})^2\right]\end{aligned} \qquad (2.61)$$

Because each summation term is independent, each term can be minimised individually:

$$\min_{\theta_{\dim(1)}} \sum_{i=1}^{N}(d_{i,\dim(1)} - \theta_{\dim(1)})^2$$
$$+ \min_{\theta_{\dim(2)}} \sum_{i=1}^{N}(d_{i,\dim(2)} - \theta_{\dim(2)})^2 + \cdots + \min_{\theta_{\dim(p)}} \sum_{i=1}^{N}(d_{i,\dim(p)} - \theta_{\dim(p)})^2$$
$$(2.62)$$

Therefore, a multivariate least squares location estimate (using a L_2 norm) is simply the mean in each dimension.

It is also interesting to consider what would happen if we replaced the least squares estimator $\rho(r) = r^2$ with the estimator function $\rho(r) = |r|$. The resulting estimator is often called the *Least Absolute Residual Estimator* (LAR). In the one-dimensional case the resulting minimisation can be written as

$$\min_{\boldsymbol{\theta}} \sum_{i=1}^{N} |d_i - \theta| \qquad (2.63)$$

where we minimise the 'sum of *absolute* residuals' rather than the sum of *squared* residuals. This minimisation will give the median of the dataset. We prove this result as follows [121]. Imagine we have 2k + 1 (one-dimensional) data points **d** and we choose θ^t to be the median of **d** plus some unknown but positive number t. All k absolute residuals $|d_i - \theta^t|$ below the true median θ will now be increased by t and all k absolute residuals below θ will be decreased by t. The net result is that there is no change in the sum of these 2k absolute residuals. However, the absolute residual corresponding to the true median $|\theta - \theta^t|$ will equal t rather than zero thus increasing the overall sum. Therefore, (2.63) is minimised when $\theta^t = \theta$.

One may well ask if the multidimensional extension of this analysis will show that the result is given by the median in each dimension in the same way that the multidimensional LS estimate ($\rho(r) = r^2$) with L_2 norm was given by the mean in each dimension. The answer is yes, if the L_1 norm is used. We begin as before, writing the equation to minimise:

$$\min_{\boldsymbol{\theta}} \sum_{i=1}^{N} ||d_{i,\text{dim}(1)} - \theta_{\text{dim}(1)}| + |d_{i,\text{dim}(2)} - \theta_{\text{dim}(2)}| + \cdots + |d_{i,\text{dim}(p)} - \theta_{\text{dim}(p)}||$$

(2.64)

Clearly, taking the absolute value of the entire sum is redundant here and can be removed. As before, each term is independent so we can write (2.64) as

$$\min_{\theta_{\text{dim}(1)}} \sum_{i=1}^{N} |d_{i,\text{dim}(1)} - \theta_{\text{dim}(1)}|$$
$$+ \min_{\theta_{\text{dim}(2)}} \sum_{i=1}^{N} |d_{i,\text{dim}(2)} - \theta_{\text{dim}(2)}| + \cdots + \min_{\theta_{\text{dim}(p)}} \sum_{i=1}^{N} |d_{i,\text{dim}(p)} - \theta_{\text{dim}(p)}|$$

(2.65)

which is the median in each dimension.

2.2.9 Properties of simple estimators

A useful estimator should be *consistent*. That is,

(a) the bias in its estimate should be zero (*unbiased*) or it should tend towards zero as the size of the dataset increases (i.e. the estimator should be *asymptotically unbiased*); and
(b) the variance of the estimate should tends towards zero as the size of the dataset increases.

Bias is defined as the difference between the true value of a parameter θ^t and expectation of the estimated parameter $E[\theta]$. That is

$$b = \theta^t - E[\theta]$$

(2.66)

Clearly, we should use estimators that either give the true value for the parameter (i.e. are unbiased), or, at worst, converge to the true value of the parameter (asymptotically unbiased). Naturally, we would also like the variance of the estimate to decrease as the sample size increases. This means that the estimated value for the parameters will approach ever more closely to the true values with each added sample.

As it turns out, the LS estimator is consistent (it is *optimal*) if the residuals have a Gaussian distribution. Remembering that the LS estimator gives parameters which lead to the residual taking on a value equal to the arithmetic mean of all residuals (Section 2.2.8), we can say that the LS estimator is consistent for residuals with a Gaussian distribution because (we do not prove these results here):

(a) the arithmetic mean will converge to the true mean (the expectation of the residuals) as the size of the dataset increases;
(b) the variance will tend to zero as the sample size increases.

2.3 Robust estimation

In the simple estimation problems considered so far, we have assumed that the residuals conform to a single Gaussian distribution. Given this condition, the LS estimate is optimal [239]. However, what happens if the residuals do not have a Gaussian distribution? What if the residuals are made up of more than one Gaussian distribution? Data points corresponding to residuals that do not conform to the assumption of a single Gaussian distribution are called *outliers* or *leverage points* (depending on where the error occurs) and it would be very useful to have estimators that can tolerate a small number of such points. As we shall see, the LS estimate is not tolerant of outliers and leverage points; however, there are a number of *robust estimators* that are.

2.3.1 Outliers and leverage points

We begin our consideration of robust estimators by considering just what can go wrong with our data to illustrate why robust estimators are necessary. Consider the now familiar case of two-dimensional linear least squares regression. Assume we have a dataset with a nominal slope (m) of 2 and intercept (c) of 1.5 corrupted with zero mean Gaussian noise of variance 0.66. In each of the plots that follow, the LS estimate of these parameters is shown above the plot. First, we consider the case where the residuals have a Gaussian distribution (no outliers) as illustrated in Figure 2.24.

As expected, with a relatively low level of Gaussian noise the estimated slope (2.096) and intercept (1.48) are close to the nominal values. Now imagine the same dataset, except that there is an error in one of the output measurements. The measurement might have been misread, there might have been a fault in the transmission of the data, or some other mistake might have occurred. Whatever the case, the data point does not conform to the assumed Gaussian distribution for residuals and hence is an outlier. The result is illustrated in Figure 2.25.

44   Motion vision: design of compact motion sensing solutions

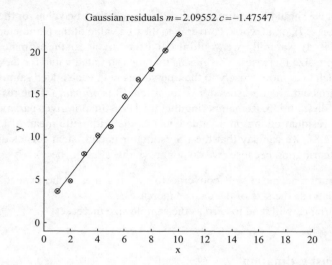

Figure 2.24   No outliers. The residuals have a Gaussian distribution

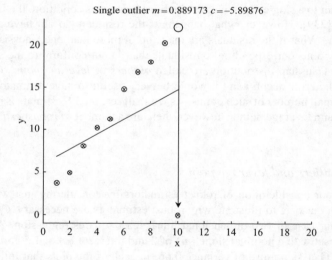

Figure 2.25   Data corrupted with a single outlier

In this example the final data point has a value of zero rather than the expected value of around 21.5. Clearly this single outlier has significantly skewed the line-of-best-fit away from its previous location and outliers will have a similar effect in location estimates. Even a single outlier could have an arbitrarily large effect on the estimate, so we say that the LS estimate is not robust.

We can explain the skew in the estimate as follows. If the line-of-best-fit had remained in its previous (Figure 2.24) position, the majority of data points will

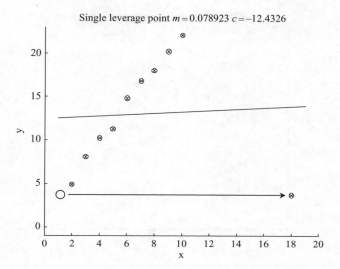

Figure 2.26   A leverage point

have small residuals; however, the outlier will have a large residual. Remember that residuals are squared before being included in the summation of (2.35), thus the outlier will have a large influence on the summation. It turns out that the summation (2.35) will be smaller overall if the line-of-best-fit moves away from the rest of the data and towards the outlier because the increased residuals for the majority of the data is counteracted by the decreased residual for the outlier.

Errors need not only occur in the output variable. Indeed, errors are more likely to occur in input variables because, in general, there may be many input variables. Data points with an error in the input variable are often called leverage points [239] rather than outliers since they can have a strong lever effect on the line-of-best-fit[27] as we can see in Figure 2.26. Note that leverage points can only occur in linear regression problems – location estimates cannot contain leverage points since they have no input variables.

Naturally, we only consider errors that move a data point away from the estimate to be outliers or leverage points. It is entirely possible for an error in a data point to move that point *closer* to the estimate. We have no way of knowing that this has occurred so we do not consider it further.

Another possibility is that errors occur in a number of data points as illustrated in Figure 2.27. There are two possible causes for multiple errors. In the simplest case, there could have been multiple measurement/transmission, etc. errors. The chance of a number of such errors occurring naturally increases with an increased number of data points and dimensions.

---

[27] The source of this leverage is explained further in Section 2.3.3.2.3.

46 *Motion vision: design of compact motion sensing solutions*

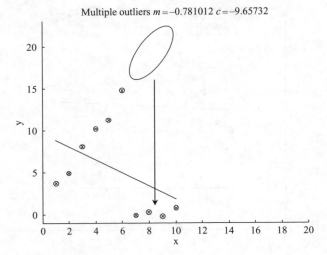

Figure 2.27   Multiple outliers

The arrangement of outliers in Figure 2.27 suggests another possibility – outliers might also have a more structured cause. For example, the system under test may change state somehow so that the model[28] that best suited the first part of the data, does not fit the second part of the data. The residuals for each section of the data may well have a Gaussian distribution, however, LS estimation only gives a correct answer if *all* residuals can be described by a *single* Gaussian distribution. Note that only output variables (outliers) can have this type of structured error since it is caused by a change in the input-output function of the system. That is, the system gives outputs corresponding to a different input-output model for different combinations of inputs (as in Figure 2.27).

In cases where the outliers have structure we may be interested in isolating each section and analysing it independently. In this chapter we only consider estimators that analyse the structure of the majority population – that is, they try to find the model that best fits half (or more) of the data. If there is no majority population (i.e. if a single model does not fit at least half of the data) then any robust estimator will fail. Recursive application of such estimators would allow analysis of each substructure in the data (assuming each forms a majority population of the remaining dataset).

Structured outliers also occur in the case of multiple linear regression and regularisation problems if the underlying assumptions (of local constancy or global smoothness) are violated. To clarify this, let us go back to the motion estimation example of Section 2.2.5. In the multiple linear regression approach we assume that the motion was constant in the local region around the pixel under analysis. But what if the pixel under analysis lies near the boundary of a moving object as in Figure 2.28?

---

[28] That is, the linear regression model, or location model.

Mathematical preliminaries 47

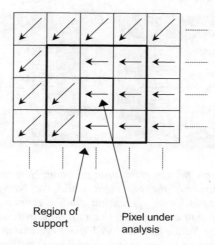

Figure 2.28   Failure of the local constancy assumption. Arrows indicate the true motion. The motion estimation problem requires us to find an estimate of this true motion

In this situation, the LS regression estimation will effectively smooth across the boundary (Figure 2.29a) – it will not recognise that some of the data belong to a different motion object. Similarly, a regularisation based solution will smooth across the boundaries because the membrane model will stretch across the motion discontinuity. This smoothing is illustrated in Figure 2.29b. As discussed earlier, the secondary motion will appear as outliers in the estimation problem since the data points corresponding to the secondary motion have a different underlying model. Leverage points can occur but they are related to errors in the input data (I_x and I_y in the case of motion estimation) and not to the secondary motion.

2.3.2   Properties of robust estimators

Above and beyond the property of consistency (which can be difficult to prove for robust estimators) robust estimators have a range of other properties. The most important of these are described below.

2.3.2.1   Breakdown point

Robustness can be measured in many ways [122, 148, 303], however, a relatively simple measure of robustness arises if we ask what fraction of the input data needs to be outliers (or leverage points) before we get an arbitrarily bad result (see Reference 239 for a more formal description). This measure is called the breakdown point and we have already seen that the LS estimator has a breakdown point of zero[29] because even

---

[29] One outlier in N data points is sufficient for the estimator to break down. The fraction 1/N tends towards zero as N increases.

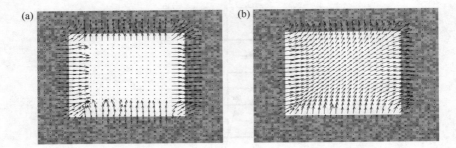

Figure 2.29 An example motion estimation problem. Here the square moves diagonally towards the top right while the background is stationary. (a) Multiple linear regression motion estimate. (b) Regularisation based motion estimate. Notice that both estimates smooth across the boundary. Notice how the local linear regression technique cannot find a solution within the square where there is not texture. The global membrane method is able to stretch its membrane across the square giving a fuller result

one outlier or leverage point can significantly skew the result. At best, an estimator will have a breakdown point of 50 per cent – if more than half the data are outliers it becomes impossible to tell which are good data and which are bad.

For linear regression problems, the breakdown point is related to both outliers and leverage points. Location estimates have no 'input' variables, hence they are only susceptible to outliers. As we shall see in Section 2.3.3.2, robust estimators also arise in the context of the membrane model if we allow the elastic link between estimates to break. In this case it does not make sense to ask about the breakdown point since the robust estimate is related to breaking the membrane (i.e. the links between estimates) and not to the data points.

2.3.2.2 Relative efficiency

The accuracy (variance) of any estimator can be no better than a lower bound known as the Cramer–Rao bound [155]. The *relative efficiency* (or *efficiency*) of an estimator is the ratio of the Cramer–Rao lower bound to the variance achieved by that estimator. It can be shown [155] that the maximum likelihood estimator (and hence the LS estimator if we assume Gaussian residuals) is the most efficient possible estimator. Improving efficiency of an estimator typically worsens its breakdown point.

2.3.2.3 Scale equivariance

An estimator is scale equivariant if scaling of the data by a constant leads to an equivalent scaling of the estimate. For example, imagine the two datasets in Figure 2.30 for which we seek a location estimate.

The dataset on the left was a generated random process with a Gaussian distribution of mean (2,2) with a variance of one. On the right we have the same data scaled

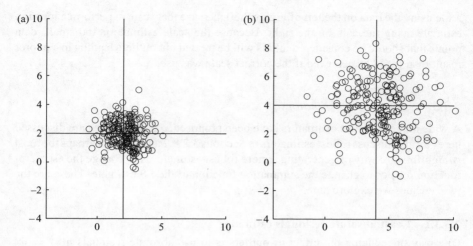

Figure 2.30 Example location estimate data. The data in (a) are equivalent to the data in (b) except for a scaling of 2. We know the true location estimate (indicated by cross lines) since we know the distribution of the data

by a factor of two thus the mean is (4,4) and the variance is two. An estimator should give a correct estimate in either case. Note that the LS estimator is scale equivariant.

As we shall see, some estimators are not scale equivariant though they can be made equivariant if the residuals are normalised by the *scale* σ (i.e. the variance). That is, instead of minimising

$$\min_{\theta} \sum_{i=1}^{N} \rho(r_i) \qquad (2.67)$$

we minimise

$$\min_{\theta} \sum_{i=1}^{N} \rho\left(\frac{r_i}{\sigma}\right) \qquad (2.68)$$

The estimation problem is now more complex because, before we can estimate our parameters, we need a *robust* estimate of the scale. If we are using a robust estimator in a real-time system, the extra complexity of calculating the scale may not be acceptable. One of the more common estimates of scale is the 1.483 times the median absolute deviation from the median (MAD) for the data. More formally, the MAD is

$$\mathrm{MAD} = \mathrm{med}\{|x_i - \mathrm{med}(\mathbf{x})|\} \qquad (2.69)$$

The factor of 1.483 appears to ensure maximum efficiency for a Gaussian distribution [216][30]. It is important we have the correct scale. Imagine we estimate

---

[30] Readers seeking further information regarding scale estimation are referred to References 122, 148, 239 and 303.

scale using the data on the left of Figure 2.30 then use that scale to perform a location estimate using the data on the right. Because the scale estimate is too small, data points that should be considered inliers will be treated out outliers leading to a poorer result than could be achieved if the correct scale was used.

2.3.3 Some robust estimators

A wide range of robust estimators have been proposed, a few of which are discussed here (we also discuss robust estimators in Sections 3.2.8.2 and 5.2.6). Perhaps the most straightforward way of generating a more robust estimator is to change the estimator function $\rho(r)$ or to change the summation function of the LS estimate. These are the two avenues we explore here.

2.3.3.1 Least absolute residuals estimator

One way of reducing the effect of outliers is to use absolute residuals $\rho(r) = |r|$ rather than the quadratic function $\rho(r) = r^2$ of the LS estimate [37]. The resulting estimator is called the *Least Absolute Residuals Estimator* or the L_1 *Estimator*[31]. This estimator is good for location estimates since it can cope with a dataset where up to 50 per cent of the elements are outliers, and it has maximum efficiency if residuals have a Laplacian distribution[32] and are scale equivariant. In linear regression problems, this estimator is resistant to outliers; however, it has a breakdown point of zero since a single leverage point can skew the estimate significantly [239]. A number of algorithms for calculating the L_1 estimate are described in [37].

2.3.3.2 M estimators and GM estimators

Estimators of the form of (2.70) are called *M estimators* since they are related to the maximum likelihood estimator (MLE)[33].

$$\min_{\theta} \sum_{i=1}^{N} \rho(r_i) \qquad (2.70)$$

The power of M estimators lies in the ability to choose the estimator function ρ so that the effect of outliers on the summation is minimised. Our earlier choice of $\rho(r) = r^2$ (the LS estimator) led to a relatively poor M estimator, since large residuals have a disproportionately large effect on the summation in (2.70). To reduce the LS estimator's sensitivity to outliers, we should use an estimator function that does not grow as quickly as the quadratic. One possible choice is to use $\rho(r) = |r|$ (the Least Absolute Residuals Estimator) since $|r| < r^2$, for $r > 1$; however, we can further reduce the effect of large residuals by choosing an estimator function that

---

[31] This refers to the fact that the absolute value is the L_1 norm.
[32] Its relative efficiency is worse than the LS estimator, however, if the residuals are Gaussian.
[33] An M estimator can be thought of as a maximum likelihood estimator where $-\log p = \rho$ and p is a density function describing the likelihood [148, 238].

saturates as residuals become larger. Some possible choices for the estimator function are illustrated in Figure 2.31.

2.3.3.2.1 Influence functions and M estimator classification

Each estimator function ρ has associated with it an *influence function* ψ (see Figure 2.31) that arises when we take the derivative of (2.70).

$$\frac{\partial}{\partial \theta_n} = \sum_{i=1}^{N} \psi(r_i) \frac{\partial r_i}{\partial \theta_n} = 0 \qquad (2.71)$$

Influence functions take their name from the fact that they indicate the influence of one additional data point on an estimate based on a large dataset. One can see that for the quadratic estimator (the LS estimator) the influence function is unbounded, hence a single additional data point can have an unbounded effect on the estimate and the LS estimator is not robust to outliers. On the other hand, the influence function of the least absolute residuals estimator is constant, indicating that an additional data point will have a bounded effect on the estimate and that the least absolute residuals estimator is robust to outliers.

M estimators are classified based on the shape of their influence function. An M estimator whose influence function rises monotonically (or saturates) is called a *Monotone M estimator* while a *Redescending M estimator* has an influence function that tends back towards zero after a cut-off point, indicating that residuals over this magnitude have a progressively smaller effect. Redescending M estimators can be further classified into soft-redescending M estimators whose influence function tends to zero asymptotically, and hard-redescending M estimators whose influence function returns to zero immediately after the cut-off point.

Redescending M estimators are of particular interest as robust estimators because they greatly reduce the effect of outliers. Any function that has a unique minimum at zero, is even, is positive definite and has an influence function that tends towards zero as the magnitude of residuals increases, (as in Figure 2.31, parts c and d) can be utilised as a redescending M estimator. The difference between these functions lies in the assumptions they make regarding the distribution of the residuals. For example, the truncated quadratic has good efficiency for Gaussian inliers and the Lorentzian is efficient for a Cauchy distribution [35]. In practice, however, real data never exactly conform to the assumed distribution [201] thus the choice of M estimator can be made more from a computational perspective (i.e. which is the simplest to implement) than from a statistical perspective.

2.3.3.2.2 M estimator properties

Redescending M estimators are not scale equivariant since they require a choice of cut-off point: scaling the data will change the cut-off point. We can make redescending M estimators scale equivariant by normalising the residuals using an estimate of the scale as in (2.68), or we can ignore the equivariance problem and choose a cut-off point to suit each dataset. This is particularly common in weak-membrane type models where the breaking point of the membrane (i.e. the cut-off point) is likely to be set heuristically by the user rather than through an automated calculation.

(a)
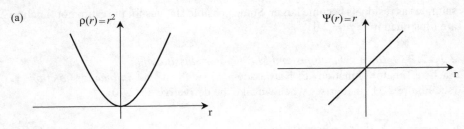
Monotone – Quadratic

(b)
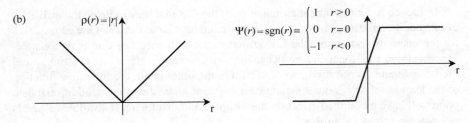
Monotone – Absolute value

(c)
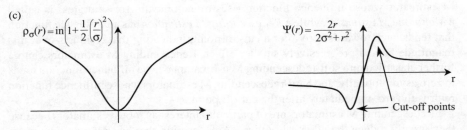
Soft redescending – Lorentzian

(d)
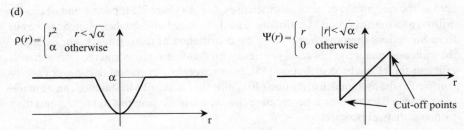
Hard redescending – Truncated quadratic

Figure 2.31   Classes of M estimators. There are two broad classes – the monotone and redescending M estimator

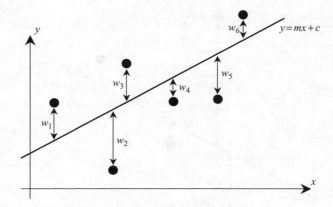

Figure 2.32 *Optimal values for m and c arise (i.e. a line-of-best-fit) when the total weight above the line (positive residuals) and below the line (negative residuals) is equal*

While redescending M estimators are resistant to outliers, leverage points will skew estimates, thus the breakdown point is 0 per cent.

2.3.3.2.3 M estimators and leverage points

To understand why M estimators are sensitive to leverage points we refer back to (2.70) where we aim to minimise the sum of some function of the residuals. To find the minimum we take the derivative of (2.70) and set it to zero. In the case of two-dimensional linear regression this gives us

$$\frac{\partial}{\partial m} = \sum_{i=1}^{N} \psi(y_i - mx_i - c)x_i = 0 \qquad (2.72)$$

$$\frac{\partial}{\partial c} = \sum_{i=1}^{N} \psi(y_i - mx_i - c) = 0 \qquad (2.73)$$

We can think of the influence function $\psi(r)$ as a function that converts the residuals into weights, hence (2.72) and (2.73) can be interpreted as a sort of balancing act (see Figure 2.32): in order for these equations to equal zero, parameters must be chosen that define a line-of-best-fit with an equal amount of weight above (positive residuals) and below (negative residuals)[34].

To make this idea clearer, let us consider the specific case of the L_1 estimator where $\rho(r) = |r|$ and hence $\psi(r) = \text{sgn}(r)$. We will only use (2.73) for the moment to avoid the complicating factor of the additional x_i term in (2.72). The L_1 influence

---

[34] In the case of location estimation, the location will be chosen so that there is an equal weight either side of the final estimated location in the direction of the derivative.

54 Motion vision: design of compact motion sensing solutions

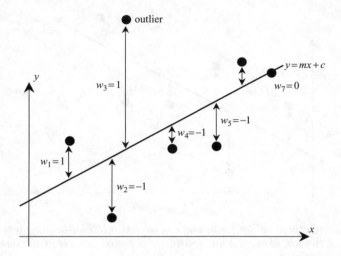

Figure 2.33 In L_1 regression a weight of one is assigned for points above the line-of-best-fit and negative one for points below the line. Therefore, outliers do not influence the result. Notice that the line-of-best-fit can pass through points giving them a zero weight

function gives all positive residuals a weight of $+1$, all negative residuals a weight of -1 and zero residuals a weight of zero (see Figure 2.33). Clearly the summation of these weights will be zero only if the line-of-best-fit is chosen so that there is an equal number of positive and negative residuals[35] (i.e. so that there is an equal number of *data points* above and below the line-of-best-fit). Outliers will be rejected because the magnitude of the outlier is not considered – we only consider only whether the point is above or below the line-of-best-fit. Notice the line of best fit will not be unique since any line that places an equal weight (equal number of points) above and below the line-of-best-fit will minimise (2.73).

The additional x_i term in (2.72) has the effect of modifying the weight specified by the influence function. In regression problems, these additional multiplicative terms will always be input variables (i.e. x_i, potential leverage points) never the output variable (i.e. y_i, a potential outlier)[36]. Therefore, we see that leverage points have an opportunity to modify the weights in a way that outliers do not: this is why M estimators are sensitive to leverage points but not to outliers.

Now let us consider how these modified weights will affect our L_1 parameter estimate. To begin with, let us assume that all the input data points x_i are relatively

---

[35] The line can also pass through any number of data points since the residual for these points will be zero.

[36] These additional multiplicative terms arise due to the differentiation of (2.70) with respect to the parameters. Because there is no parameter associated with (2.31) with the input variable (d_0 in the general case, or y in our specific two-dimensional example) the input variable will not appear as a multiplier.

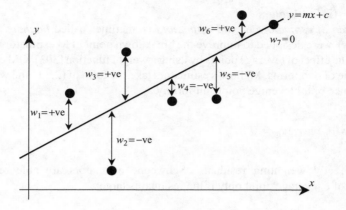

Figure 2.34 *In L_1 regression (indeed in any regression) the weights are sometimes modified by the input parameters, however even in this case the final estate balances the weight above and below the line-of-best-fit. The only way a data point (where $x \neq 0$) can have a zero weight is if it falls on the line-of-best-fit*

evenly distributed. The weights will no longer be $w = \psi(r) = \pm 1$ (or 0), they will be $w = \psi(r)x_i = \pm x_i$ (or 0) – that is points above the line-of-best-fit will have a positive weight, points below the line-of-best-fit will have a negative weight and points on the line-of-best-fit will have a weight of zero. As before, the value for m will be the value that gives an equal weight above and below the line-of-best-fit (see Figure 2.34).

The weight of a leverage point[37] can be significantly different to the weights of the other data points – many data points may be needed to balance the weight of the leverage point and this will cause the line-of-best-fit to be skewed from where we expect it. In L_1 estimation the line-of-best-fit can actually pass through the leverage point because it can be impossible to balance the weight of the outlier. Consider the data points $x_1 = 1$, $x_2 = 2$, $x_3 = 3$, $x_4 = 6$, $x_5 = 50$ – clearly there is no way the other data points can balance the weight of x_5, therefore the only way to achieve the balance we seek is for the line-of-best-fit to pass through x_5, thereby making its weight zero. However, even if the line-of-best-fit passes through this point a problem still remains – there is no way to balance the remaining weights and we see that L_1 regression can be degenerate [37].

Of course, our discussion here has been focused on the simple case of two-dimensional L_1 regression, however, the ideas presented here can easily be extended to other M estimators and to higher dimensions.

---

[37] This can also occur if the input data values are not evenly distributed about the mean, however, the effect will not be as strong.

2.3.3.2.4 GM estimators
Generalised M estimators or *GM estimators* (sometimes called *bounded influence estimators*) were designed to improve the breakdown point of M estimators by counteracting the effect of leverage points by using a weight function [303]. GM estimators take on one of two forms. Mallow's estimators take the form of (2.74) and will reduce the influence of all leverage points uniformly:

$$\sum_{i=1}^{N} w(d_n) \psi(r_i) d_n = 0 \qquad (2.74)$$

By directly weighting residuals, Schweppe estimators are able reduce the influence of a leverage point only if the residual is large:

$$\sum_{i=1}^{N} w(d_n) \psi\left(\frac{r_i}{w(d_n)}\right) d_n = 0 \qquad (2.75)$$

Even with this improvement, the breakdown point of GM estimators remains proportional to p^{-1} (where p is the number of parameters) at best [239].

2.3.3.2.5 M estimators and the membrane model
M estimators also arise in membrane problems. Often we would like to allow the membrane to model discontinuities in the estimates (e.g. the boundary between two different moving objects) which can be achieved by 'breaking' at locations where the difference between neighbouring estimates is large (over a cut-off point). The resulting model is called the weak membrane model [35, 36], which is formed by adding what is called a line process to the smoothness constraint as follows:

$$\sum_{\text{image}} \left(E_d^2 + \lambda^2 E_s^2 (l - 1) + \alpha l\right) \qquad (2.76)$$

Here l is the binary line process. If l is zero then the smoothness constraint is used and if l is one the smoothness constraint is replaced with α. The connection between motion estimates will break if $\alpha < \lambda^2 E_s^2$ because during minimisation l will be set to the value that gives the minimum sum. To simplify the minimisation, the line field can be eliminated (it is 'minimised away' – see Figure 2.35) yielding

$$\sum_{\text{image}} \left(E_d^2 + \rho_{\alpha,\lambda}(E_s)\right) \qquad (2.77)$$

where

$$\rho_{\alpha,\lambda}(r) = \begin{cases} \lambda^2 r^2 & \lambda^2 r^2 < \alpha \\ \alpha & \text{otherwise} \end{cases} \qquad (2.78)$$

The resulting function ρ is the truncated quadratic M estimator that we saw earlier, thus we see that M estimators do appear in membrane problems. The effect is that the

Mathematical preliminaries 57

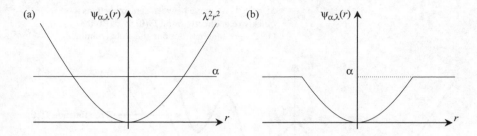

Figure 2.35 The truncated quadratic: (a) explicitly shows the line process, (b) shows the truncated quadratic after minimising the line process away

membrane will break in regions where neighbouring estimates differ by more than the cut-off point allowing rejection of outliers caused where the underlying model has changed (as we saw in Figure 2.28). The smoothing effect of the membrane between these breaks will reduce estimation noise. What this does not take into account is outliers and leverage points caused by *incorrect* input data. Black [35] points out that we could achieve some robustness to outliers and leverage points by applying a robust estimator to the data constraint as well as the smoothness constraint. In this way, the link between the data and the estimate can be broken if the data is giving a result that is clearly inconsistent with the membrane, allowing smoothness of the membrane to be maintained.

2.3.3.2.6 Solving M estimator problems
Solving for the parameters when using a redescending (G)M estimator is difficult because the minimisation is non-convex. If we use direct descent, steepest descent (Section 2.2.6.1) or any other naïve minimisation technique (e.g. Iteratively Reweighted Least Squares – see Section 5.2.6.3) it is unlikely that we will find the global minimum; rather we will only find a local minimum. One relatively simple way of finding a global minimum is via the Graduated Non-Convexity (GNC) algorithm of Blake and Zisserman [36]. GNC is a *deterministic annealing* technique that uses a series of approximations $f^{0...N}(\boldsymbol{\theta})$[38] to the function being minimised $f^0(\boldsymbol{\theta})$. The term annealing is used in analogy to the physical annealing process where the temperature of the material is gradually lowered so that the final material is in its lowest energy state. Similarly, in GNC we gradually reduce a temperature parameter (the exponent) until we find the global minimum of the function under consideration.

GNC begins (see Figure 2.36) with a convex approximation of the function $f^{max}(\boldsymbol{\theta})$, for which a minimum is found using one of the standard minimisation techniques. Because the function is convex, the unique, global minimum will be found.

---

[38] Note that the exponent does not necessarily increase in integer steps.

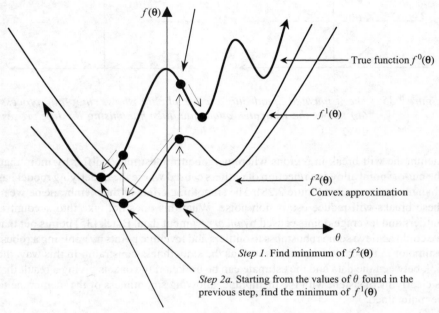

Figure 2.36 *An example of Graduated Non-Convexity. In this example there are two approximations to the true function*

A new approximation $f^{\max -1}(\theta)$[39] closer to the true function is then chosen and the minimum of this new function is found. The minimisation process begins using the value for θ that corresponds to the location of the minimum of the previous approximation with the presumption that location of the minimum for the previous approximation will be close to the location of the minimum for the current approximation. This process of updating the approximation and minimising is repeated until we reach the true function $f^0(\theta)$ (or some arbitrarily close approximation to the true function $f^{\min}(\theta)$). Note that GNC is not guaranteed to work in all cases, though proof does exist that it works for the membrane model [36]. In cases where GNC does not function, techniques such as Mean Field Annealing and Simulated Annealing may be more appropriate [191, 282, 313].

---

[39] Note that here max −1 means the next closer approximation; it does not (necessarily) mean that the next temperature should be one less than the previous temperature.

It is important that the temperature parameter is not reduced too quickly. For example, consider step 2b in Figure 2.36. Here we can see that if we had jumped straight from $f^2(\boldsymbol{\theta})$ to $f^0(\boldsymbol{\theta})$ then the GNC algorithm will not find the global minimum. Blake and Zisserman [36] show that GNC will find the true minimum if the temperature parameter is varied continuously towards zero, though they claim that an update schedule of the form

$$t \to \frac{t}{P} \tag{2.79}$$

where P is no larger than two should work well in most cases. On the other hand, Black [35] claims that two steps of the temperature parameter (a convex approximation and the true function) are generally enough to get a good solution. In practice, some experimentation may be necessary to find the best update schedule for a particular application. This will be a trade off between better accuracy (i.e. lower P) and faster convergence (i.e. higher P).

Construction of the approximations is straightforward – it can be done by modifying the estimator functions rather than the entire function $f^0(\boldsymbol{\theta})$. In their original formulation, Blake and Zisserman used the truncated quadratic as their estimator function (since their estimator function comes from the weak membrane model) and constructed approximations as follows:

$$\rho^T(r) = \begin{cases} \lambda r^2 & |r| < q \\ \alpha - \dfrac{c(|r|-s)^2}{2} & q \leq |r| < s \\ \alpha & \text{otherwise} \end{cases} \tag{2.80}$$

$$c = \frac{1}{4T}, \qquad s^2 = \alpha\left(\frac{2}{c} + \frac{1}{\lambda^2}\right), \qquad q = \frac{\alpha}{\lambda^2 s}$$

This approximation inserts a quadratic arc into the estimator function as illustrated in Figure 2.37a. The GNC algorithm starts with $T = 1$ and gradually proceeds to $T = 0$ (or some small number) at which point the approximation will be identical to the original truncated quadratic of (2.78). This approximation is relatively complex and simpler approximations are possible. For example, Li [191] suggests using the truncated quadratic and setting the parameter α high enough to ensure that all residuals fall within the cut-off point. The parameter α is then gradually reduced until it reaches the desired cut-off point. Similarly, Black [35] uses the Lorentzian (Figure 2.37b) and begins the GNC process with the σ parameter set high enough to encompass all residuals and gradually lowers σ until the desired cut-off point is reached.

2.3.3.3 Least median of squares

The *Least Median of Squares Estimator* (LMedS) is based on the observation that the summation typical of M estimators is not the only possible way of combining residuals. Rather than minimising the sum as in (2.70), the LMedS estimator

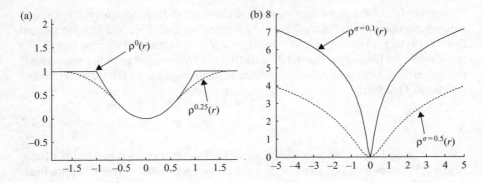

Figure 2.37 Example approximations of ρ: (a) shows the Blake and Zisserman approximations to the truncated quadratic with the temperature parameter set to 0 and 0.25; (b) shows the Lorentzian function with $\sigma = 0.1$ and $\sigma = 0.5$

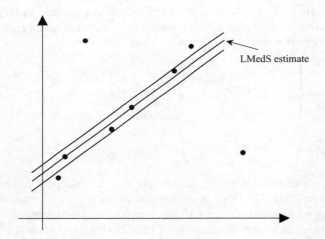

Figure 2.38 LMedS linear regression gives a result that lies in the middle of the narrowest band (measured vertically) that contains half the data

minimises the median of the squared residuals.

$$\min_{\theta} \; \text{median}_i \; r_i^2 \tag{2.81}$$

LMedS estimator can be thought of as a type of mode estimator [239] in that it will give a result that lies near the highest density region in a location estimate. In a linear regression problem, LMedS gives a line that lies in the middle of the narrowest possible band that contains half the data [201] (see Figure 2.38).

LMedS has a breakdown point of 50 per cent and is scale equivariant so there is no need to estimate a scale parameter separately. The drawback of LMedS is that it has low efficiency and it is difficult to compute. Typically, one must use stochastic sampling of the data to find a solution (see References 8, 202 and 240 for examples of how to solve for the LMedS estimate); however, in the case of one-dimensional location estimation, a simple solution technique exists. Given an ordered set of data points $\vec{\mathbf{d}}$ with N elements we take the pairwise differences as follows:

$$\vec{d}_h - \vec{d}_1, \vec{d}_{h+1} - \vec{d}_2, \ldots, \vec{d}_N - \vec{d}_{N-h+1} \tag{2.82}$$

The mid-point of the smallest difference is the LMedS location estimate. In principle, the choice of $\rho(r) = r^2$ is not necessary. For example, $\rho(r) = |r|$ could be used [201], however, this can cause problems if N is small and even [201, 239].

2.3.3.4 Least trimmed squares
Rather than minimising the sum of all the residuals, the *Least Trimmed Squares* (LTS) estimator seeks to minimise the sum of the smallest h squared residuals:

$$\min_{\theta} \sum_{i=1}^{h} (\vec{r}_i^2) \tag{2.83}$$

Here $\vec{\mathbf{r}}^2$ is the ordered set of squared residuals (note that the residuals are squared first, then ordered). The LTS estimator works by completely discarding the largest residuals, ensuring they cannot influence the estimate. The LTS estimator has a breakdown point of 50 per cent if $h = \frac{1}{2}N$, it is scale equivariant and has better efficiency than the LMedS estimator, though the fact that the estimate is always based on a fixed number of data points means the efficiency is lower than that of M estimators. As with the LMedS, the LTS estimate is difficult to compute (see References 8 and 232 for examples of how to compute the LTS estimate), except in the case of a one-dimensional location estimate. Given the ordered dataset $\vec{\mathbf{d}}$, the LTS estimate can be calculated by computing the following means:

$$\begin{aligned}
\bar{d}^1 &= \frac{1}{h} \sum_{i=1}^{h} \vec{d}_i \\
\bar{d}^2 &= \frac{1}{h} \sum_{i=2}^{h+1} \vec{d}_i \\
&\vdots \\
\bar{d}^{N-h+1} &= \frac{1}{h} \sum_{i=N-h+1}^{N} \vec{d}_i
\end{aligned} \tag{2.84}$$

The LTS location estimate corresponds to the mean $\bar{d}^n$ that has the smallest sum of squares which is calculated as follows:

$$SQ^1 = \sum_{i=1}^{h}\{\vec{d}_i - \vec{d}^1\}^2$$

$$SQ^2 = \sum_{i=2}^{h+1}\{\vec{d}_i - \vec{d}^2\}^2 \qquad (2.85)$$

$$\vdots$$

$$SQ^{N-h+1} = \sum_{i=n-h+1}^{N}\{\vec{d}_i - \vec{d}^{N-h+1}\}^2$$

Variations of the LTS estimator have also been proposed, such as the weighted scheme [220].

2.3.3.5 Other estimators

There exists a range of robust estimators beyond the ones presented here including RANSAC (which gives an estimate corresponding to the centre of a band width user-defined width that encompasses the densest part of the data) [91, 201], Minpran [216] and the estimators discussed in Section 5.2.6. Readers interested in other estimators are referred to References 122, 148, 201, 216, 239, 267 and 303.

Chapter 3
Motion estimation

The Sense of Movement: A Sixth Sense?
Berthoz [32]

If an autonomous vehicle is to navigate through dynamic environments, it is vital that it has a sense of the motion in that environment. Given the constraint of sensors rigidly attached to a vehicle[40], this chapter explores the methods available to estimate environmental motion and considers a number of important issues in this field including the brightness constancy assumption, data structure and robustness. Beyond consideration of algorithms, this chapter also examines a range of hardware implementations of motion estimation.

Motion 'sensing' has been widely studied both in computational contexts [25, 28, 174, 204, 269, 278] and biological contexts [40, 66, 128, 129, 185, 215, 246, 296, 302]. Experiments conducted by Exner as far back as 1875 showed the ability of the human visual system to detect motion as a fundamental quantity [297]. In this experiment, the timing of a series of electrical spark generators could be controlled. With correct timing, observers would see a single moving spark rather than a series of individual sparks. Computational studies of motion sensing have appeared more recently (most prominently since the 1970s) as computing power has increased.

Since that time, a vast number techniques (e.g. gradient, frequency and token based methods) and frameworks (e.g. regularisation, robust statistics, Markov random fields and mixture models) have been proposed for the estimation of motion. The resulting motion estimates have been used in applications as diverse as video compression, human–machine interfaces, robotic navigation, surveillance, weather analysis, biomedical problems, video understanding and military applications [204, 241]. In the autonomous navigation context, motion information has been

---

[40] Hence, excluding any 'active vision' approaches where the camera orientation can be changed.

used in ways that seem to mirror the processes that occur in biological systems[41]. For example, motion data have been used to detect obstacles [53, 242, 273] and to estimate properties such as the time-to-collision [60, 78], egomotion [50, 218] and depth [88, 130, 291].

One of the earliest hints that motion information may be important for navigation in biological systems came from Helmholtz [132] who observed that motion could act as a cue for depth perception. More recently, Gibson suggested that the focus of expansion (the point from which motion seems to emanate) provides 'the chief sensory guide for locomotion in space' [105]. Indeed, it has been shown that motion information plays a role in visual tasks as diverse as control of eye movements, depth perception, segmentation and estimation of egomotion and time-to-collision [215]. Evidence for the use of motion information is not limited to humans. For example, birds and flies use the time-to-collision parameter to trigger landing behaviour [208] and frogs respond to the motion of small dark objects [288] assumed to be prey.

The beauty of using vision to understand the environment is twofold. First, our own experience shows that vision can work extremely well in allowing us to understand and interact with our environment. This fact in itself drives the vast array of work devoted to various aspects of machine vision. Second, vision is passive. It does not send a signal into the environment that may interfere with other sensing processes. Despite this, it should be remembered that biological solutions to motion estimation (and sensing in general) may not be the best solutions in the computational domain [297]. Therefore, in this book we look to biological vision for broad inspiration but not for implementation details. Researchers have come to understand that a single type of sensor is not ideal under all conditions and that different types of data can be complementary, giving a more complete or more accurate description of the environment when combined [118]. Therefore, we consider motion both from a visual perspective and from the perspective of data fusion.

3.1 The motion estimation problem

Before considering methods of motion estimation, the nature of the motion estimation problem should briefly be considered. Motion is a geometric property of the environment inasmuch as the effect of motion is a change in the relative position of objects. As such, motion cannot be measured directly; rather, it must be inferred from the changing patterns in a time-varying data source [174].

When a camera views a scene, the three-dimensional relative motion in the environment is projected onto the camera's sensing surface producing a two-dimensional (translational) *apparent motion* field (Figure 3.1). In the simplest case, the problem

---

[41] We do not mean to imply that calculations are performed in a manner similar to that which occurs in biological systems. We simply state that both biological and computational systems seem to extract similar properties from the environment.

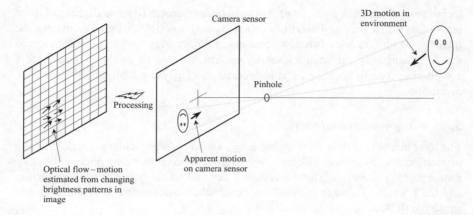

Figure 3.1 Three-dimensional motion; apparent motion and optical flow using a pinhole camera model

of *visual* motion estimation is to approximate this two-dimensional apparent motion field based on the changes in intensity seen by a single camera. Such an approximation of the apparent motion field is known as *optical flow* (or *optic flow*). Often this problem is extended from the estimation of the apparent motion to the estimation of some *parametric motion* model [271] that is closer to the original three-dimensional motion. Similarly, the general motion estimation problem is to estimate the motion (projected or parameterised) given one, or more, spatiotemporally varying data sources.

In this text we use the term *motion estimation* to mean any estimate of motion, be that an estimate of the apparent motion, or some parametric motion model. The term 'optical flow estimation' is used only when considering the approximation of the apparent motion. Furthermore, optical flow can be measured as a velocity[42] or displacement[43] though the two representations are equivalent if *temporal aliasing* (see Section 4.1) is avoided – a condition that cannot always be guaranteed. We use the term 'motion' as a substitute for both 'velocity' or 'displacement'. The meaning is clear from the context. Finally, the term 'three-dimensional velocity' is used to differentiate between optical flow and the velocity of an object in the three-dimensional environment.

3.2 Visual motion estimation

Broadly speaking, there are three classes of visual motion estimation algorithms: *gradient based* techniques which operate using image derivatives, *frequency domain* techniques which analyse the image sequence in the Fourier domain and *token based*

---

[42] Measured in pixels/frame – represented by (u, v).
[43] Measured in pixels – represented using $(\delta x, \delta y)$.

techniques which track some image token between frames. To some degree, all these techniques utilise the *Brightness Constancy Assumption* (BCA); they assume that the intensity of light reflected from a point on an object does not change over time so that all changes in the image intensity pattern are due to motion. Thus, before considering specific techniques, it is necessary to analyse the brightness constancy assumption.

3.2.1 Brightness constancy

The light intensity (brightness) captured by a camera at a particular pixel is generally proportional to the amount of light reflected from the corresponding point in the environment [257]. How much light is reflected depends on the *reflectance* of the surface and the prevailing *illumination*. Changes in either illumination or reflectance will cause the BCA to fail.

A simple model [73] of reflectance (see Figure 3.2) defines the total amount of light reflected from a surface as the sum of the *specular* and *Lambertian* reflection. Lambertian surfaces scatter light in all directions equally, hence the reflectance is dependent only on the angle of incidence and distance to the source: it is independent of the viewing angle. Specular surfaces produce mirror-like reflections where the angle of incidence is equal to the angle of reflection.

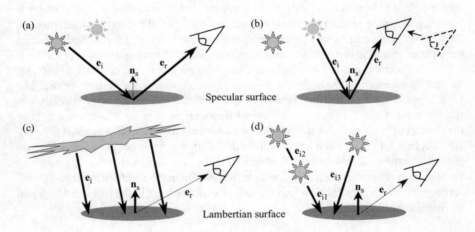

Figure 3.2 *Specular and Lambertian reflection. Parts (a) and (b) show specular surfaces where the angle between surface normal n_s and the incident ray e_i is equal to the angle between n_s and the reflected ray e_r. If the observer or the surface moves, the radiance of a given point on the surface will necessarily change. (d) For Lambertian surfaces the reflected radiance e_r depends on the distance to the light source (e.g. e_{i1}, e_{i2}) and the angle e_i to n_s. (c) If the light source is very diffuse or distant then the angle e_i to n_s will not change with observer or surface motion. In all cases we must assume a stationary light source (adapted from Reference 73)*

Using this simple reflectance model, a stationary, constant light source[44] and a single object, it is necessary to ask under which *motion* conditions the BCA holds. For specular surfaces, the BCA fails under most conditions. Any motion (except rotation of the camera) will change the brightness of a point because the brightness depends on the viewing angle. For a Lambertian surface and a diffuse light source (e.g. sunlight on a cloudy day) brightness will remain stable unless the object rotates, in which case any brightness variation in the object caused by shading will not move with the object. This shading will be fairly soft (if present at all) due to the flat lighting conditions so the problem is minimal.

Between the extremes of specular surfaces and Lambertian surfaces in diffuse light are Lambertian surfaces with a nearby light source, or a point light source (e.g. sunlight on a bright day or a single light bulb). Under these conditions, the angle of incidence or the distance to the light source is likely to change with object motion and, hence, so will the reflected light. Furthermore, any object with a varying surface normal will exhibit strong shading that can skew motion estimates. Shadows are particularly problematic under these conditions since they will be quite dense and can cause the appearance of apparent motion where there is none. If other objects are allowed into the environment the situation becomes more difficult since one object's shadows affect the brightness of other objects. The relationship between reflectance and motion (as illustrated in Figure 3.2) is summarised in Table 3.1.

Specular reflections can cause another problem. For a specular surface, visual motion is not that of the surface itself, but that of the reflected object. If a surface has both specular and Lambertian properties then the motion both of the surface and of reflected objects might be visible at the same point causing confusion for motion estimation techniques. Similar issues also occur for transparent or translucent surfaces. A number of techniques have been devised to deal with such *transparent motions* including the use of frequency domain methods that allow filtering of multiple velocities at a single point (e.g. [68, 69, 93]), and the use of mixture models [152]. However, the majority of motion estimation techniques assume that transparent motion is rare and attempt to find the single most dominant motion for each point given some set of assumptions.

Another important mode of failure of the BCA is that of *occlusion*. At motion boundaries previously invisible regions appear (disocclusion) or existing regions disappear (occlusion). Such regions do not correspond to any region in the previous (disocclusion) or subsequent (occlusion) frame, hence the BCA fails.

Given this discussion, one may well wonder whether it is a little dangerous to use the brightness constancy assumption. The best that can be done is to assume that the BCA usually holds true and to use one, or a combination of the methods listed

---

[44] Illumination conditions can also cause failure of the BCA. For example, a moving light source implies a changing incident angle, which leads to a change in the amount of light reflected from both specular and Lambertian surfaces. Naturally, changes in intensity of the light source will also cause the BCA to fail.

Table 3.1 Relationship between radiance and motion for a single surface (adapted from Reference 73)

Reflectance	Camera motion	Object motion	Radiance
Specular	Stationary	Translation	Variable
Specular	Stationary	Rotation	Variable
Specular	Translation	Stationary	Variable
Specular	Rotation	Stationary	Constant
Lambertian, point	Stationary	Translation	Variable
Lambertian, point	Stationary	Rotation	Variable
Lambertian, point	Translation	Stationary	Constant
Lambertian, point	Rotation	Stationary	Constant
Lambertian, diffuse	Stationary	Translation	Constant
Lambertian, diffuse	Stationary	Rotation	Variable (soft shadows)
Lambertian, diffuse	Translation	Stationary	Constant
Lambertian, diffuse	Rotation	Stationary	Constant

below, to ensure the effect of any failure is minimised. While these techniques can help, none will be 100 per cent effective in all situations.

- Robust statistical techniques to reject patently incorrect data (Section 3.2.8)
- *Illumination flattening* [257]. This method is an effective way of removing variations in lighting with low spatial frequency. It can be applied as a preprocessing step in any motion estimation algorithm; however it is rarely used in practice due to its high computational complexity.
- Prefiltering with a high pass or bandpass filter can minimise the effect of illumination change by removing DC components from images [77, 124].
- Use of models that explicitly allow brightness variation [65, 217]. These are usually used in conjunction with gradient based techniques; hence discussion is postponed until they are considered.
- One can normalise the mean or the variance of a pair of images (or image blocks) to reduce the effect of illumination variation [51]. Normalisation works well for block matching methods because the structure in the block is maintained. With gradient based techniques, care must be taken to ensure the normalisation does not distort the relationship between spatial and temporal derivatives.
- Brightness invariant tokens [16]. Use of edge or corner features has the benefit that these features are largely invariant under illumination changes.
- Phase based techniques. Frequency domain motion estimation techniques that use spatiotemporal phase are relatively insensitive to illumination change [25].

Later, it is shown that the combination of visual and range information provides another means of reducing the effect for one form of brightness variation – shadows.

3.2.2 Background subtraction and surveillance

One application of motion estimation is in surveillance applications where the camera is stationary. In such applications, *image subtraction* and *accumulated difference images* [245] are perhaps the most obvious methods for motion detection, though naïve implementations are often problematic due to the failure of the BCA (especially shadows [189, 229]) and an inability to handle occlusions (e.g. [116, 162]) and camera motion. Further, such methods do not provide velocity information. These problems have been dealt with in a number of ways including robust background subtraction (e.g. [67]) and the use of optical flow (e.g. [117]). Because our application domain (autonomous navigation) implies the use of a moving camera, these techniques are not considered further here.

3.2.3 Gradient based motion estimation

Gradient based motion estimation explicitly begins with the brightness constancy assumption [134]. Mathematically, this can be written as

$$\frac{dI(x, y, t)}{dt} = 0 \tag{3.1}$$

where $I(x, y, t)$ is the image intensity function over space and time. Using the chain rule for differentiation we obtain the total derivative

$$uI_x + vI_y + I_t = 0 \tag{3.2}$$

where (u, v) is the temporal derivative of position and represents the optical flow (as opposed to the apparent velocity represented using (u, v)). Coordinates have been omitted for clarity and subscripts indicate partial derivatives with respect to the subscript variable. Equation (3.2) is often presented in dot product form:

$$\nabla I \cdot (u, v)^T + I_t = 0 \tag{3.3}$$

Noise, failure of the BCA and other errors mean that (3.3) rarely equals zero exactly, hence in the simplest case, solution methods generally aim to find optical flow that minimises the square of this constraint. The optical flow parallel to the intensity gradient is often called the *normal flow* and can be determined using (3.4).

$$\frac{|I_t|}{\sqrt{I_x^2 + I_y^2}} = \frac{|I_t|}{|\nabla I|} \tag{3.4}$$

Equation (3.2) is known as the Optical Flow Constraint Equation (OFCE) and it relates optical flow to intensity derivatives. Note that the OFCE is underconstrained: it only constrains the velocity at a point to lie on a line parallel to the intensity gradient. To obtain a unique solution for motion at a pixel, further constraints must be applied. This is a mathematical manifestation of the *Aperture Problem* (Figure 3.3) where the structure of data in a given region of analysis is insufficient to compute the entire motion.

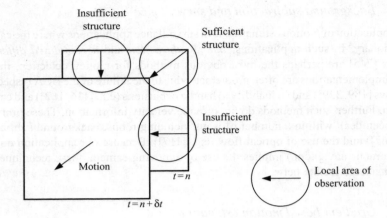

Figure 3.3   *The aperture problem. This name arises from the lack of structure in the local analysis aperture. If image structure is one-dimensional in a given region then motion can only be determined parallel to the intensity gradient. In this example only partial motion information is available at the edges while complete motion is available at the corner where there is 2D structure*

3.2.3.1 Data structure

The accuracy of all motion estimation methods, no matter the data source, depends on the presence of sufficient structure in the data. In visual motion estimation, the effect of a lack of structure is often explained using the example of a textureless[45] rotating sphere [134]. This sphere will have some shading due to lighting conditions, however, this shading will not move with the sphere, so optical flow will be zero despite the fact that the apparent motion field is obviously non-zero.

In regions of no texture, image derivatives I_x, I_y, I_t will all be zero, hence no estimate of motion can be made. Regions of limited structure can also be problematic as was seen with the aperture problem. In this case, there is only structure in a single dimension: motion cannot be estimated perpendicular to this since there are no visual cues to motion in that direction. To minimise this problem the spatial gradient should change from point to point[46]. However, this cannot be guaranteed in practice because featureless regions (ones with little or no spatial gradient) and one-dimensional structures (edges) do occur (e.g. walls).

While sufficient structure is necessary, excessive structure can also be problematic. Our derivation of the OFCE assumes $I(x, y, t)$ is differentiable; however, natural images are rarely continuous and the image must be discretised before

---

[45] Structure in visual data is commonly known as texture.
[46] In other words, if a two-dimensional motion optical flow is required, the data structure should be two dimensional.

being processed. This issue is considered in more detail when we consider aliasing (Section 4.1): for now we assume that sufficient smoothing has been applied to images so that aliasing effects are minimised. Given this discussion, we see that a trade off is required. Sufficient structure should be present to allow unambiguous motion estimation, however, the structure should not be so high as to cause aliasing [71].

3.2.3.2 Constraints for gradient based techniques

As stated, (3.2) is ill-posed[47]. It does not give a unique solution at each pixel hence additional constraints are necessary. Two classes of constraint have been proposed. The first is the local constraint (the regression approach) where an overdetermined set of equations is used to solve for motion. The second class of constraint is the global constraint (the regularisation approach) where some heuristic relationship is enforced over the entire image.

3.2.3.2.1 Local constraints

Local (or regression) methods build a collection of OFCEs in order to find a unique motion estimate. The set of equations can be formed in a number of ways including the use of multispectral images (e.g. the red, green and blue channels of a colour video stream [108]), or from filtered versions of the original sequence (e.g. application of multiple filters [58, 299]). However, the most common approach is credited to Lucas and Kanade [197] who assume motion to be constant over some small region of support R and solve the resulting set of equations using weighted least squares.

$$\sum_R W^2(uI_x + vI_y + I_t)^2 \tag{3.5}$$

Here, W is a window function giving less weight to residuals further from the centre of R. We can find expressions for u and v by minimising (3.5) with respect to u and v.

$$u = \frac{be - cd}{b^2 - ad} \qquad v = \frac{bc - ae}{b^2 - ad} \tag{3.6}$$

where

$$a = \sum_R I_x^2 W^2 \qquad b = \sum_R I_x I_y W^2 \qquad c = -\sum_R I_x I_t W^2$$
$$d = \sum_R I_y^2 W^2 \qquad e = -\sum_R I_y I_t W^2 \tag{3.7}$$

One possible extension to this approach is to use the eigenvalues of the least squares matrix as a confidence measure[48] [254]. This allows estimates from regions where the solution is poorly conditioned to be rejected. In local approaches, a larger region of support will lead to better motion estimates (due to greater effective

---

[47] An ill-posed problem is one where a solution may not exist, may not be unique or does not depend continuously on the data [31].

[48] A confidence measure attempts to indicate how well the measured flow matches apparent motion.

smoothing); however, the larger the region, the greater the risk that multiple motions may be present. Sources of error for local motion estimation methods have been considered in more detail [160].

3.2.3.2.2 Global constraints

Global (or regularisation) approaches overcome ill-posedness by applying some *a priori*, heuristic knowledge to the entire problem. Rather than assuming optical flow is constant over some small area (as in local regression), one might assume that neighbouring points in the image will usually move in a similar manner, hence optical flow may be constrained to change smoothly across the entire image. The nature of such smoothness constraints was discussed [258] with the conclusion that there are in fact four basic smoothness constraints from which all others can be derived. However, the classic example of a global method was presented by Horn and Schunck [134]. They added the Laplacian of the optical flow to the OFCE as a regularisation term, leading to what is sometimes called a membrane model [36]. The resulting equation (3.9) represents the energy stored in the membrane. Motion is estimated by minimising that energy[49].

$$\sum_{image} (uI_x + vI_y + I_t)^2 + \lambda^2 (u_x^2 + u_y^2 + v_x^2 + v_y^2) \tag{3.8}$$

Or alternatively,

$$\sum_{image} (E_d^2 + \lambda^2 E_s^2) \tag{3.9}$$

where the data term E_d corresponds to the OFCE and the smoothness term E_s corresponds to the Laplacian of optical flow. Figure 3.4 illustrates the membrane model schematically. Neighbouring flow estimates are related by an elastic link (hence the term 'membrane'), the strength of which is determined by the regularisation parameter λ. Higher λ leads to a smoother result and, conversely, lower λ gives a result more faithful to the underlying data.

To solve (3.9) for apparent velocity, it must be minimised with respect to u and v. This minimisation can be performed directly by setting the u and v derivatives to zero [134]; however, convergence may be quicker if an alternative minimisation strategy (a gradient descent method such as successive over-relaxation [36]) is used. Horn and Schunck showed that the direct minimisation approach gives the following pair of iterative update functions [134]

$$\begin{aligned} u^{n+1} &= \bar{u}^n - \frac{I_x(\bar{u}^n I_x + \bar{v}^n I_y + I_t)}{\lambda^2 + I_x^2 + I_y^2} \\ v^{n+1} &= \bar{v}^n - \frac{I_y(\bar{u}^n I_x + \bar{v}^n I_y + I_t)}{\lambda^2 + I_x^2 + I_y^2} \end{aligned} \tag{3.10}$$

---

[49] That is, motion estimates are calculated by simultaneously ensuring both a minimal OFCE and a minimal departure from smoothness.

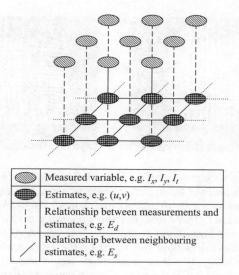

Figure 3.4   Schematic view of a global constraint (adapted from Reference 96)

Here superscripts refer to the iteration number and $(\bar{u}, \bar{v})$ are the local averages of velocity (excluding the current pixel). Because global constraints require iterative solutions, they incur a significantly higher computational cost compared to local methods. The advantage of global methods is that they have the potential to provide more accurate motion estimates since estimates can propagate from regions of low structure to regions of high structure. For example, Figure 3.5a shows that the local method is affected by the aperture problem at the edges of the square and is unable to propagate any result to the squares interior. In contrast, the global method (Figure 3.5b) is less affected by the aperture problem and motion has been propagated to the square's interior. While the propagation is not complete, the result is certainly closer to reality than the local constraint provides.

3.2.3.3   Derivative estimation

Accurate derivative estimates are vital when attempting to obtain accurate motion estimates from methods based on the OFCE. On a uniform sampling grid, derivatives can be estimated up to an arbitrary accuracy using the equation [101]

$$I'(n) = \frac{1}{h}\left(\Delta I(n) - \frac{1}{2}\Delta^2 I(n) + \frac{1}{3}\Delta^3 I(n) - \cdots \pm \frac{1}{n}\Delta^n I(n)\right) \tag{3.11}$$

where it is assumed that I(n) is a single-dimensional (image) function, and h is the spacing between elements of I(n)[50]. Δ is a difference function; the *forward difference*

---

[50] Here, we assume spacing to be one pixel. Hence h = 1 or h = ζ depending on whether derivative is taken with respect to pixel address or physical pixel position. ζ is the physical inter-pixel spacing.

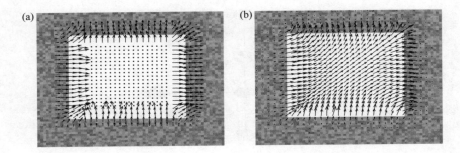

Figure 3.5 Dealing with regions of low structure. In this example the background is static and the featureless square moves with velocity (1,1). (a) is the flow generated by the Lucas and Kanade algorithm (9 × 9 rectangular window). There is significant smoothing across motion boundaries due to the large window, however, the effects of insufficient structure are clear. (b) is flow generated by the Horn and Schunck algorithm ($\lambda^2 = 100$, 500 iterations). Once again there is smoothing across motion boundaries, however, motion estimates have propagated inside the white square, minimising error for this region of low structure

Δ_f is defined as

$$\Delta_f I(n) = I(n+1) - I(n) \tag{3.12}$$

Higher order forward differences are a straightforward extension of this, for example

$$\Delta_f^2 I(n) = \Delta_f(\Delta_f I(n)) = \Delta_f I(n+1) - \Delta_f I(n) \tag{3.13}$$

Using only the first order term of (3.11) the *first-order forward-difference* (FOFD) approximation of the derivative is obtained:

$$I'(n) \approx \frac{1}{h}\Delta_f I(n) + \varepsilon = \frac{1}{h}[I(n+1) - I(n)] + \varepsilon \tag{3.14}$$

In this approximation, the error term ε is O(h). Rewriting this equation as a convolution mask (and assuming h = 1 pixel) gives

$$[-1 \quad 1] \tag{3.15}$$

Using a second order approximation reduces the error to O(h^2); however, the resulting estimate requires three elements of data from the input sequence. A simpler estimate of the derivative with similar error can be obtained using a combination of forward- and backward-difference operations. Using the *backwards-difference* operator

$$\Delta_b I(n) = I(n) - I(n-1) \tag{3.16}$$

the *first-order backward-difference* (FOBD) estimate of the derivative is obtained:

$$I'(n) \approx \frac{1}{h}\Delta_b I(n) + \varepsilon = \frac{1}{h}[I(n) - I(n-1)] + \varepsilon \tag{3.17}$$

To obtain a smoothed estimate of the derivative, take the average of (3.14) and (3.17) giving

$$I'(n) \approx \frac{I(n+1) - I(n-1)}{2h} \qquad (3.18)$$

Equation (3.18) is known as the *first-order central-difference* (FOCD) estimate of the derivative since it involves the data elements either side of the element currently under consideration. Because second order terms cancel when taking the average, this estimate has an error of order h^2. The equivalent convolution mask (assuming $h = 1$) for this estimator is

$$[-0.5 \quad 0 \quad 0.5] \qquad (3.19)$$

While higher order estimators can be used to reduce the error it should be remembered that derivatives are only valid in regions where the underlying dataset $I(n)$ is continuous. Natural datasets often contain discontinuities (e.g. edges in visual data) hence erroneous derivatives are to be expected. This becomes more of a problem with higher order derivative estimators since they use more elements from the dataset $I(n)$ and consequently will spread errors due to discontinuities further. Smoothing reduces the error caused by discontinuities, however, it also makes edges more difficult to localise.

The trade off between accuracy and error spread has been handled in a number of ways. Typical of many authors, Barron *et al.* [25] presmooth the image sequence with a spatiotemporal Gaussian filter to smooth discontinuities, then apply a five-tap central-differences filter (3.20) to estimate spatiotemporal derivatives:

$$\frac{1}{12}[-1 \quad 8 \quad 0 \quad -8 \quad 1] \qquad (3.20)$$

This derivative estimator is the most accurate of a set of three estimators [226], with error $O(h^4/30)$. The remaining two estimators have slightly larger error: $O(h^4/5)$ and $O(h^4/20)$ respectively.

$$\frac{1}{12}[-25 \quad 48 \quad -36 \quad 16 \quad -3]$$
$$\frac{1}{12}[-3 \quad -10 \quad 18 \quad -6 \quad 1] \qquad (3.21)$$

Simoncelli [253] includes smoothing in the derivative operator since 'it is easier to design the derivative of a prefilter than it is to design a prefilter and a separate derivative filter'. Using this approach he generates two-, three- and five-tap filters for derivative estimation (3.22):

$$\begin{array}{c} [-1 \quad 1] \\ [-0.4414 \quad 0 \quad 0.4414] \\ [-0.1081 \quad -0.2699 \quad 0 \quad 0.2699 \quad 0.1081] \end{array} \qquad (3.22)$$

The three-tap estimator is simply a scaled version of the FOCD estimator (3.19). This scaling has no effect on the motion estimates since all derivatives were scaled equally. Common edge detection masks such as the Sobel and Prewitt (3.23)[51] filters also provide a degree of smoothing via the weighted spatial average of three first-order central-difference estimates.

$$\underbrace{\frac{1}{4}\begin{bmatrix}-1 & 0 & 1\\ -2 & 0 & 2\\ -1 & 0 & 1\end{bmatrix}}_{\text{Sobel}} \qquad \underbrace{\frac{1}{3}\begin{bmatrix}-1 & 0 & 1\\ -1 & 0 & 1\\ -1 & 0 & 1\end{bmatrix}}_{\text{Prewitt}} \tag{3.23}$$

The derivative operator can also smooth in the temporal domain as Horn and Schunck [134] demonstrated. They calculated derivatives in $2 \times 2 \times 2$ spatiotemporal blocks essentially taking the average of four FOFD derivative estimates. For example, the derivative with respect to y is calculated as

$$I_y \approx \frac{1}{4}\left\{\begin{array}{l}[I(x,y,t) - I(x,y+1,t)] + [I(x+1,y,t) - I(x+1,y+1,t)] \\ +[I(x,y,t+1) - I(x,y+1,t+1)] \\ +[I(x+1,y,t+1) - I(x+1,y+1,t+1)]\end{array}\right\} \tag{3.24}$$

A performance comparison of these estimators is left until Section 5.2.3 where an assessment of the estimators in terms of the specific motion estimation algorithm developed in this book is conducted.

3.2.4 Displaced frame difference

The *Displaced Frame Difference* (DFD) is an alternate way of stating the brightness constancy assumption. Rather than stating that the derivative of intensity is zero, we begin by stating that the brightness of a point does not change as the point moves:

$$I(x,y,t) = I(x + \delta x, y + \delta y, t + \delta t) \tag{3.25}$$

Noise, failure of the BCA and other factors mean that this relationship rarely holds exactly. In the simplest case[52], (3.25) is rewritten as the DFD[53]:

$$\text{DFD}(x,y,\delta x,\delta y,t) = I(x + \delta x, y + \delta y, t + \delta t) - I(x,y,t) \tag{3.26}$$

---

[51] Convolution masks for derivatives in the x direction are shown. Masks for y and t derivatives can be obtained by rotating these masks appropriately.
[52] More complex cases are considered in the next section.
[53] We use the term DFD to refer to the displacement of any subset of the image, however, strictly speaking term DFD refers to a global displacement of the image. If subblocks of the image are used as regions of support, this is called the Displaced Block Difference and for individual pixels the term Displaced Pixel Difference is sometimes used [174].

Table 3.2 Applicability of the DFD and OFCE methods

	Small δt	Large δt
OFCE	Directly applicable.	Spatial subsampling (scale space – Section 4.4) required to avoid temporal aliasing.
DFD	Requires image upsampling to measure subpixel displacements.	Interpolation required to determine non-integer displacements.

Assuming $I(x, y, t)$ is differentiable and taking the Taylor expansion of $I(x + \delta x, y + \delta y, t + \delta t)$ in (3.26) about point (x, y, t) gives

$$\text{DFD}(x, y, \delta x, \delta y, t) = \delta x \frac{\partial I(x, y, t)}{\partial x} + \delta y \frac{\partial I(x, y, t)}{\partial y} + \delta t \frac{\partial I(x, y, t)}{\partial t} \quad (3.27)$$

where it has been assumed that $I(x, y, t)$ is locally linear, hence second and higher order terms are negligible. Assuming the DFD is zero, dividing through by δt and taking the limit as $\delta t \to 0$ gives

$$u I_x + v I_y + I_t = 0 \quad (3.28)$$

which is the OFCE (3.2) thus the DFD and the OFCE are equivalent[54] for small δt. For large δt, displacements may be large which can lead to temporal aliasing (see Chapter 4) in which case the derivatives in the OFCE fail. Conversely, for small δt displacements will be small (<1 pixel), so upsampling of the image is required to estimate motion using the DFD (e.g. [58]). A summary of the applicability of OFCE and DFD methods is presented in Table 3.2.

There are three ways of solving for optical flow using the DFD [278]. Each method seeks to find a value for optical flow that gives a minimum aggregate DFD over some region as follows:

$$\min_{(\delta x, \delta y) \in R_s} \sum_{(x,y) \in R} \rho(\text{DFD}(x, y, \delta x, \delta y, t)) \quad (3.29)$$

Here ρ is an estimator function, R is the region of support, and R_s is the set of candidate displacements. This aggregation reduces sensitivity to noise [271] by smoothing the results over a region of space. Using a large region of support gives more effective noise reduction, however, using larger regions increases the likelihood of their being more than one motion within that region. Typically $\rho(r) = r^2$ which leads to a least squares problem, though it is common to use $\rho(r) = \text{abs}(r)$ since this estimator

---

[54] Notice also that correlation/block based motion estimation algorithms that use a local sum of the DFD become equivalent to the Lucas and Kanade gradient based motion estimation algorithm.

(the least absolute residuals estimator) can give superior results while reducing computational complexity [271]. However, we saw in Chapter 2 that these estimators are not robust, hence it may be more appropriate to choose a robust estimator function for $\rho(r)$. We discuss this idea more in Section 3.2.8.

The first means of solving for optical flow using the DFD is the classic block matching strategy, where the displaced frame difference becomes the displaced block difference (DBD – see Section 3.2.6). A second way of obtaining optical flow from the DFD is to assume $\delta t = 1$ (frame) and solve the set of equations generated by (3.27) over the region of support R. This assumes small motions so that the derivatives are valid and hence is effectively the same as the Lucas and Kanade solution to the OFCE.

The final means of using the DFD to calculate flow is to assume the DFD equals zero and solve (3.27) using what is referred to as a *pel* recursive strategy [222, 278]. Pel recursive algorithms are causal – they operate one pixel at a time and only have access to a small number of *previously processed* pixels (i.e. the region of support is a queue of the n most recent motion estimates). They aim to minimise the square of (3.27) via a procedure such as the steepest descent algorithm [230] (see Section 2.2.6.1).

The steepest descent method is an iterative means of finding the parameters (θ) that minimise a function. In this algorithm, the result of the current iteration (θ^n) is based on a prediction related to the previous result ($\hat{\theta}^{n-1}$) minus an update term related to the gradient of the function being minimised. This algorithm essentially steps us 'down the slope' of the function in steps of size λ.

$$\theta^n = \hat{\theta}^{n-1} - \lambda \nabla_\theta \rho(f(\theta^{n-1})) \qquad (3.30)$$

In this case the function ρ must be convex so that the minimisation is not stuck in a local minimum. For pel recursive motion estimation, θ is the optical flow ($\delta x, \delta y$) and $f(\theta)$ becomes the square of (3.27). The update term effectively enforces the brightness constancy assumption while the prediction term enforces smoothness since it is possible to use the motion estimate at the previous pixel as the initial prediction for current pixel. Note that the Horn and Schunck algorithm takes on a similar form, however, it is not pel recursive since it is not causal (it can use pixels 'ahead' of the current pixel). A straightforward implementation of the pel recursive method is known as the *Netravali–Robbin algorithm* [278] while the *Walker–Rao algorithm* is a slightly modified version of this algorithm that uses an adaptive step size and an update term that is set to zero if the DFD using the current motion estimate is 'near enough' to zero or if the spatial gradient is too low. In Reference 278 it is claimed that the adaptive step size of the Walker–Rao algorithm allows convergence in as few as five iterations.

Rather than assuming motion to be smooth in a local area (as was implicitly done by taking the local velocity average as the prediction term for (3.30)) it is possible to follow Lucas and Kanade's approach and assume that motion is constant within some causal region of support R.

$$\theta^n = \hat{\theta}^{n-1} - \lambda \sum_{(x,y) \in R} \nabla_\theta \rho(\mathrm{DFD}(x, y, \theta^{n-1}, t)) \qquad (3.31)$$

This can be solved using a Netravali–Robbin type approach or by using Wiener estimation to minimise the update term in (3.31) [127, 278].

3.2.5 Variations of the OFCE

In this section a number of variations and alternatives to the classic OFCE are considered. We begin with an alternative proposed by Schunck [248] (3.32). He argued that a term explicitly modelling optical flow divergence[55] should be added to the OFCE leading to the following variant of the OFCE:

$$I\nabla \cdot (u, v) + \nabla I \cdot (u, v) + I_t = 0 \qquad (3.32)$$

This variant is rarely used since divergence due to relative motion towards or away from the camera is usually quite small (see Section 4.2). The divergence does grow to infinity at the edges of rotating objects, however, this effect tends to be very localised.

While (3.32) extends the OFCE to model motions more complex than simple translation, a more controlled way of achieving this is to explicitly model motion using a parametric model. Such models can be designed with specific assumptions regarding object motion (e.g. *translational*, *affine* or *rigid motion*) [15, 271] and can be designed to take into account *image formation models* (e.g. *orthographic*, *weak perspective*, *projective* models) [304]. For example, affine motion under orthographic projection can be modelled using six parameters (a_1, a_2, a_3, b_1, b_2, b_3) as in (3.33). This affine model models all motion as linear transformations of the image. Explicit modelling of acceleration can also be useful [271].

$$\begin{bmatrix} u \\ v \end{bmatrix} = \begin{bmatrix} a_1 & a_2 \\ b_1 & b_2 \end{bmatrix} \begin{bmatrix} x \\ y \end{bmatrix} + \begin{bmatrix} a_3 \\ b_3 \end{bmatrix} \qquad (3.33)$$

A fundamental limitation of any model using purely visual data is that motion at a given point can only be estimated up to a scale factor equal to the depth at that point. This will become clearer when the perspective rigid body motion model is discussed in the following chapter.

Gupta developed another alternative to the OFCE [113, 114]. He reformulated the OFCE in integral form by integrating the OFCE and applying Gauss's divergence theorem to eliminate intensity derivatives, leading to the surface and volume integrals in (3.34). The advantage of this form is that it does not depend on numerical differentiation of intensity. Gupta fits an affine motion model to his constraint in order to overcome the aperture problem:

$$\int_s u I \, dy \, dt + \int_s v I \, dx \, dt + \int_v I(u_x + v_y) \, dx \, dy \, dt + \int_s I \, dx \, dy = 0 \qquad (3.34)$$

Modifications to the OFCE can also be designed to allow some degree of departure from the BCA. For example, one might assume constancy of the brightness gradient

---

[55] As occurs for objects approaching the camera, or for rotating objects.

rather constancy of brightness itself [289]. That is

$$\frac{d\nabla I}{dt} = 0 \tag{3.35}$$

This assumption places stronger constraints on allowable motion than the BCA since it does not allow rotation or dilation, which change the intensity gradient. In conditions where this assumption holds, optical flow is found using

$$\begin{bmatrix} \frac{\partial^2 I}{\partial x^2} & \frac{\partial^2 I}{\partial x \partial y} \\ \frac{\partial^2 I}{\partial x \partial y} & \frac{\partial^2 I}{\partial y^2} \end{bmatrix} \begin{bmatrix} u \\ v \end{bmatrix} + \frac{\partial \nabla I}{dt} = 0 \tag{3.36}$$

This equation requires the use of 2nd derivatives which may be difficult to compute due to the high pass nature of derivative filters; however, a number of implementations using this constraint have appeared in the literature [25, 308] and report good results. Local constraints are not required to solve for optical flow using this formulation, however, the resulting motion estimates often needs smoothing [308]. Because of this, Treves and Konrad [280] extended this general method to use the concept of displaced gradient difference so that smoothing could be performed via regularisation.

Cornelius and Kanade [65] proposed an alternative way of dealing with brightness variation (due to illumination changes, shadows, etc.). They added simple additive offset C to the OFCE representing brightness change for each image pixel:

$$uI_x + vI_y + I_t = C \tag{3.37}$$

They solve for both velocity and the offset using a global method where smoothness is enforced for both the optical flow field and the brightness change field. Extending this, Negahdaripour [217] suggests a generalised brightness change model where deviation from brightness constancy[56] is modelled as a linear transformation:

$$M(x, y, t)I(x, y, t) + C(x, y, t) = I(x + \delta x, y + \delta y, t + \delta t) \tag{3.38}$$

This leads to a constraint equation similar to (3.37), except that the offset C is now a linear transformation of intensity. Since M and C can vary from pixel to pixel they are called the multiplicative and additive brightness offset fields respectively:

$$uI_x + vI_y + I_t = M_t I + C_t \tag{3.39}$$

Negahdaripour [217] solves this equation for velocity and for M_t and C_t (temporal derivatives of the M and C fields respectively) using a local approach. Mukawa [209] reaches a similar constraint through the use of an explicit optical model incorporating light sources. While this constraint can be used to explain the effect of spatiotemporally varying illumination, the effect is clearer if we reformulate this constraint slightly. Rather than modelling brightness change as a linear variation in the DFD

---

[56] Here, brightness constancy is modelled as the Displaced Frame Difference – this is discussed in Section 3.2.4.

it can be modelled directly by assuming measured intensity is linearly related to the 'true' intensity – that is, the intensity that would be measured if the BCA held true [115]. We write this as

$$I = MI + C \qquad (3.40)$$

where I is the measured intensity and I is the 'true' intensity. The spatiotemporal derivatives of intensity become

$$I_x = MI_x + M_x I + C_x$$
$$I_y = MI_y + M_y I + C_y \qquad (3.41)$$
$$I_t = MI_t + M_t I + C_t$$

and these derivatives can be substituted into the OFCE yielding

$$u(MI_x + M_x I + C_x) + v(MI_y + M_y I + C_y) + (MI_t + M_t I + C_t) = 0 \quad (3.42)$$

Rewriting (3.42) in a form similar to (3.37) and (3.39), where brightness variation is modelled as an offset to the OFCE, yields

$$uI_x + vI_y + I_t = \frac{-u(M_x I + C_x) - v(M_y I + C_y) - (M_t I + C_t)}{M} \qquad (3.43)$$

This form shows more clearly the effect of an illumination gradient. Consider the motion of a rotating textured sphere in an environment where lighting causes strong shading of the spheres surface. In this case the temporal derivatives of the M and C fields are zero, however, the spatial derivatives are non-zero and this leads to a velocity dependent offset in the velocity. This makes sense – if the sphere were to rotate faster, points on its surface would move through a greater section of the illumination gradient causing the illumination gradient to have a larger effect on the measured velocity.

3.2.6 Token based motion estimation

In *token based motion estimation* methods, image tokens selected from one frame are matched to tokens in a subsequent frame to measure the apparent displacement between frames. Such methods have three key components: selection, search and matching. The selection method defines which tokens will be used, the search method defines how the search for matching tokens is undertaken, and the matching method gives a measure of how similar a candidate is to the template token. Token based methods are unable to measure subpixel motion directly since features can only be localised with pixel accuracy. If subpixel accuracy is required interpolation is used [107]. Two token based techniques are considered here – block based and feature based.

Feature based methods rely on the extraction of 'interesting' features such as points or edges (see Reference 92 for a range of possibilities) and subsequent tracking of the features and segmentation of the resulting trajectories. Since they make an explicit high level decision regarding 'interest', feature based methods run against the philosophy of extracting motion as directly as possible from the data. For this reason these algorithms are only briefly considered here. Authors who use feature based methods often argue that methods based on dense optical flow calculation implicitly rely on feature points (e.g. the need for 'sufficient' structure to avoid the aperture problem) and fail where there is no texture. The implication is that it is a waste of time trying to determine motion away from feature points. This ignores the fact that images with sufficient structure to allow application of feature point methods generally have sufficient structure for dense optical flow estimates over a broad part of the image. Regularisation based methods have the advantage of being able to smooth motion estimates across small untextured regions.

If appropriate features are used, feature based methods have the potential advantage of stability under brightness change however the use of correlation and explicit functions of intensity as an initial indication of match in many algorithms reduces this advantage [256, 285]. This is usually compensated by feature tracking (e.g. [149]), which allows prediction of a feature's future location thus leading to a reduction in the search space and the potential for match ambiguity. This tracking algorithm must be robust to the appearance and disappearance of feature points whether caused by failure of the feature extractor or by motion. Feature based methods have the added advantage that they do not operate on the entire image; hence they have the potential to operate very quickly.

Smith [256] describes the ASSET-2 system where feature tracking is performed in the 2D image domain allowing real-time motion estimation and segmentation (25 Hz). Rather than using a simple 2D motion model, Torr [285] uses the fundamental matrix and the trifocal tensor to impose a rigid motion model. Wiles [304] uses a rigid motion assumption together with a hierarchy of camera models (this hierarchy is formulated both for arbitrary motion for ground plane motion) where the most appropriate camera model for a given scenario is chosen dynamically. This allows for graceful degradation – if a full solution cannot be obtained, a partial solution is given.

Block based[57] (or *correlation based*) motion estimation methods are similar to feature based approaches except that tokens are not chosen for their interest – simple rectangular blocks of pixels are used instead. Block based methods take each token in the first frame and a search for a matching token in a subsequent frame (see Figure 3.6), however block based methods tend not to explicitly track tokens over time since the blocks have no specific meaning in the context of the image as do 'features'.

There are two common approaches for selecting tokens in block based approaches. To generate a motion estimate for each pixel, a token can be selected centred on each pixel (Figure 3.7a); however, the computational cost of this approach is relatively high. The alternative is to divide the image into what are called macroblocks. Each macroblock is used as a token and the motion estimate determined for this token is assigned to each pixel in the macroblock (Figure 3.7b). This approach is often used

---

[57] Note that the phase correlation approach described in the next section also falls into this category.

Motion estimation 83

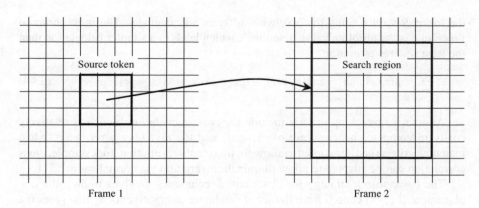

Figure 3.6   Feature and block based motion estimation take a source token (a block is illustrated here though any token can be used) from the first frame and attempt to find a matching token in a subsequent frame. To reduce computational complexity the search region is generally limited to a subset of the image. The distance between the centre of the source token and the centre of the best match in the subsequent frame defines the motion

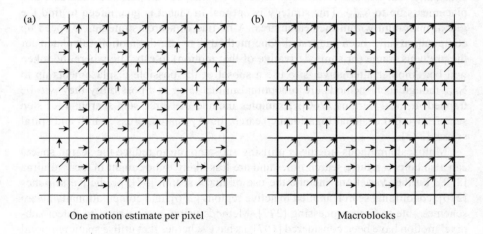

Figure 3.7   Choosing tokens in block based motion estimation. The two usual strategies are to (a) take a token centred on every pixel in the image leading to a motion estimate that can vary from pixel to pixel or (b) to divide the image into macroblocks. Each macroblock is treated as a single token and the motion estimated for the token is assigned to each pixel in the macroblock

in video coding systems (the most prolific user of block based motion estimation techniques), where speed is often critical [84].

The degree of match between a pair of tokens is usually measured using the Displaced Block Difference (DBD) (3.44). The DBD sums the function $\rho(r)$ over the token R with r being the pixelwise difference in intensity. The most common function ρ is the absolute value function[58], which leads to a simpler calculation than the least squares estimator:

$$\sum_{(x,y)\in R} \rho(r) = \sum_{(x,y)\in R} \rho(I(x,y,t) - I(x+\delta x, y+\delta y, t+\delta t)) \quad (3.44)$$

Alternative matching criteria include the *cross correlation* (which is relatively costly to compute, hence is not often used) and the *matching pixel count*. Most matching criteria can be adapted to allow for illumination variation, for example, cross correlation can be adapted to allow illumination variation via normalisation [51].

The basic search strategy for block based techniques involves finding the displacement $(\delta x, \delta y)$ chosen from the set of candidate displacements R_s that gives the minimal DBD:

$$\min_{(\delta x, \delta y) \in R_s} \sum_{(x,y)\in R} \rho\{I(x,y,t) - I(x+\delta x, y+\delta y, t+\delta t)\} \quad (3.45)$$

Usually the maximum displacement $\delta_{\max}$ in R_s is limited based on the maximum expected displacement, thus reducing computational complexity and match ambiguity. The *full search* strategy arises if R_s includes all possible candidate displacements up to $\delta_{\max}$. This strategy is optimal in that it is guaranteed to find the global minimum within the search area. A myriad of search strategies to speed up computation have been suggested. One method is to use the motion calculated for the previous frame (or some derivative of that motion) to estimate where the token will be. Another alternative is to use a subset of the possible displacements up to $\delta_{\max}$ though such methods are suboptimal in that there is a possibility they will be trapped in local minima [260]. Examples include the three step search, four step search, two-dimensional logarithmic search, binary search, cross search and spiral search [287].

Multiscale methods (methods utilising subsampling to reduce the search space) can also help to reduce search times and are less likely to be caught in local minima [260]. Further variations include the use of fuzzy search [57], predictive searches [276], conditional search applied to active regions [236], and computationally aware schemes able to limit processing [277]. Methods using interpolation to detect sub-pixel motion have been considered [107], as have schemes that utilise spatiotemporal correlation [54] and schemes that use both block matching and global smoothness leading to algorithms similar in concept to the membrane model [29, 287].

One limitation of block based schemes is that they are unable to deal with rotation and deformations such as scaling since they implicitly assume locally constant motion. A possible mechanism to alleviate this problem is to use a deformable mesh to search in a higher-dimensional space [13, 278]. For example, one may use dimensional affine

---

[58] The resulting summation is often called the 'Sum of Absolute Differences' or SAD. If the square function is used the summation is called the 'Sum of Squared Differences' or SSD.

space rather than the standard two-dimensional translational space. Of course, such techniques greatly increase computational complexity.

3.2.7 Frequency domain motion estimation

The category of *frequency domain motion estimation* encompasses all techniques that utilise a frequency domain representation of an image sequence. Evidence that mammalian visual systems contain neurons acting as spatiotemporal bandpass filters [293, 298, 300] has driven much of the research on these methods. Frequency domain techniques can give accurate motion estimates [25], however, they tend to be computationally expensive since a large array of filters is required to sample the frequency domain properties of an image sequence.

Like most authors (e.g. [205]), we begin by considering the simplest case where two images (or image regions) I_1 and I_2 are related by a global displacement

$$I_1(x, y) = I_2(x - ut, y - vt) \tag{3.46}$$

This is a variant of the brightness constancy assumption (3.25). In the Fourier domain, displacement becomes phase shift via the shifting property of the Fourier transform. Denoting the Fourier transform of I as $\hat{I}$ we can write [205]

$$\hat{I}_1(\omega_x, \omega_y) = e^{-j2\pi(u\omega_x + v\omega_y)t} \hat{I}_2(\omega_x, \omega_y) \tag{3.47}$$

Here (ω_x, ω_y) are the 2D spatial frequencies. Writing the phase difference explicitly we have

$$\arg\{\hat{I}_1(\omega_x, \omega_y)\} - \arg\{\hat{I}_2(\omega_x, \omega_y)\} = -2\pi(u\omega_x + v\omega_y)t \tag{3.48}$$

which allows solution for velocity in a manner similar to the local constraint for gradient based motion estimation. One may sample phase at a number of frequencies and solve the resulting equations, however, phase unwrapping can be problematic. Approximate solutions of (3.48) are also possible [307]. Another possible solution is to use the phase correlation function. In the Fourier domain, correlation becomes multiplication and the normalised cross-power spectrum can be written as [278]

$$\hat{c}(\omega_x, \omega_y) = \frac{\hat{I}_1(\omega_x, \omega_y)\hat{I}_2^*(\omega_x, \omega_y)}{|\hat{I}_1(\omega_x, \omega_y)\hat{I}_2^*(\omega_x, \omega_y)|} \tag{3.49}$$

where the superscript * indicates complex conjugate. Substituting (3.47) into (3.49) and taking the inverse Fourier transform gives the phase correlation function

$$c(x, y) = \delta(x - ut, y - vt) \tag{3.50}$$

Thus, velocity can be determined by searching for an impulse in the phase correlation function. If more than one motion occurs in the region under consideration, there will be one impulse for each motion, hence this technique can be applied where transparent or multiple motion occurs. Issues surrounding the use of this equation (boundary effects, spectra leakage etc.) have been discussed [174, 278]. Frequency domain techniques can also explicitly include the temporal domain if the static image function I_2 is given a temporal parameter:

$$I_2(x, y, t) = I_2(x, y, 0) = I_2(x, y) \qquad \forall t \in \delta t \tag{3.51}$$

This is an image that does not change over the time interval δt. Assuming velocity is constant over this period the following result is obtained:

$$I_1(x, y, t) = I_2(x - ut, y - vt, t) \tag{3.52}$$

The 3D Fourier transform of (3.52) can be obtained from (3.51) and (3.47) [245]:

$$\hat{I}_1(\omega_x, \omega_y, \omega_t) = \int_{-\infty}^{\infty} \{e^{-j2\pi(u\omega_x + v\omega_y)t} \hat{I}_2(\omega_x, \omega_y)\} e^{-j2\pi\omega_t t} dt$$

$$= \hat{I}_2(\omega_x, \omega_y) \int_{-\infty}^{\infty} e^{-j2\pi(u\omega_x + v\omega_y + \omega_t)t} dt$$

$$= \hat{I}_2(\omega_x, \omega_y) \delta(u\omega_x + v\omega_y + \omega_t) \tag{3.53}$$

Thus the spatiotemporal Fourier transform is only non-zero in a plane defined by

$$u\omega_x + v\omega_y + \omega_t = 0 \tag{3.54}$$

This is an analogue of the OFCE in the Fourier domain (indeed, this result can be obtained directly from the OFCE [253]) and can be solved for optical flow by using appropriate filters to sample frequencies and solving the over-determined set of equations (e.g. [253]). When using such an approach special care must be taken to negate a linear filter's contrast dependence [28, 253]. Variations of this method are often presented as models of biological visual systems [205, 298].

Methods using zero crossing techniques have also been considered as frequency based (or more specifically, *phase based*) motion estimation techniques since 'zero crossings can be viewed as level-phase crossings' [25]. Phase based techniques are more robust to failure of the BCA than other methods since phase remains relatively unperturbed by changes in brightness. For example, Perrone [228] uses an array of closely spaced second derivative operators to identify the location of intensity edges and tracks these edges over time to estimate optical flow. Another phase based method is that of Fleet and Jepson [93] who use what is essentially a phase constancy assumption rather than a brightness constancy assumption to determine normal flow. The normal flow is then integrated over each local region using a linear motion model so that the total optical flow is obtained.

As with the gradient based techniques, numerous variations of the basic frequency domain techniques have been suggested. For example, Simoncelli [253] extends the basic method of (3.54) to create an algorithm that presents motion estimates as a probability distribution of likely motions rather than as simple motion vectors. Another variation is that of Ogata and Sato [223] who use a two-stage algorithm where a frequency domain motion estimate (based on the model of

Watson and Ahumada [298]) provides a rough initial approximation of motion. This is then refined by a feature tracking based method that uses this initial estimate to reduce match ambiguity and reduce the computation required to find correspondences. Justification for use of a two-stage algorithm is given from experiments on the human visual system that suggest that separate short-range motion estimation (i.e. small temporal support, using brightness pattern directly) and long-range motion estimation (i.e. iconic matching) coexist in the brain [66]. A similar approach is suggested by Konrad [174] who suggests the use of the phase correlation function as a preprocessing step giving the likely motions within a region. Correlation based techniques then use these initial estimates to reduce the search space and refine the result using an interpolated search.

Simoncelli [253] shows that under certain conditions, frequency domain techniques are in fact related to gradient based techniques with local constraints. Thus, in general, a lack of spatial structure affects both classes of motion estimators similarly.

3.2.8 Multiple motions

The motion estimation techniques discussed so far have made strong assumptions regarding motion: motion is either globally smooth, or locally constant with respect to some motion model. In most practical situations these assumptions are violated since environments generally contain *multiple independently moving objects*, or objects at different depths that induce different flow due to camera motion. Because objects usually move coherently, the assumptions of smoothness or local constancy will hold piecewise through the image[59], however, the occlusion and disocclusion that occurs at motion boundaries will cause both the brightness constancy assumption and image derivatives (if used) to fail. Thus the motion estimation process could be improved by smoothing over the largest possible area (as defined by motion boundaries) without smoothing across motion boundaries. This section briefly reviews some of the techniques that have been proposed to solve this problem. While the majority of this presentation assumes estimation of optical flow, many of the techniques can be extended for use with alternate (higher-dimensional) motion models.

3.2.8.1 Preserving discontinuities

A naïve approach to finding piecewise smooth motion is to attempt segmenting the optical flow directly. While authors have considered this approach (e.g. using histograms of normal flow [264]) there is a distinct 'chicken and egg' [135] aspect to the problem of segmenting optical flow. To be able to segment optical flow, the optical flow must not be smoothed across motion boundaries, but how can one prevent smoothing across motion boundaries if their location is unknown?

---

[59] While we consider globally smooth motion and piecewise smooth motion (also referred to as smoothness in layers) here, another class of motion known as optical snow can also occur. Such motion is characterised by the presence of motion discontinuities everywhere, much like one might see with snow or in very cluttered environments [184].

If prevention of smoothing across motion boundaries is more desirable than explicit modelling of boundaries, a number of *discontinuity preserving* motion estimation algorithms are available. In the domain of global smoothness constraints, Nagel [214] proposed a constraint where smoothing is suppressed across steep intensity gradients:

$$\sum_{\text{image}} (uI_x + vI_y + I_t)^2 + \frac{\beta^2}{|\nabla I|^2 + 2\lambda} \times \begin{bmatrix} (u_x I_y - u_y I_x)^2 + (v_x I_y - v_y I_x)^2 \\ + \lambda(u_x^2 + u_y^2 + v_x^2 + v_y^2) \end{bmatrix}$$

(3.55)

where λ is the regularisation parameter as before (which controls the degree of smoothing) and β allows us to control the relative attenuation of smoothing. An obvious limitation of any such approach is that, for the most part, the presence of an intensity edge does not imply the presence of a motion discontinuity; therefore the motion will tend to be oversegmented. However, the presence of an intensity edge could be used as a confirmation of the presence of a motion edge computed using other means (e.g. [175, 282]). Rather than preventing smoothing of motion estimates across intensity edges, Deriche, Kornprobst and Aubert [72] replace the isotropic smoothness term in Horn and Schunck's algorithm with an anisotropic smoothing term that does not smooth along the motion gradient. Proesmans *et al.* [234] compute flow using both a global gradient based technique and a block matching motion estimation technique and detect motion edges by checking for inconsistencies between the estimates. This inconsistency is then fed back to adjust the regularisation parameter of the gradient based algorithm to reduce smoothing across detected boundaries.

There are also a number of ways to preserve discontinuities when using local methods. For example, one might use the quality of the fit in a regression algorithm to detect discontinuities [104]. Alternatively, the size or shape of the region of interest can be changed so that the region of interest conforms to motion boundaries and hence does not smooth across them [261]. Another approach with an adaptive region of support based on Adaptive Structure Tensors has been presented [195]. Schunck proposed an alternative local method called Constraint Line Clustering [247]. The goal is to find 'the tightest cluster containing roughly half of intersections' of lines defined by the OFCE (called constraint lines) in the region of support. This cluster will correspond to the most prominent motion within the region of support. To simplify the search for a cluster, the search for intersections is performed along the line defined by the central element of the region of support. Of course, this makes the assumption that the constraint line is reasonably accurate and that it corresponds to the majority population (i.e. the most prominent motion). Nesi *et al.* [219] also use the concept of the constraint line, however, they transform from lines in velocity space (u, v) to points in (m, c) space where m $= -I_x/I_y$ and c $= -I_t/I_y$. In the (m, c) domain, the problem of clustering is transformed into a search for a line on which a maximum number of points reside and this line can be determined using a combinatorial Hough transform.

3.2.8.2 Robust estimators

The primary limitation of the techniques discussed so far is that they do not explicitly model motion discontinuities and hence cannot provide the locations of motion discontinuities directly. Perhaps the simplest way to model motion discontinuities is through the membrane model associated with global smoothness constraints (Figure 3.4). As formulated, the link between neighbouring motion estimates is infinitely flexible, but what if this link is allowed to break if stretched too far, turning the membrane into a weak membrane [36]? To do this a line process is added to (3.9) yielding

$$\sum_{\text{image}} (E_d^2 + \lambda^2 E_s^2 (l-1) + \alpha l) \tag{3.56}$$

Here l is the binary line process. If l is zero then the smoothness constraint is used and if l is one the smoothness constraint is replaced with α. The connection between motion estimates will break if $\alpha < \lambda^2 E_s^2$ because during minimisation l will be set to the value that gives the minimum sum. To simplify the minimisation, the line field can be eliminated in advance (Figure 3.8) yielding

$$\sum_{\text{image}} (E_d^2 + \rho_{\alpha,\lambda}(E_s)) \tag{3.57}$$

where

$$\rho_{\alpha,\lambda}(r) = \begin{cases} \lambda^2 r^2 & \lambda^2 r^2 < \alpha \\ \alpha & \text{otherwise} \end{cases} \tag{3.58}$$

The function ρ, known as the truncated quadratic, is a member of a family of robust estimators known as M estimators. The M estimator has been applied with great effect to optical flow estimation problem. For example, Black [35] uses the Lorentzian as an M estimator and formulates both the smoothness term and the data term of the Horn and Schunck algorithm using robust estimators. This allows breaks in the membrane where the data indicate a motion discontinuity and breaks from the data where the data appear to be grossly incorrect (e.g. due to failure of the BCA due to occlusion, etc.). Black also applied the M estimator to block based motion estimation. Gupta [113]

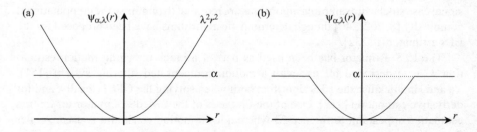

Figure 3.8 The truncated quadratic. (a) explicitly shows the line process, (b) shows the truncated quadratic after minimising the line process away

uses the truncated quadratic and shows that one must normalise the intensity gradient in the image to ensure motion edges are located consistently. Both solve the resulting non-convex minimisation using the Graduated Non-Convexity algorithm.

Just as there are many variants of the M estimator, there are many classes of robust estimations beyond the M estimator [239]. In the motion estimation literature three robust estimators are particularly common: the M estimator, Least Median of Squares (LMedS) and Least Trimmed Squares (LTS). An analysis [268] of these estimators showed that, when noise properties (standard deviation) are unknown, the LTS estimator provides the best results.

The LTS and the LMedS estimators (both introduced by Rousseeuw and Leroy [239]) have gained popularity in recent times. Rather than minimising the sum of all squared residuals, the LTS estimator minimises the sum of the smallest h squared residuals thus ensuring that outliers do not contaminate the estimate:

$$\min_{\theta} \sum_{i=1}^{h} (\vec{r}_i^2) \qquad (3.59)$$

Here $\vec{r}_i^2$ is the ordered set of squared residuals (squaring occurs first, then ordering). The LMedS estimator replaces the summation with the median operator. Thus, rather than finding the parameters that minimise the sum of squared residuals, we find the parameters that minimise the median of the squared residuals:

$$\min_{\theta} \operatorname*{median}_{i} r_i^2 \qquad (3.60)$$

Both estimators have a breakdown point of 50 per cent (in the case of the LTS algorithm, this occurs for $h = N/2$); however, the LTS estimator is generally considered to be superior since it has higher relative efficiency (i.e. lower variance). In practice, the LTS and LMedS estimators are used to detect outliers, then a simple weighted least squares estimate (with weights set to exclude outliers) is used to further improve efficiency [239]. This is vital since, for example, the LTS estimator bases its estimate on exactly h residuals, not on the entire set of inliers. Both the LTS and LMedS estimators are scale equivariant. While simple solution methods for both the LTS and LMedS are available for the one-dimensional case, in the multidimensional case stochastic sampling algorithms are required to minimise the computational complexity [8, 202, 240] though deterministic algorithms have been proposed for the LTS estimator [8, 232].

The LTS estimator has been used as part of a block matching motion estimation scheme designed for use with a motion compensated filtering system [151]. Ye and Haralick use the LTS algorithm both when solving the OFCE locally, and for derivative estimation [312]. One of the first uses of the LMedS estimator for motion estimation appears in Reference 23 where a local motion constraint is used. As per Rousseeuw and Leroy's original formation of the LMedS algorithm, outliers are first detected and removed from the region of support using LMedS then a motion estimate calculated with the remaining data using a least squares estimator. This is then extended to use a total least squares[60] method (see Section 2.2.1). Another method using the LMedS estimator together with a local approach appears in Reference 224. Here, motion is fitted to an affine model and a block shifting technique allows motion to be accurately determined in regions with up to three motions thus overcoming the requirement of robust estimators for there to be a single motion occupying at least half (for a breakdown point of 50 per cent) of the region of support. Qian and Mitchie [235] fit a rigid motion model to the optical flow and solve for motion parameters using the LMedS estimator. They then use a k-means algorithm to segment the resulting motion estimates. Finally, we consider Reference 251 where the LMedS estimator is used with a global smoothness assumption to estimate motion. This technique fuses *Maximum A Posteriori* (MAP) estimation[61] (and the Markov Random Field) with the LMedS estimator leading to a Robust Reweighted MAP motion estimation scheme.

3.2.8.3 Markov random fields

The *Markov Random Field* (MRF) is a Bayesian method closely associated with the membrane model; however, rather than being constructed mechanically (i.e. via a membrane) it is constructed probabilistically. Essentially, an MRF is a collection of random variables $\mathbf{R} = (\mathbf{R}_1, \ldots, \mathbf{R}_m)$ defined on a set of sites $\mathbf{S}$ [191]. Each random variable $\mathbf{R}_i$ can take a value φ_j from the set of labels $\mathbf{L}$. The two key properties of MRFs are

- Positivity. All realisations of the field (i.e the joint probability (3.61)) have a non-zero probability (Φ is the solution space).

$$p_\mathbf{R}(\mathbf{R}_1 = \varphi_i, \ldots, \mathbf{R}_m = \varphi_m) = p_\mathbf{R}(\varphi) > 0 \qquad \forall \varphi \in \Phi \qquad (3.61)$$

- Markovianity. The probability of a particular realisation of a random variable is only dependent on the neighbouring random variables. In (3.62), the index $\mathbf{S} - \{i\}$ is the set difference and R is the set of sites neighbouring site i. Typically the first order neighbourhood (i.e. the four immediate neighbours) is used as R:

$$p_\mathbf{R}(\varphi_i | \varphi_\mathbf{S} - \{i\}) = p_\mathbf{R}(\varphi_i | \varphi_j j \in R) \qquad (3.62)$$

MRFs are generally solved by finding a set of parameters that maximise the joint probability given the observed data (this is the MAP estimate), however, it is not obvious how the joint probability $p_\mathbf{R}(\varphi_i)$ can be derived from the local conditional probabilities (3.62) [191]. This problem was solved when the equivalence between the MRF and the *Gibbs Random Field* was established. This allows us to specify $p_\mathbf{R}(\varphi_i)$ as a *Gibbs distribution*:

$$p_\mathbf{R}(\varphi) = \frac{1}{Z} e^{-(U(\varphi)/T)} \qquad (3.63)$$

---

[60] TLS has been used for motion estimation (e.g. [299]).
[61] MAP maximises a conditional joint probability function making it a mode seeking estimator.

where

$$Z = \sum_{\Phi} e^{-(U(\varphi)/T)} \qquad (3.64)$$

Z is a normalisation constant known as the partition function where the summation is over all realisations of the MRF, T is the temperature (a term taken from statistical mechanics) and $U(\varphi)$ is an energy function of the form

$$U(\varphi) = \sum V_c(\varphi) \qquad (3.65)$$

Here V_c is the clique potential function that describes how neighbouring sites interact. Thus we can define the joint probability of an MRF by specifying its *clique potential functions* and the joint probability can be maximised by minimising $U(\varphi)$. The similarity between the membrane model and the MRF arises from (3.62) which is analogous to the smoothness constraint that led to the membrane model. Despite this similarity, Stiller makes it clear [269] that the regularisation approach is fundamentally different to Bayesian methods. In the Bayesian approach all terms must be probability distributions and the final form of $p_R(\varphi_i)$ (and hence the clique potential function) is defined by Bayes rule (e.g. [271, 282] see Section 2.1.11). Explicitly defining probability distributions allows one to verify assumptions experimentally. Put another way, the function of an MRF is not described in terms of residual minimisation (or, in the case of discontinuity detection, in terms of outlier rejection) but in terms of probability maximisation. On the other hand regularisation based approaches add a smoothness term to the OFCE to make the motion estimation problem well-posed. Regardless of this distinction, the minimisation problems that arise from both approaches can be quite similar.

When estimating *velocity* or *displacement field* (i.e. an MRF describing velocity or displacement) it is assumed that the set of sites **S** on which the motion field is defined corresponds to pixels in an image. One can model discontinuities by adding an additional set of sites between pixels called the *line field*. This line field interacts with the motion field in the clique potential function in much the same way as the line process introduced in the truncated quadratic M estimator. It switches off the link between neighbouring sites if there is a large difference in motion estimates. For example, Tian and Shah [282] use the OFCE together with the line field approach and arrive at a clique potential function analogous to (3.57).

As was the case with robust estimators, the resulting minimisation is non-convex[62]. Tian and Shah solve this using a mean field technique and arrive at a simple set of iterative update equations[63] that do not require construction of approximations of the estimator function [100], as does the Graduated Non-Convexity algorithm. The result is an iterative algorithm that simplifies to the Horn and Schunck algorithm if the line field is eliminated. Tian and Shah assume that motion discontinuities are likely to

---

[62] Excellent discussions of minimisation procedures are available [191, 278].

[63] These equations simplify to the Horn and Schunck update equations if the line field is assumed to be zero everywhere.

correspond with intensity edges, hence the control parameter for the line field is set to make formation of motion edges easier at intensity edges. It is interesting to note that in this algorithm the sites in the line field are not binary, but rather can take on any value in the interval from zero to one thus avoiding a hard decision on boundary locations.

Rather than using line fields, Iu [150, 278] uses a threshold to exclude neighbours from participating in smoothing. The energy function relating neighbouring motion estimates is written as

$$U(d) = \sum_{i \in \text{image}} \left\{ \frac{1}{N_h} \sum_{j \in R} \delta_j (d(i) - d(j))^2 \right\} \quad (3.66)$$

where i and j are two-dimensional position variables and

$$\delta_j = \begin{cases} 1 & (d(i) - d(j))^2 < \text{Threshold} \\ 0 & \text{otherwise} \end{cases}$$
$$N_h = \sum_j \delta_j \quad (3.67)$$

The effect of this smoothness term is that smoothness is only imposed for neighbours whose difference is less than a threshold and this threshold is either a constant or varies so that each site has the same number of cliques $(d(i)-d(j))$ included in the summation over the region of support R. A solution is found using simulated annealing.

Since a number of fields can easily be combined in an MRF formulation, it is common to see extensions of the basic weak membrane model. For example, motion discontinuities are considered more likely if they lie on an intensity edge[64], and if they vary smoothly. Explicit modelling of occlusion is also common. The classic example of such a formulation is that of Konrad and Dubois [175]. They begin by defining a structural model (the BCA), then extend this to allow for noise leading to the observation model (the Displaced Pixel Difference – i.e. pixelwise DFD). A displacement field model is then developed that assumes piecewise smooth motion through the use of a line field which is biased to prefer smooth discontinuities and discontinuities corresponding to intensity edges. A MAP solution of the resulting energy function is found using simulated annealing. This model is extended to include a smoothly varying occlusion field [76]. Simulated annealing is again used to minimise the energy function.

In a similar approach, Zhang and Hanauer [313] use an additional segmentation field to label regions where the BCA has failed due to occlusion and disocclusion. The idea of labelling occlusion sites is similar to that of Black [35] where a robust estimation is applied to both the smoothness and data terms. Zhang solves the resulting

---

[64] Here, the dependence is explicitly included in the motion model. In Tian and Shah's algorithm, this constraint was imposed by externally manipulating a discontinuity formation control variable.

energy function using the Mean Field approach arguing that it provides better results than the greedy, *Iterated Conditional Modes* (ICM) algorithm that is fast, but can get stuck in local minima and generates results more quickly than the stochastic Simulated Annealing algorithm. Stiller [269] examined the statistical characteristics of a number of image sequences and determined that the zero-mean generalised Gaussian distribution better represents the distribution of the DFD than the usual stationary, spatially uncorrelated, Gaussian assumption. Rather than use a line process to describe motion discontinuities, Stiller uses a label field (defined on the same sites as the motion field) to define regions of similar motion. This approach allows regions of occlusion to be determined by the motion of labelled regions rather than being modelled as an additional field. Stiller's motion model favours spatially and temporally smooth motion with smoothly moving discontinuities preferably located at intensity edges. The resulting energy function is solved in a multiscale framework helping the ICM solution algorithm to avoid local minima.

Other interesting MRF based motion estimation algorithms include that of Bober and Kittler who use a Hough transform approach to generate an initial estimate of the motion field and its discontinuities, then use an MRF to perform the final smoothing and segmentation [38]. A solution to the MRF is found using a supercoupling approach [39]. Heitz and Bouthemy use a similar approach [131]. They first compute optical flow using both a gradient based method, and an edge tracking method. The assumption here is that the techniques are complementary, since gradient based methods will fail at discontinuities (both in motion and in intensity) while feature (edge) tracking methods operate well in these regions. These motion estimates are weighted by a confidence measure and combined using a coupled MRF. A line field that takes labels from the set $-1, 0$ or 1 is used with 0 indicating no discontinuity while the values ± 1 indicate which side of an edge corresponds to the occluding object; again, smooth motion discontinuities at intensity edges are encouraged. An excellent discussion of MRF based motion estimation is available [278].

3.2.8.4 Mixture models

Finally, we present a brief introduction to the use of *mixture models* for motion estimation. Robust estimation methods (and MRF methods) give the motion corresponding to the majority population (or to the most likely motion – the mode) under the assumption that there is a majority population. This works well if the region of support is smaller than smallest moving region. If one explicitly wishes to know all the motions within a given region of support without the restriction of a majority population, the concept of mixture models can be used. Such methods are ideal where transparent motion or fragmented occlusion exists or where there are multiple motions such that there is not a majority population. The key to mixture models is that they do not assume a piecewise smooth model, rather they assume a layered model with each layer corresponding to a different motion model. For example, consider the one-dimensional motion estimate in Figure 3.9.

A traditional approach (e.g. non-robust local regression) will smooth across the discontinuities as in Figure 3.9a while a robust method would assume piecewise

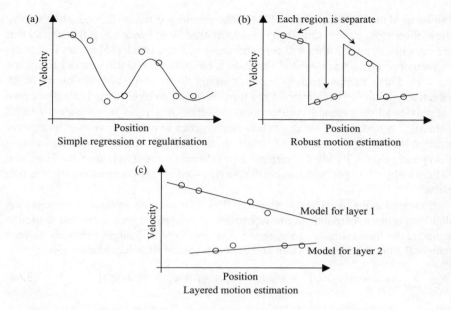

Figure 3.9 Illustration of motion estimation techniques. Simple regression or regularisation will smooth across motion discontinuities. Robust methods allow a piecewise smooth motion field. Layered representations assign each estimate to a particular motion model (adapted from Reference 294)

smooth motion as in Figure 3.9b. Methods based on a mixture model presume some number of models exists (this number may be determined automatically) and assign each estimate a label indicating which model it best fits [301] (Figure 3.9c). Mixture model formulations are often solved using the expectation minimisation algorithm (eg. [153, 301]), however, other methods such as the k-means algorithm can also be applied [278, 294].

3.3 Temporal integration

These robust motion estimation methods are designed to integrate motion estimates over the largest possible area to reduce noise without integrating data across motion boundaries. What has not been considered so far is the possibility of extending integration to include the temporal dimension. *Temporal integration* has two primary benefits, it improves motion estimates over time, and it can reduce computational requirements. It has also been suggested that temporal integration can model certain features of the human visual process [35].

Computational benefits arise from temporal integration if one uses a motion estimation algorithm that requires iterative minimisation. Rather than starting minimisation from some arbitrary configuration (e.g. zero flow), the previous flow estimate can be used as a starting point[65] [211]. The previous estimate is arguably closer to the motion at the current time than is a zero motion field; hence one might expect that the minimisation algorithm will converge more quickly. Black [35] argues that faster convergence can be achieved if one uses a constant acceleration model to update the flow. Further gains are achieved by warping the resulting optical flow by itself, effectively giving a predication of the flow for the upcoming frame [35]. Black also formulates an incremental minimisation algorithm where his minimisation control parameter is updated allowing gradual convergence to a final motion estimate over time and reducing the number of iterations required for a given frame. Rowekamp [241] takes this to its logical extreme by performing one iteration of the Horn and Schunck algorithm per frame, using the previous optical flow estimate as the 'starting point'.

Potential gains in accuracy of optical flow estimates are obvious. By integrating data over a spatiotemporal volume rather than over a spatial area, more data is used to generate the final estimate. To integrate data over time, one might explicitly enforce temporal smoothness in the Horn and Schunck optical flow formulation (3.8):

$$\sum_{\text{image}} \{(uI_x + vI_y + I_t)^2 + \lambda^2(u_x^2 + u_y^2 + u_t^2 + v_x^2 + v_y^2 + v_t^2)\} \tag{3.68}$$

Chin *et al.* solve this formation using Kalman filters [59]. Similarly, Black [35] adds a robust term to his formulation linking measured and predicted optical flow and Stiller [269] assumes temporally smooth motion estimates in his MRF based formulation.

The advantages of temporal integration are also available to token based motion estimation techniques. If a particular motion feature is tracked over an extended time, a model of its motion can be formed. Using this model, the future location of that feature can be predicted, hence reducing the search space necessary in the matching process. This reduction in the search space also reduces the chance of ambiguous matches thus improving motion estimation results [29, 304].

3.4 Alternate motion estimation techniques

It is only natural to assume a camera will be used in the task of motion estimation; our eyes and brains perform that task so well. However, it is interesting to ask whether other sensory modalities or combinations of modalities can be used for the task. Here a number of possible alternatives are considered.

Rather than use a single camera to determine motion, some authors have considered the possibility of using a stereo pair of cameras for the motion estimation task. This is quite natural since motion estimation and visual stereo are similar problems.

---

[65] This assumes that the motion at each pixel remains constant – quite a reasonable assumption if motion is less than one pixel per frame.

Measurement of motion essentially involves the search for matching points in a pair of frames separated in time. In visual stereo, disparity is measured by searching for matchings points in a pair of frames separated in space.

Sudhir *et al.* [275] consider the cooperative fusion of the stereo and motion estimation tasks via a Markov Random Field formalism. In their approach, optical flow is constrained to be similar in corresponding points of the stereo image pair and discontinuities in the optical flow indicate potential discontinuities or occlusion in disparity and vice versa. They also use a 'round-about' compatibility constraint to reduce the disparity search space. This is based on the premise that if point p_{left} in the left image corresponds to point p_{right} in the right image and these points are mapped to p_{left}' and p_{right}' in the subsequent frame by the optical flow, then p_{left}' and p_{right}' will also be corresponding points.

Wang and Duncan [295] use a combination of range and stereo to determine 3D motion. They use the rigid body motion model (Section 4.2) to relate optical flow to 3D motion then use stereo disparity (where available) to eliminate the depth term. Where stereo disparity is not available, an approximate method for 3D motion estimation is used. A multilevel iterative solution scheme allows them to segment the environment into coherently moving regions. Franke and Heinrich [95] also use disparity to remove the need for explicit depth information, though they use the equations of perspective projection rather than a rigid body motion model. This allows them to define an expected 'disparity weighted velocity' (based on known camera motion) which is compared to the measured velocity to locate independently moving objects. A third means of eliminating the need for the depth term of the rigid body motion equations is presented in Reference 21. Here, both the optical flow and disparity calculations are written in terms of the rigid body equation with the solution for depth from the disparity calculation used to eliminate depth from the optical flow calculation.

While using vision to estimate motion is common, motion can, in principle, be determined using any appropriate time-varying data source. One alternative is the use of range data though one must be careful to avoid motion distortion if a scanning sensor is used [255]. In the simplest case, motion can be estimated using range data by segmenting the data into different regions and tracking the motion of those regions as was done by Ewald and Willhoeft [83] in the 1D case. In Reference 311, the optical flow constraint equation is extended to a range flow constraint equation (3.69), thus allowing estimation of 3D velocity from range:

$$UZ_X + VZ_Y + W + Z_t = 0 \tag{3.69}$$

Here (U,V,W) is the three-dimensional range velocity and Z is the depth. Subscripts indicate partial derivates with respect to the world coordinate system. This is extended [56] to use specially defined invariant features rather than range directly. In References 26 and 263 range and optical flows[66] are directly combined using,

---

[66] Optical flow is redefined so that derivatives are with respect to the 3D world coordinate system rather than the 2D camera coordinate system thus allowing optical flows and range flows to be equated.

in the first case, a local constraint, and in the second case, a global constraint that is solved iteratively.

Typically, cameras sample on a rectangular sampling grid, however, Bolduc and Levine [41] consider the use of *space variant sampling schemes* designed to reduce data rates for a given field of view. Such cameras are often inspired by the human retina, which is composed of a high resolution fovea and a low resolution periphery [293]. Tistarelli and Sandini [284] show that polar and log-polar sampling significantly simplifies calculation of parameters such as time to collision and relative depth, and Tunley and Young [286] give results for extraction of first order flow properties from simulated log polar data. Boluda *et al.* [42] show that the combination of a log-polar image sensor and a differential motion estimation technique can automatically discard camera translation along the optical axis. Rather than using a space variant pixel arrangement, Gao *et al.* [99] consider the use of a wide angle, high distortion lens and reformulate the Lucas and Kanade optical flow algorithm to correctly operate with such a lens. Finally, in Reference 82 a foveated sensor based on rectangular sampling designed for tracking applications is considered.

Structured light techniques use a camera to measure the deformation of a known pattern of light projected into the environment using laser. Based on this deformation (and the change in deformation) the shape and motion of objects in the environment can be deduced. For example, a camera can detect the deformation of a laser stripe to determine relative motion between a vehicle and an object [207]. In Reference 315 a system for tracking an object using an laser stripe and camera on a robotic arm is described. Another alternative is the use of *laser speckle images*. When a laser shines on an optically rough surface, a speckle pattern arises. This pattern can be detected by a camera and processed to give the motion of the illuminated surface [137]. This concept has been implemented using an application specific integrated circuit [136]. Motion estimation in situations where camera motion (or zoom) can lead to blurring caused by defocus has been considered [213]. Here, the aim is to extract both affine motion and degree of defocus simultaneously so that motion estimates are more accurate. Finally, Rekleitis [237] considers frequency domain analysis of a single blurred image as a means of motion estimation. Motion blur causes an easily detectable ripple in the log power spectrum and velocity can be measured from the period of the ripple.

3.5 Motion estimation hardware

Thus far the focus has been on algorithmic aspects of motion estimation; however, an important feature of this work is implementation of a motion estimation algorithm in semicustom digital hardware to create a motion sensor. Therefore, a range of custom and semicustom implementations of motion estimation are now considered. This discussion excludes software solutions based on Digital Signal Processor (DSP) or PC hardware.

The *Vision Chip* [206] is an approach to vision processing where the imaging and (usually analogue) processing components are incorporated into a single device.

These devices tend to be of low resolution[67], difficult to design, and, in the case of analogue devices, sensitive to process variation [206]. However, the advantages of compactness and low power dissipation means that there continues to be significant interest in the design of vision chips.

A recent example of a vision chip is that of Yamada and Soga [310] who implemented a 10×2 pixel motion sensor based on the tracking of intensity edges which are located by the ratio of intensity of neighbouring pixels. A sensor based on a similar principle is presented in Reference 81. In this case both 1×9 and 5×5 pixel sensors are developed that detect intensity edges using zero-crossing operator. Harrison and Koch [124] use a Reichardt motion detector (often used as a model of biological vision), which correlates a pixel's value and a delayed version of its neighbour. A pair of such correlations is then subtracted to yield a direction sensitive response. Moore and Koch [208] use a one-dimensional variant of the optical flow constraint equation to solve for velocity while in Reference 82 a foveated motion estimation chip is described where a central, high density array of pixels (the fovea) is used to detect motion direction and a less dense peripheral region is used to detect target edges. This sensor is used to track motion with the peripheral region generating saccades (large motions of the camera to centre a target in the image) and the foveal region smoothly pursuing the target once it has been acquired.

Kramer *et al.* [177] developed a one-dimensional sensor to detect motion discontinuities by comparing the outputs of neighbouring velocity sensors. They also created a one-dimensional sensor able to segment a scene into regions of coherent motion through the use of a non-linear resistive network. MRF-based motion estimation using analogue resistive networks has been considered [198] and in Reference 188 a hybrid analogue and digital binary block correlator is proposed showing a significant size reduction over an equivalent digital circuit without loss of processing performance. Analogue neuroprocessors have also been used for motion estimation. Further reviews of vision chip design are available [79, 206, 243].

Beyond the vision chip, motion estimation is performed using camera and separate processing hardware. The use of custom digital logic (in the form of an ASIC) can allow motion estimation at extremely high frame rates as demonstrated by Rowekamp [241] whose pipelined ROFlow sensor can generate a field of 128×128 optical flow estimates (based on the Horn and Schunck algorithm) at 1500 fps given sufficient memory and camera bandwidth. Experimental results show operation at 50 fps. Adorni *et al.* [7] use a massively parallel SIMD processor known as PAPRICA, specially designed for processing two-dimensional data. They implement an optical flow estimation algorithm that performs correlation with a one pixel region of support followed by postprocessing to remove noise due to poor matches. An implementation of this algorithm using a 16×16 array of processors is able to produce flow estimates at 26 k/s [111] (with a search region of ± 8 pixels) and simulations show that up 2 M flow estimates per second could be generated using a linear array of

---

[67] Vision chips can also suffer from a low fill factor (ratio of light sensitive silicon to the total sensor area) since processing elements are generally interspersed with the light sensors.

128 processors. Newer generation PAPRICA hardware [47] increases this to 12.5 M flow estimates per second assuming an array of 256 processing elements.

Many authors have also considered custom digital implementation of block matching algorithms in the context of video coding. For example, Chen [57] uses a modified fuzzy block matching algorithm with a more regular data flow than the standard algorithm to produce real-time motion estimates for video with resolution of 720×480 pixels, at 30 Hz. A complete VLSI module for full search block based motion estimation has been described [84]. When clocked at 105 MHz, this module can generate 704×576 motion estimates at 30 Hz using a block size of 32 pixels and search region of ± 8 pixels. Both these implementations use non-overlapping macroblocks blocks, leading to a somewhat fragmented motion estimate. Kuhn [181] gives a more general discussion of block based motion estimation hardware considering the various design trade offs especially in the light of the MPEG-4 video coding standard that allows arbitrarily shaped blocks for motion estimation. Finally, Stöffler and Farber [272] use commercially available correlation chips (designed for use in MPEG compression) to calculate the optical flow at 25 fps for navigation purposes [273]. They compute a confidence criterion in parallel with the motion estimation so that poor flow estimates can be discarded resulting in a sparse flow field.

3.6 Proposed motion sensor

Based on the preceding discussion we now elaborate the design and implementation of our intelligent motion sensor. Our goal is to build a sensor able to estimate *relative translational three-dimensional ground plane motion* for the purpose of *real-time autonomous navigation*. We begin by developing a *dynamic scale space* based on the nature of the real-time constraints in the autonomous navigation context. We then use a simple fusion of *visual* and *range* information to provide an initial estimate which is refined using a variant of the *Least Trimmed Squared Estimator*. Since environmental *segmentation occurs in the range domain* rather than in the motion domain, the resulting algorithm is *extremely simple*. Its accuracy is illustrated both in simulation and using a prototype sensor.

Part 2
Algorithm development

Chapter 4
Real-time motion processing

If one aims at describing the structure of unknown real-world signals, then a multi-scale representation of data is of crucial importance.

Lindeberg [194]

The meaning of *real-time* depends on the application, however, it usually implies that processing is performed quickly with respect to the dynamics of the system under consideration. When estimating motion for autonomous navigation, there are two constraints that define the meaning of 'real-time'. Both constraints derive from the need to accurately capture environmental dynamics. The first limit is related to the imaging system and the second is imposed by the navigation problem.

- Temporal image sampling must occur sufficiently quickly to avoid *temporal aliasing*.
- Motion processing must proceed with an update rate and latency that ensures navigation decisions are made using the true current state of the environment.

In this section it will be shown how scale space can be used to avoid temporal aliasing and that the inclusion of range information in scale space leads to a *dynamic scale space* which removes the overhead associated with more traditional approaches.

4.1 Frequency domain analysis of image motion

To avoid errors caused by temporal aliasing, it is important to consider the frequency domain characteristics of an image sequence (spatial and temporal sampling are considered). While we focus on image data here, this analysis applies equally to any spatiotemporal data source.

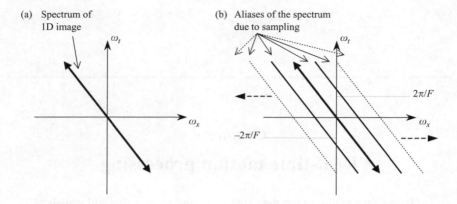

Figure 4.1 The power spectrum (b) of a moving 1D image (a) is confined to a line defined by (4.3). Temporal sampling of the image sequence causes the power spectrum to be repeated at intervals of the sampling frequency. In this case sampling rate is sufficient to prevent temporal aliasing for the given velocity (slope of the line)

A camera's spatial sampling rate is defined by the spacing of its sensing elements. If the frequency of the texture projected onto the sensing elements[68] exceeds the *Nyquist frequency*[69] aliasing occurs and manifests itself as jagged edges or incorrectly rendered detail (e.g. small details are missed all together). By using the equations of perspective projection (4.8), it is possible to determine an upper limit on the allowable spatial frequencies in the environment. However, natural environments contain numerous discontinuities in texture (edges) and are replete with fine detail (e.g. distant objects), all of which cause spatial aliasing so such an analysis would be moot – spatial aliasing is inevitable. In this work, the effect of spatial aliasing is minimised by appropriately filtering (smoothing) the images.

Temporal aliasing occurs when images are not captured at a sufficient rate relative to environmental dynamics. A common manifestation of temporal aliasing is the 'propeller effect' where a propeller seen on a television seems to repeatedly reverse direction as it accelerates. For gradient and frequency based motion estimation algorithms, temporal aliasing is particularly problematic. Gradient based algorithms often use small filter kernels to compute derivatives (Section 3.2.3.3) and any motion greater than the scale of the filter will lead to erroneous derivatives. The effect of aliasing is most clearly seen in the frequency domain where temporal sampling leads to replication of the spectrum (Figure 4.1) with obvious ramifications for frequency domain motion estimation techniques. This replication can also be problematic for feature matching algorithms, though the periodic image features needed to induce ambiguity

---

[68] The camera optics will determine the finest texture that can be projected onto the imaging surface.

[69] 'A band-limited continuous time signal, with highest frequency B Hertz can be uniquely recovered from its samples provided that the sampling rate is $\geq 2B$ sample/second' [233].

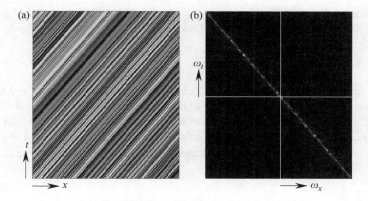

Figure 4.2 Frequency domain analysis. The left portion of the diagram shows how the 1D pattern moves over time. The right diagram illustrates the corresponding power spectrum. Note that the power spectrum lies on a line perpendicular to the iso-intensity lines

are rare in practice. In this sense, features matching techniques have an advantage over other methods.

The concept of temporal aliasing is formalised by applying the analysis of Section 3.2.7 to a single-dimensional image. This better matches the motion estimation algorithm developed later. In this case the following image function applies:

$$I_1(x,t) = I_2(x - ut, t) \tag{4.1}$$

As this one-dimensional pattern translates, it generates a striped pattern as in Figure 4.2. Fourier analysis of this function gives

$$\hat{I}_1(\omega_x, \omega_t) = \hat{I}_2(\omega_x)\delta(u\omega_x + \omega_t) \tag{4.2}$$

Equation (4.2) can be interpreted as the Fourier transform of the 1D image projected onto the line defined by (4.3) (Figure 4.2)

$$u\omega_x = -\omega_t \tag{4.3}$$

Under the assumption of pixel pitch ζ (m) and frame interval F (s) the Nyquist theorem can be applied to determine the maximum allowable spatial and temporal frequencies [308]. This gives to an upper allowable spatial frequency of

$$|\omega_x| < \frac{\pi}{\zeta} \tag{4.4}$$

This spatial frequency is measured in the image plane. Similarly, the upper allowable temporal frequency is given by

$$|\omega_t| < \frac{\pi}{F} \tag{4.5}$$

Combining (4.3) and (4.5), it is possible to determine the maximum of the horizontal component of optical flow for a given frame interval F.

$$|u| < \frac{\pi}{|\omega_x|F} \qquad (4.6)$$

The upper-bound on flow is a function both of the frame interval and the spatial frequency. Indeed, areas of low spatial frequency allow measurement of higher optical flow. However, as discussed earlier, spatial aliasing is inevitable in natural image sequences. This implies that spatial frequency components near the upper-bound of $\omega_x = \pi/\zeta$ will occur often; hence in the worst case, the maximum optical flow is

$$|u| < \frac{\zeta}{F} \qquad (4.7)$$

We can see that for practical purposes, the maximum allowable flow is one pixel per frame. We confirm this result later when we consider the accuracy of derivative estimators (Section 5.2.3).

4.2 Rigid body motion and the pinhole camera model

Before considering frequency domain issues further, the *pinhole camera model* used to approximate image formation in our camera is introduced (Figure 4.3). Two coordinate systems are defined: the camera-centred coordinate system (X, Y, Z) and the image coordinate system (x, y) with the units for both systems being metres. The image coordinate system lies on a plane called the image plane (or the imaging surface, or the sensor surface) that is parallel to the XY plane and distance f (focal length) behind it. Under this model, a point $\mathbf{P}(X, Y, Z)$ in the camera-centred coordinate system maps to a point $\mathbf{p}(x, y)$ on the image plane via the equations of perspective

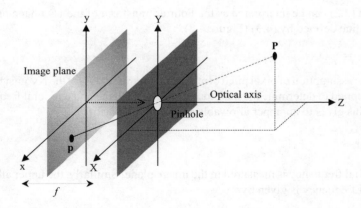

Figure 4.3   Pinhole camera geometry

projection [204]:

$$x = \frac{fX}{Z} \qquad y = \frac{fY}{Z} \qquad (4.8)$$

At this point we introduce the fundamental formula [204] which relates a point's velocity $\mathbf{P}'(X', Y', Z')$ to its position $\mathbf{P}(X, Y, Z)$ and its relative[70] three-dimensional translational velocity $\mathbf{T} = (T_X, T_Y, T_Z)$ and rotational velocity[71] $\mathbf{\Omega} = (\Omega_X, \Omega_Y, \Omega_Z)$ under the assumption of rigid (non-deforming) motion.

$$\mathbf{P}' = \mathbf{\Omega} \times \mathbf{P} + \mathbf{T} \qquad (4.9)$$

Combining (4.8) and (4.9) we arrive at the perspective rigid body motion model[72]:

$$\begin{aligned} u &= -\frac{1}{Z}(xT_Z + fT_X) + \frac{xy\Omega_X}{f} - \left(f + \frac{x^2}{f}\right)\Omega_Y - y\Omega_Z \\ v &= -\frac{1}{Z}(yT_Z + fT_Y) + \left(f + \frac{y^2}{f}\right)\Omega_X - \frac{xy\Omega_Y}{f} + x\Omega_Z \end{aligned} \qquad (4.10)$$

This model describes the relationship between position on the image plane (x, y), depth (Z), velocity ($\mathbf{T}, \mathbf{\Omega}$) and apparent motion (u, v). While simpler models exist (e.g. affine, weak perspective, etc. [304]) this model makes fewer assumptions regarding the environment and most accurately describes the imaging process (though it does not include lens distortions). The calculation of apparent motion has, thus far, used units of metres for distance and seconds for time, giving us a natural means to describe the motion of objects in the environment. However, the goal here is to estimate three-dimensional motion, thus optical flow and apparent motion must be assumed to be equivalent, and the apparent motion must be converted to the units of optical flow (pixels/frame). This is achieved by multiplying apparent velocity by a factor ψ.

$$\psi = \frac{F}{\zeta} \qquad (4.11)$$

Recall that F and ζ were introduced earlier as the frame interval and pixel spacing, respectively. Since apparent velocity and optical flow are now considered to be equal, these terms will be used interchangeably. The next step is to simplify the rigid body motion model. This is done by assuming that objects are constrained to move on a smooth ground plane; a reasonable assumption in many autonomous navigation applications. Under these conditions, $T_Y = \Omega_X = \Omega_Z = 0$ and (4.10) reduces to the

---

[70] This velocity takes into account both ego- and object motion.
[71] This rotation is about the camera axes.
[72] This model relates object velocity to apparent velocity. Different equations are used to relate object displacement to apparent displacement, though the equations are equivalent if temporal aliasing is avoided [278].

well known rigid ground plane motion model [304]

$$u = \psi\left(-\frac{1}{Z}(xT_Z + fT_X) - \left(f + \frac{x^2}{f}\right)\Omega_Y\right)$$
$$v = \psi\left(-\frac{yT_Z}{Z} - \frac{xy\Omega_Y}{f}\right) \tag{4.12}$$

This can be further simplified by assuming that there is little rotation about the Y axis (i.e. $\Omega_Y = 0$). Any rotation that does occur is assumed to have a large radius and to strongly resemble horizontal translation (a reasonable assumption in the on-road environment where the majority of motion is approximately in a straight line). This leaves

$$u = \psi\left(-\frac{1}{Z}(xT_Z + fT_X)\right) \tag{4.13a}$$

$$v = \psi\left(-\frac{yT_Z}{Z}\right) \tag{4.13b}$$

A final simplification eliminates the need to compute vertical flow. The vertical component of optical flow is only non-zero for ground plane motion if there is relative motion towards (or away) from the camera, in which case a diverging (or contracting) flow pattern emerges. This diverging pattern is not used to extract T_Z since the divergence tends to be small[73]. Instead, T_Z is calculated more accurately using range data as discussed in the next chapter. The consequence of this is that calculation of the vertical component of flow is redundant. The final motion model assumes that all motions can be modelled as simple horizontal translation.

$$u = -\frac{F}{\zeta}\frac{fU_X}{Z} \tag{4.14}$$

T_X is replaced with U_X since this velocity parameter no longer encodes the three-dimensional translation in the X axis, rather it encodes the velocity projected onto the X axis. Comparing the forms of (4.13a) and (4.14) it can be seen that

$$U_X = \frac{x}{f}T_Z + T_X \tag{4.15}$$

Thus, T_X and U_X will coincide only in image regions where the relative velocity $T_Z = 0$. That is, they will coincide where an object moves parallel to the X axis if the camera is stationary [74]. As we shall see, this is not an overly restrictive model for our application: it provides sufficient information to allow estimation of relative three-dimensional, translational, ground plane motion which is sufficient for navigation in a dynamic environment.

---

[73] Especially if we use the techniques described later in this section to avoid temporal aliasing.

[74] In principal, T_X and U_X could also coincide if the environment is stationary and the camera translates in the **X** axis, however, vehicle kinematics usually do not allow such translational motion.

Table 4.1   Operational parameters for motion estimation

Name	Symbol and units	Value
Image resolution	(x_r, y_r) pixels	512×32
Focal length	f (metres)	4.8e−3
Pixel pitch	ζ (metres)	12.5e−6
Minimum distance	D_{min} (metres)	0.25
Maximum distance	D_{max} (metres)	10
Maximum velocity	V_{max} (metres/s)	0.1
Safety margin (D_{min}/V_{max})	S (s)	2.5

4.3   Linking temporal aliasing to the safety margin

Now that a relationship between three-dimensional relative velocity and optical flow has been established, the next important question is: What three-dimensional velocity does the maximum optical flow of 1 pixel/frame correspond to? Before this question can be answered, it is necessary to define appropriate operational parameters.

The first three parameters listed in Table 4.1 are the intrinsic parameters of the camera system used in this work and the final five parameters describe the conditions under which the motion estimation sensor is expected to operate, including minimum object range, maximum object range and maximum velocity[75]. Importantly, a parameter called the *safety margin* is also defined[76]. This parameter mirrors the practice of drivers maintaining a two second[77] minimum stopping time to the vehicle in front. For additional safety a slightly larger safety margin is used, though the kinematics of the test vehicles does not necessitate this. In general, the safety margin should be chosen based on vehicle kinematics (i.e. how quickly can the vehicle stop or steer around an object) and system reaction time. If an object falls closer than this safety margin our motion estimation algorithm cannot guarantee that temporal aliasing will be avoided so alternate sensing systems must take over. Such a situation would generally be considered an emergency and one might reasonably expect that sensors specially optimised for use in close quarters will take over[78] so that special obstacle avoidance and harm minimisation procedures can be applied (e.g. [89]).

While the motion estimation model (4.14) uses a single parameter that is the projection of all motion onto the X axis, in reality objects translate in the XZ plane (4.15),

---

[75] These values are defined with respect to the ICSL CAMR Cooperative Autonomous Vehicles [292].

[76] A similar parameter (*lookahead*) is defined by Hancock [123] in his work on laser intensity based obstacle detection.

[77] This time is determined by average human reaction time and typical vehicle kinematics.

[78] This changeover process may be controlled by a higher-level data fusion [118] or planning module [270].

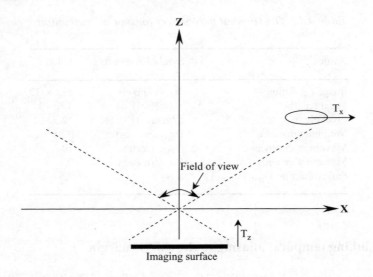

Figure 4.4 Worst case apparent motion

hence it is important to consider both three-dimensional ground plane velocities T_X and T_Z when determining the worst case apparent motion. Using the parameters in Table 4.1 together with (4.14) and (4.15) one can easily conclude that:

- The closer the object, the higher the apparent velocity. Thus, worst case apparent velocity will occur for $Z = D_{min}$.
- For motion in the Z axis, the highest apparent velocity occurs at image edges where $x = (x_r/2) \times \zeta = 3.2$ mm. Apparent velocity will be zero at the centre of the image for such translations.
- Motion in the X axis will always produce a higher apparent velocity than motion in the Z axis since $\forall x \mid f > x$.

Remember that the sensor is mounted at the front of a vehicle and vehicle kinematics [212] do not allow egomotion in the X direction. Therefore, the worst case apparent motion will occur with the camera moving forwards and an object near the edge of the scene captured by the camera moving along the X axis, out of view of the camera (Figure 4.4)[79]. If both the camera and object move at the maximum velocity V_{max}, (4.13a) can be rewritten as

$$u_{\text{worst case}} = 1 = \frac{8 \times 10^{-3} F V_{max}}{D_{min} \zeta} \quad (4.16)$$

---

[79] We could equally consider the camera to be stationary and the object to have a velocity $T_x = 10$ cm/s, and $T_z = 10$ cm/s.

Rewriting (4.16) gives a relationship describing the minimum frame rate required to avoid temporal aliasing in the worst case:

$$\text{FPS}_{\text{worst case}} = \frac{8 \times 10^{-3}}{S\zeta} \tag{4.17}$$

Here the ratio $V_{\text{max}}/D_{\text{min}}$ has been replaced with our safety margin S. From this, it can be deduced that a frame rate of approximately 256 fps is required to avoid temporal aliasing in all situations in our environment.

4.4 Scale space

The next issue to consider is the practicality of attempting to build a system capable of such a high frame rate. From a technical perspective, it may be feasible, though obtaining a camera and creating a processing system able to operate at this rate would be costly. More serious difficulties arise when considering practical issues: if a system is built that is able to detect worst case motion, will the system be able to accurately detect slower motion? Consider this: if object velocity falls from 10 cm/s to 5 cm/s then apparent motion halves. Also, if the objects are twice as far from the camera (50 cm) apparent motion will halve again. If the camera is stationary there is no global motion and apparent velocity falls again. Temporal derivatives generally fall with apparent motion which, in turn, makes motion estimation more difficult.

The need to deal with a large range of apparent motion values has long been recognised in the motion research community [80]. Equation (4.16) shows that for a given set of environmental dynamics, there are two parameters that can be tuned to avoid temporal aliasing: frame rate and pixel pitch. Since the frame rate is fixed for the majority of vision systems[80], researchers have manipulated the pixel pitch ζ by subsampling the image sequence leading to the *hierarchical* or *scale space*[81] approach to motion processing. Not only does scale space allow detection of a wider range of velocities, it can also increase computational efficiency by reducing the number of pixels that must be processed.

The use of scale space is not limited to motion estimation – it is a generically applicable technique that can be used whenever interesting image features may appear over a range of scales in an image [6, 161, 194] and there is evidence that it is used in the human visual system [6]. Scale space techniques take each image in the sequence, filters it and decimates it a number of times to generate an *image pyramid* as shown in Figure 4.5. The filtering operation performs two roles. Primarily, its role is to remove high frequency components so that spatial aliasing is minimised at coarser scales. In this case, Gaussian filters are usually used leading to a Gaussian scale space.

---

[80] This is true in that the temporal sampling rate cannot easily be increased. It can be decreased via temporal subsampling, however, there is little computation benefit in doing so unless object motion is very small (e.g. [190]).

[81] Other common terms for this concept are multiscale and multiresolution.

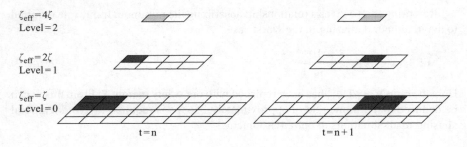

Figure 4.5 Scale space motion estimation

However, it may be argued that filtering can also be used to highlight particular image features. For example, higher frequency image components (i.e. textured regions) generate larger derivatives and hence more accurate motion estimates. A high pass spatial filter will also remove any DC component minimising the effect of illumination changes. In such a situation a Laplacian, rather than Gaussian, scale space may be more appropriate [194].

The benefit of the scale space approach is obvious. At each higher (coarser) level or scale[82] of the pyramid, ζ is effectively doubled, hence the apparent motion is halved. Referring again to Figure 4.5, the four pixel/frame apparent motion evident in the original image, becomes a 1 pixel/frame motion at the 2nd scale, and temporal aliasing is avoided. Using the convention that the bottom level of the image pyramid (i.e. the original image) is level 0, the relationship between ζ and the level of the pyramid is

$$\zeta_{\text{eff}} = 2^{\text{level}} \zeta \tag{4.18}$$

Thus, we can determine the frame rate required at any given level of the pyramid by using the effective pixel pitch in (4.17):

$$\text{FPS}_{\text{worst case}} = \frac{8 \times 10^{-3}}{S \zeta_{\text{eff}}} \tag{4.19}$$

A 512 × 32 pixel image can reasonably be subsampled by a factor of 16 leading to a 32 × 32 pixel image[83], thus at the highest level of the pyramid, the frame rate requirement falls to 16 fps, which is for more manageable than the 256 fps required for level 0. While this discussion has only considered uniform subsampling, non-uniform multiresolution methods based on space variant 'meshes' [13, 61] and quadtrees [27] have also been proposed.

Difficulties with a complete scale space scheme arise when trying to integrate data between levels. One approach is the *coarse to fine* strategy [13, 35, 253] which

---

[82] The words 'level' and 'scale' are used interchangeably when referring to levels of the image pyramid.
[83] In this work, subsampling occurs only in the X direction. Each image line is effectively treated as a separate image. This is explained more fully in Section 5.2.2.

begins with motion being computed at the coarsest scale. This motion is then projected down one level of the pyramid and is used to partially register the current and previous images[84] effectively removing any motion greater than one pixel per frame from this level. The motion estimation algorithm is then applied at this new level and the procedure is repeated until the lowest level of the pyramid is reached. The final motion estimate is then the sum of the estimates at each level. A similar approach is presented by Xu [308] who, rather than explicitly registering images, derives a multiresolution optical flow constraint equation using a modified derivative operator. This operator uses the motion estimate from the higher level as an offset, effectively performing partial registration during the derivative calculation. These methods are computationally very costly due to the need for warping and the need to repeatedly apply the motion estimation algorithm. Error is also a problem for such methods since any error at the coarser scales will affect finer scales. Warping, which involves interpolation, will introduce error of its own. Simoncelli [252, 253] shows that both error sources can be dealt with using a Kalman filter based course to fine scheme where errors are explicitly modelled.

Black [35] suggests a flow through strategy where the coarsest level giving a motion estimate higher that 0.5 pixels/frame determines a pixel's motion estimate. Quad-tree scale space strategies based on a similar premise have also been proposed [27, 179]. The advantage of these approaches is that they prevent the propagation of error and, if non-global motion estimation techniques are used, some computational advantage can be gained over processing at a single scale. However, since estimates are not refined at each level, the absolute error in the motion estimates will be higher at higher scales (though all estimates will have the same relative error).

In contrast to these 'top-down' approaches, methods where calculation can move both up and down the image pyramid have also been proposed. For example, if successive over-relaxation (SOR) is used to minimise the energy function of a global gradient based motion estimation algorithm one can apply Terzopoulos' method [36, 279]. In this method, the minimisation process is switched between levels automatically based on the spectral composition of the error signal (e.g. the data term in (3.9)). Some authors claim as much as a 100-fold reduction in the convergence time for the SOR algorithm using this method [36] since it allows low frequency data to be propagated more quickly at higher levels of the pyramid.

4.5 Dynamic scale space

One possible means of avoiding the overhead of a complete scale space implementation is to utilise the availability of range information to dynamically select a single, most appropriate spatial scale for the current environment [165, 166, 170]. Because

---

[84] This partial registration is achieved by using the estimated flow to warp the previous image towards to current image.

Table 4.2 Scale crossover points

Depth (m)	Level	Image width (pixels)
0.25	4	32
0.5	3	64
1	2	128
2	1	256
4	0	512

navigation is the primary goal of this investigation, it is assumed that the nearest object is of most relevance since it is the one that is most likely to cause a collision in the short term. Humans often use this as a navigation strategy; a graphic demonstration being on very crowded roads where personal experience shows that amid the chaos, collisions are (usually) avoided by simply avoiding the vehicles directly in front.

Dynamic choice of scale has been considered before (e.g. Niessen *et al.* [221] chose a spatial (ζ_{eff}) and temporal (FPS) scale to maximise the robustness of a motion estimate), however, the choice of scale has always been made using purely visual measures leading to a large overhead since all levels of the pyramid must be available in order to choose which level is the most appropriate. Dynamic scale space is unique in that it uses depth to focus the system's attention to a particular object (i.e. scale) and hence avoid temporal aliasing. This occurs with no overhead since correct subsampling can be performed on the input data stream before it is processed.

Using (4.16) under the assumption that processing occurs at the minimum requirement of 16 Hz, it is possible to calculate the depths at which scale changes must occur (Table 4.2). For example, if the current minimum depth is 4 m or more, no subsampling is required, however, for 2 m $\leq$ depth $<$ 4 m the image must be subsampled at a rate of 2 to avoid temporal aliasing. Note that scale changes always occur when the *worst case* apparent motion falls to 0.5 pixels/frame.

4.6 Issues surrounding a dynamic scale space

There are a number of issues that must be considered when discussing the use of dynamic scale space. These include the tendency to miss particular objects, edge localisation accuracy, motion estimation error and computational complexity. Each of these issues is considered now.

First, using a single scale implies that small, distant, or slow moving objects may be missed altogether. It is natural to concentrate on the nearest object to the exclusion of all others since that is the object most likely to be involved in a collision in the short term. As such, this algorithm is tuned to detect a range of velocities based on assumptions about camera and object dynamics. If an object is moving so slowly that the algorithm cannot detect its motion, it can reasonably be considered a static obstacle from the perspective of our sensor. This is accepted as a reasonable consequence of this approach and it is argued that if information regarding such objects is necessary it should be obtained from a sensor specially designed for that task.

It may be useful to design a dynamic scale space where the scale is chosen using a combination of minimum range and minimum resulting optical flow. Thus, if a scale is chosen and little or no motion is detected it is possible to move down the image pyramid until a reasonable apparent velocity is found. Also, temporal subsampling could be applied to 'amplify' the apparent motion. The danger with such schemes is that apparent velocity may change suddenly leading to a failure in motion. These considerations are left for future work.

As shall be seen later, dynamic scale space is similar to the flow-through methods since, regardless of the scale which is currently in use, the result is eventually projected back to the lowest scale. This means that error does have a dependence on scale, however, as is shown in Section 7.2.6.2.2, this dependence is swamped by other factors.

It is interesting to note that dynamic scale space has a slightly perverse effect with regards to computational complexity. While this entire section has been devoted to analysis and avoidance of the 'problem' of temporal aliasing, it appears that temporal aliasing is algorithmically quite simple to avoid. All that is required is spatial subsampling. The result of subsampling is that larger motions actually require less processing because they are detected at a higher level of the image pyramid. Conversely, small motions are more difficult to measure since they require a finer resolution, which entails a larger image and hence more computation. In fact, the processing rate of a system using dynamic scale space[85] is determined by the largest image size it may need to process, not by the need to avoid temporal aliasing.

In a similar vein, it is important to ask whether 'fast enough to avoid temporal aliasing' (i.e. 16 Hz in this case) is equivalent to 'fast enough for reasonable navigation'. Arguably, it is fast enough because the dynamic scale space is configured relative to the dynamics of the environment. In particular, if the safety margin is chosen correctly then the motion estimation system will provide motion data quickly enough to make plans to avoid the nearest object. Put another way, generating motion estimates more quickly will not *usually* generate new data; it will simply generate an interpolated version of the data. The ability to deal with situations where changes occur rapidly has as much to do with vehicle dynamics as it does with sensing. There is no point 'over-sensing'. Of course, latency must be minimised so that old information is not used for motion planning, however, latency is as much a function of the implementation as it is of the algorithm. Our algorithm has a three frame latency if latency is measured as per Chapter 1.

---

[85] This is not the frame rate required to avoid temporal aliasing – it is the maximum frame rate at which a particular implementation of our algorithm is capable of operating.

Chapter 5
Motion estimation for autonomous navigation

By the time you perceived the shape of another car crashing into you, it would be too late for both you and your visual system.

Watanabe [297]

The byline for this chapter eloquently summarises our approach to motion estimation: it must happen quickly and it must happen without regard to any ancillary description of the environment. To avoid a moving object, all that is required is to know its location and approximate velocity – knowing what the object is, its colour or its features is not necessary in this process. We treat motion as a fundamental quantity that is measured as directly as possible.

This naturally leads to the gradient based methods, where the only intermediate description of the visible world consists of the intensity derivatives. Alternative motion estimation techniques require more complex intermediate descriptors: frequency domain techniques require sets of filter banks or Fourier transforms, and token based methods require explicit extraction of some type of structure from the world.

This chapter draws together gradient based motion estimation, the rigid body motion model, dynamic scale space and a set of environmental assumptions to create a simple motion estimation algorithm. The resulting algorithm determines quickly the piecewise projection of relative three-dimensional translational motion onto the camera's X axis.

5.1 Assumptions, requirements and principles

To simplify the rigid body motion model, and to define operational parameters, a number of assumptions have already been made. Before arriving at the final algorithm a number of additional assumptions and requirements must be defined. This section considers the principles upon which the algorithm is built and analyses the impact of all assumptions and requirements.

5.1.1 Application

Our intention is to build an intelligent sensor that allows autonomous vehicles to navigate in dynamic environments, hence the algorithm chosen to estimate motion will need to be implemented in 'real-time', as described in the previous chapter. To help achieve this aim, we will implement the algorithm in semi-custom digital hardware, allowing complete control over all design aspects, and provide the ability to exploit parallelism. The upshot of using digital hardware is that the algorithms selected must be suited to implementation in such technology. Deterministic algorithms are preferred over algorithms requiring stochastic sampling, and complex operations (like sine, cosine, exponential, etc.) are avoided in favour of addition and subtraction operations and perhaps a small number of multiplication and division operations. Algorithms that require tuning of parameters (such as the regularisation parameter in Horn and Schunck's algorithm or the cut-off point of an M estimator) should also be avoided since optimal tuning tends to depend on the nature of the input data. In short, we seek simplicity.

As well as operating in real-time, the sensor must give an output that is useful for navigation in an environment containing dynamic and static objects. Equation (4.15) shows that a combination of visual motion information (projection of motion onto the X axis) and range information can be used to assign a relative three-dimensional, translational ground plane velocity to each image point. Arguably, this is sufficient to make a short term navigation plan that will avoid both static and dynamic objects and it improves upon existing navigation methods that use visual motion to avoid only static objects (e.g. [63, 73, 178]). Our algorithm measures relative velocity – it does not attempt to remove the effect of egomotion. In Section 5.3 we show that relative motion can be used directly to determine the likelihood of collision.

5.1.2 Data sources

While a camera is the obvious sensor choice for motion estimation, a single camera describes the environment as a spatiotemporal distribution of *brightness* and provides only a two-dimensional sense of the environment. A single camera does not *directly* give a sense of depth. While it is certainly possible to make various motion estimates directly from visual data, this lack of depth information raises an important question. Since motion is a geometric property of the environment (motion is all about change in position after all), could better motion estimates be made if depth data were available? Would a simpler motion estimation algorithm arise? There is evidence that depth information (as determined by stereopsis) does play a role in motion estimation in the human brain [185] so this idea is worth following. Indeed, we saw in Chapter 3 that the complementary information found in vision and range can be exploited for motion estimation.

Both the mammalian brain and a number of proposed motion estimation algorithms (e.g. [26]) exploit the availability of dense range information[86], however, such data are difficult to obtain at the same rate and resolution as an image. While experimental sensors able to generate dense range data quickly do exist (e.g. Beraldin et al. [30], Farid and Simoncelli [86]), the most obvious means of obtaining dense range data is via visual stereo. This would be ideal since it allows us to build a completely passive sensing system that closely resembles human vision. However, visual estimation of depth can be error prone and it adds a significant degree of complexity to a motion estimation system.

Active sensors give a more direct way of obtaining range information, however, they have the side effect of 'polluting' the environment with artificial signals leading to the possibility of interference between sensors. The first active sensors that spring to mind are radar and ultrasonic ranging devices. Unfortunately, both are difficult to use in conjunction with visual data. Their resolution is too low and ultrasonic sensors are not suitable for use with automobiles since wind interferes strongly with the acoustic signal. The upcoming field of terahertz imaging [225] may overcome these issues, however, at present, the best way of obtaining high resolution range data is via a scanning laser range sensor. Not only do such sensors provide accurate range information, the localised nature of a laser reduces the likelihood of interference. Such sensors can scan a 270° field of view with 0.25° angular resolution at a rate of 10 Hz to generate one-dimensional range scans: higher rates are possible at lower resolutions [141]. A 10 Hz frame rate is certainly compatible with many camera systems, but how can a two-dimensional image be integrated with one-dimensional range data?

Indeed, why not just use range data alone? The concept of range flow [262] shows that gradient based motion estimation can be applied to range data, as can token based techniques [141]. However, the task here is to extend a sense of geometry to visual motion estimation by fusing range and vision [118]. This fusion is not performed to improve the reliability of the motion estimate [26]; rather, range and vision are fused to obtain data that are not available when these data sources are treated separately and to simplify motion estimation. For example, (4.13) shows that, without range, velocity is ambiguous – a slow moving nearby object may appear to have the same velocity as a distant fast moving object. Range information also enables the use of an efficient dynamic scale space to avoid temporal aliasing.

Thus, a sense of geometry allows us to go beyond the usual task of simply estimating apparent motion and allows measurement of the three-dimensional translational velocity of objects. This sense of geometry also gives a simple means of segmenting the environment. Rather than using complex formulations to segment a scene in the visual or motion domain, segmentation occurs in the range domain. This is more natural than segmentation using visual information because the extent of an object (or a portion of an object) is necessarily a geometric property. Range discontinuities nearly always indicate object edges especially in

---

[86] Sampled roughly as densely as the corresponding image.

the autonomous navigation field where objects tend to be well separated. Visual edges only sometimes correspond to the physical edge of an object, and motion discontinuities are somewhat difficult to determine from the motion information itself.

So, to answer the earlier question, how to integrate two-dimensional range data with a one-dimensional range? It is first assumed that the camera's X axis and the scan axis of the range sensing device are parallel and as close together as possible, thus ensuring the range scan is made within the field of view of the camera. Second, range and visual information are registered. That is, one range measurement is mapped (via interpolation) onto each image column. In doing so an assumption is made that each column contains a planar surface parallel to the camera sensor and covering the entire extent of the column. To help ensure that assumption holds, relatively narrow images are used – only 32 pixels tall (giving an approximately 4.6° vertical field of view). Using a narrow image also reduces the amount of data that must be processed. Of course, this assumption may fail – there may be two regions of distinctly different depth within the same column[87]. To overcome this, it is assumed that at least half of the image column contains data corresponding to the measured depth and a robust average estimator is used to discard the remainder.

An important issue is the accuracy of range segmentation. Since the algorithm developed here maps one-dimensional range data to image columns, range data are subsampled along with the image. Furthermore, because segmentation is performed using range data alone, edge localisation accuracy in the image plane will decrease as subsampling increases (see top of Figure 5.1). For example, at level one of the pyramid, edge localisation accuracy is ± 1 pixel, while at the highest level it is effectively ± 8 pixels. However, the degree of subsampling is directly related to object distance, hence the error in the estimated edge location in the camera coordinate system (X, Y, Z) does not increase for the object currently focused upon[88]. Referring to the bottom of Figure 5.1, at level one error is ± 1 pixel at a maximum range of 4 m. This translates to a positional error of approximately 1 cm. At scale level four the error increases to ± 8 pixels, however, the maximum range is only 0.5 m which corresponds to a positional error of approximately 1 cm as before.

One further assumption is made regarding the sensors in this system. They are attached rigidly to the vehicle. The use of a pan/tilt system or other devices to move the sensors independently of the vehicle is beyond the scope of this work; however, such active vision systems certainly have the potential to improve motion estimation. Not only could they be used to minimise 'motion noise' as discussed in Section 5.1, but experiments with the human visual system [106] show that eye movement (tracking of objects) greatly extends the detectable range of motions indicating that similar improvements may be attainable when appropriate hardware is used.

---

[87] Small deviations from the measured range are acceptable.

[88] It will, of course, increase for more distant objects, however, this is not too serious since in general the nearest object poses the largest danger, hence its edges must be located most accurately.

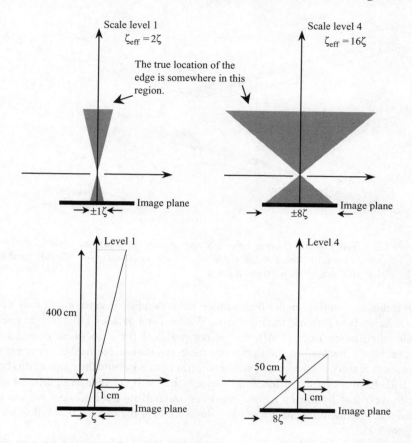

Figure 5.1 Edge localisation accuracy. Although the error in edge localisation will increase in the image plane at higher levels of the image pyramid (top of figure), the localisation accuracy in the camera coordinate system remains constant since subsampling only increases as objects come closer to the camera. The bottom section of the diagram shows the worst case error at the centre of the image

5.1.3 Motion

The first assumption we made about the world was that of rigid motion (motion without deformation) on a smooth ground plane. This assumption is reasonable under laboratory conditions, however, real roads have bumps and hills that affect motion estimation in different ways. High frequency deviations from smoothness (bumps and holes) cause the vehicle and hence the sensor to pitch and roll, quickly inducing 'motion noise' into the image and range data. This motion noise will be of high frequency (potentially many pixels per frame) so temporal image derivatives will be severely affected. The algorithm presented here has a very poor response to

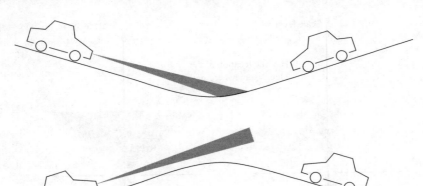

Figure 5.2   Nearing the bottom of a hill our sensor may not be able to detect an imminent threat since its field of view is so narrow. Similar problems will occur as we near a crest

high frequency motion: under high motion noise conditions zero output may occur. The solution is to stabilise the input data. While visual data lend themselves to electronic stabilisation (e.g. [154]), this is not practical for a one-dimensional range scanner hence mechanical stabilisation is the best solution. No further comment will be made here since the design of a stabilisation system is beyond the scope of this book.

Hills have a slightly different influence – they will not induce significant noise due to their low frequency nature; however, the narrow field of view we use may mean that imminent threats are not visible as we near the bottom of a hill or a crest (Figure 5.2).

If the safety margin has been chosen correctly, this is only an issue when a vehicle is within its safety margin range of the dip/crest. Unfortunately, it is difficult to detect and correct for this situation. One might imagine that a tilt mechanism could assist in the case of a dip[89] although it remains difficult to determine that the sensors are actually looking into a dip rather than at valid data. The dynamic scale space will enter 'emergency mode' as the base of the dip nears since the safety margin is violated. This is a reasonable response in the absence of a tilt mechanism. There is no simple solution for the case of a crest – vehicles on the other side of the hill remain invisible regardless of sensor tilt and entering 'emergency mode' on the basis of a 'lack' of range data is inappropriate since this is generally a safe situation. If the sensor is used in a road environment, it must rely on road engineers to design roads with gradients that allow the maintenance of a safety margin[90].

[89] The IBEO range sensor has the ability to scan in four vertical planes to compensate for such situations.
[90] Highways are indeed designed using a parameter called the *crest stopping sight distance* that relates the radius of the crest to object and observer height such that the object will be visible some minimum time before impact [123].

To further simplify the rigid body motion equations it is assumed that all objects turn with a large radius so that the turns can be approximated by translations. This is a reasonable assumption since the turning radius for the majority of vehicles (and other objects in the on-road environment) is quite large, especially while in motion.

It is also assumed that only the projection of velocity onto the X axis needs to be retrieved. The rationale here is that, on a ground plane (and in the absence of camera pitch caused by hills and bumps), Y axis (vertical) apparent motion only occurs as objects approach (or recede from) the camera. The resulting expansion (or contraction) in the image is equally encoded in the horizontal and vertical components so extracting vertical motion would be redundant. Furthermore, such motion tends to induce small apparent motions making motion estimation difficult. It is argued that motion towards or away from the sensors is best measured using range data. Since motion is guaranteed to be less than 1 pixel per frame through the use of dynamic scale space, a simple differencing operation will yield good estimates of this motion. Thus, no attempt is made to measure rate of approach using visual data. Our lack of interest in vertical motion further strengthens our use of a narrow image – a tall image is unnecessary when considering only horizontal motion of objects touching the ground.

The final assumption regarding motion in the environment was the safety margin defined in Table 4.1. This parameter is defined with respect to vehicle dynamics and road conditions and is used to determine when scale changes must occur.

5.1.4 Environment

We make three assumptions regarding the general environment. The first is an assumption of relatively flat illumination. While the combination of range and visual information makes this algorithm more robust to illumination variation than the basic Horn and Schunck and Lucas and Kanade algorithms (especially to shadows), illumination changes still can cause problems such as skewing of the temporal derivative in cases of a global illumination change, and a phase dependent offset in the motion estimate as is shown in Section 5.2.4.

The second environmental assumption was hinted at earlier. Segmentation of the environment is based on range information, not visual or motion information. In this work, the environment is not divided in terms of 'objects' but in terms of spatially continuous regions called blobs. A blob may correspond to part of a single object, or it may be composed of a group of closely clustered objects. From a navigation perspective, provided collisions are avoided with all blobs, the mapping from blobs to objects is irrelevant. The algorithm fulfils the principle of treating motion as a primitive quantity since the only assumption regarding blobs is that they move rigidly. No reasoning or calculation occurs at the blob level – blobs simply define coherently moving regions. Note that the left edge of the first blob is always the first image column and the right edge of the final blob is always the last image column.

The final assumptions are related to dynamic scale space. First, the system focuses on the nearest blob whether it moves or not. This may make it impossible to detect the motion of more distant blobs; however, it is the closest blob that we are most likely to collide with in the short term, hence we focus attention on that. Second, if an object comes closer than a predefined safety margin it is impossible to guarantee that motion estimates will be correct, so the system will enter 'emergency mode' where alternate sensors and navigation strategies take control. This is not an unusual assumption – control while cruising on a highway is different to the control required in close quarters or if velocity can be extremely variable [200].

5.2 The motion estimation algorithm

This section considers each constituent component of the algorithm – from the inputs and outputs, to the one-dimensional constraint equation, to some practical issues regarding derivative estimation and dynamic scale space and a deterministic robust averaging estimator based on the Least Trimmed Squares estimator [239]. Finally, the performance of the algorithm is assessed using data generated by a specially designed simulation package thus allowing us to ensure the algorithm operates as expected and to allow testing of the algorithm's performance before going to the time and expense of hardware implementation.

As was explained in the introduction, it is not possible to directly compare our algorithm to the work of others because our algorithm performs a very specific task that is different to the tasks performed by other algorithms. In summary, our motion estimation algorithm combines range and visual information to give a one-dimensional motion estimate in units of (centi)metres/second. Perhaps the most similar algorithm is that of Argyros *et al.* [20] who also used robust techniques (the LMedS estimator) and range information (via stereo vision). Their algorithm, however, is far more complex than ours, and gives a two-dimensional motion estimate which includes an estimate of translation and rotation for each point in the image. Many of the techniques used in the field of 'Vision Chips' [206] (e.g. the Reichardt detector [124], see Section 3.5) can also superficially resemble our algorithm since they usually give a one-dimensional result. However, the details of these techniques differ significantly from our algorithm, especially in the use of robust techniques and integration of data over image columns, and they do not use range data.

5.2.1 Inputs and outputs

The algorithm sources data from both a camera and a laser range scanner. Preprocessing of camera data involves subsampling to the appropriate scale of the dynamic scale space, then filtering to minimise high spatial frequency content. Preprocessing of range data begins with conversion from polar to rectangular coordinates, then data is interpolated onto the (subsampled) camera coordinate system with one range measurement provided for each image column. Images can range in resolution from 512×32 pixel to 32×32 pixels depending on the scale. The algorithm outputs a single motion estimate U_x (projection of motion onto the X axis) for each image column, and the interpolated range data thus allowing the full three-dimensional, translational ground plane motion to be extracted for each image column using (4.15).

5.2.2 Constraint equation

Given this input data, a means of estimating motion is required. Since an estimate of the relative three-dimensional translational ground plane motion (m/s) is desired, it is necessary to incorporate the rigid body motion model into the relationship. For the purpose of this design, all environment motion is assumed to occur on a smooth ground plane motion with large radius turns. Furthermore, only the projection of the velocity onto the X axis is of concern. This allows us to use (4.14) (repeated here as (5.1)) as a model for apparent velocity:

$$u = -\frac{F}{\zeta_{\text{eff}}} \frac{fU_X}{Z} \qquad (5.1)$$

As only the horizontal component of motion is used, the optical flow constraint equation (OFCE) (3.2) can also be simplified. The motion sought is single dimensional, thus rather than treating the input image as a single array, it is treated as 32 separate one-dimensional vectors. This (in addition to discussion in Section 5.1.3) means that there is no vertical motion so vertical optical flow v can be eliminated from the OFCE leaving

$$u = -\frac{I_t}{I_x} \qquad (5.2)$$

This is, in fact, the 1D normal flow. While theoretically there are cases (such as in Figure 5.3) where this normal flow may not be related to the horizontal motion of an object, the ground plane motion assumption ensures that this is rare. If it is assumed that optical flow is a good estimate of the apparent motion, (5.1) and (5.2) can be combined to give the constraint equation, (5.3):

$$U_X = \frac{ZI_t}{I_x} \frac{\zeta_{\text{eff}}}{fF} \qquad (5.3)$$

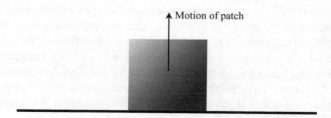

Figure 5.3   *For an image patch whose texture is at an angle, any upward motion will induce a normal flow parallel to the ground plane much as the barber pole illusion generates the appearance of vertical motion when the motion is in fact horizontal*

This relatively simple constraint equation can be solved directly at each pixel without further constraints or assumptions. As only a single dimension is considered, the aperture problem has disappeared. The final implementation uses a slightly modified version of this equation:

$$U_X = \frac{ZI_t}{I_x} \frac{\zeta_{\text{eff}}}{fF} \frac{SU_1}{SU_2} \tag{5.4}$$

Here SU_1 and SU_2 are scaling factors used to ensure sufficient numerical accuracy. This is discussed in more detail in Section 7.2.6.2.2.

5.2.3 Derivative estimation – practicalities

Given this constraint equation, the next issue to consider is which of the derivative estimators from Section 3.2.3.3 is best suited to this application. This decision is made based on the complexity and accuracy of the estimator. More complex (larger) derivative estimators do theoretically give better results, however, they leave a larger band of erroneous estimates around discontinuities and they are more costly to calculate.

We follow Simoncelli's [253] method of analysis; that is, we use the ability of derivative estimates to solve for optical flow as a measure of accuracy. Since the motion estimation problem is single dimensional, only single-dimensional estimators are tested (including a 1D variant of the Horn and Schunck estimator). Test data for this experiment are a sequence of sinusoids (of the form of (5.5)) with varying velocity and spatial frequencies

$$\text{round}[\sin(2\pi(x + v_s t)f_s)] \tag{5.5}$$

where f_s is the spatial frequency of the sinusoid[91], x is a pixel location, v_s is the velocity of the sinusoid in pixels/frame and t is time (measured in frames). To ensure this reflects the imaging process as accurately as possible, the sinusoid is scaled and quantised to fit in the integer range 0–255, corresponding to 8 bit intensity resolution available from most cameras. Optical flow estimates are also quantised so that they are represented using a nine-bit signed value, reflecting the representation used in the FPGA implementation. Optical flow is calculated directly from (5.2) using each of the derivative estimators[92]. To ensure that comparisons between various sinusoids remain valid, data are not prefiltered since this would alter the spatiotemporal characteristics of the sinusoid. Furthermore, flow estimates are made as nearly as possible to the same relative point in the waveform (i.e. zero-crossing, peak, etc.) no matter the spatial frequency. Results are shown in Figures 5.4 and 5.5 for five derivative estimators: FOFD, FOCD, Barron's four point central difference, Simoncelli's four point central difference and a modified Horn and Schunck estimator where the average of two (rather than four) FOFDs is used.

---

[91] Spatial frequency must be less than 0.5 cycles/pixel to avoid aliasing.
[92] Estimators are appropriately rotated when calculating I_t.

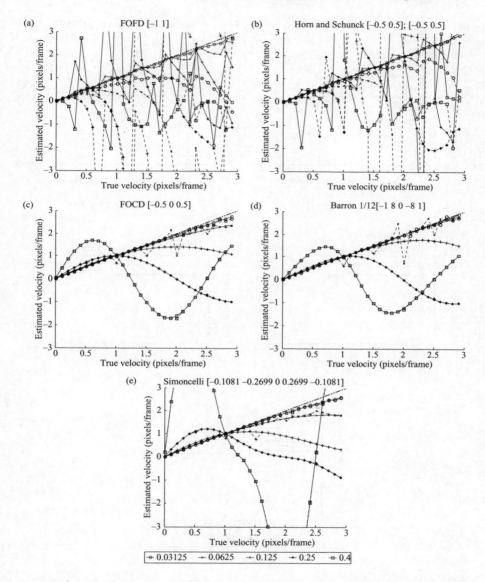

Figure 5.4 Estimated optical flow for the (a) FOFD, (b) the Horn and Schunck estimator, (c) Barron's 4 point estimator, (d) FOCD and (e) Simoncelli's 4 point estimator. Each plot shows motion estimates for a range of spatial frequencies as indicated by the legend at the bottom. Optical flow estimates were made at a point corresponding to 0° phase (solid lines) and 45° phase (dotted lines).

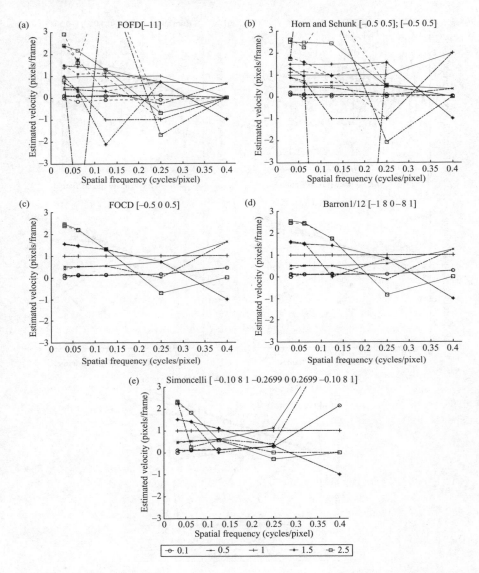

Figure 5.5 *Flow estimates as a function of spatial frequency. Each of our derivative estimators is tested to see how estimated optical flow varies with spatial frequency. This is tested at a number of frequencies as indicated in the legend at the bottom. Flow estimates were made at 0° (solid), 90° (dashed) 45° (dash-dot) phase. Note that as we near the aliasing limit of 0.5 cycles/frame all estimators perform poorly. Once again, the forward-difference estimators under-perform the central-difference estimators. Best estimates come from Barron's estimator and the FOCD which give very similar results*

Note that whilst we have followed Simoncelli's [253] general approach to testing of derivatives, the method used here differs from his in the following important ways:

- It uses a single-dimensional optical flow estimate. Consequently there is no need to test for rotation invariance [85]. This also leads to a divergence between the results obtained here and those of Simoncelli; however, the final conclusion remains the same – larger estimators do tend to provide better derivative estimates.
- A larger number of derivative estimators are tested.
- Sensitivity to extremely small optical flows is tested.
- A wider range of velocities and spatial frequencies are tested to give a better understanding of the characteristics of each estimator.
- Motion estimation is performed at more than one point. Motion estimates are made at the zero crossing, peak and the midpoint to compare how the relative location at which motion estimates are made affects the result. Ideally, results should be insensitive to phase.

In Figure 5.4 the velocity of the sinusoid is varied and flow estimates made at each velocity. This is repeated at a number of spatial frequencies for each estimator. Solid plots correspond to flow estimates made to a zero crossing (0° phase) and the dotted plots represent flow estimates made at 45° phase allowing some analysis of how sensitive derivative estimators are to phase. What is most striking is that the FOFD (Figure 5.4a) and Horn and Schunck estimators (Figure 5.4b) both give poor flow estimates, and both have a very strong dependence on phase. The FOFD estimate of flow is roughly correct at very low spatial frequency and velocity, however, beyond this the flow estimate fails completely. The Horn and Schunck estimator performs marginally better, especially at low spatial frequencies, however, the calculated optical flow remains poor at higher frequencies.

The Barron (Figure 5.4c), FOCD (Figure 5.4d) and Simoncelli (Figure 5.4e) estimators perform far better, giving excellent results when spatial frequency is low, with relatively little sensitivity to phase for optical flow below 1 pixel/frame. Above this, the quantisation of the sinusoid begins having some effect – if the sinusoid was not quantised both the zero-crossing and mid-point estimate of optical are identical for all three derivative estimators. Over all, Barron's derivative estimator gives the best flow estimates.

These plots also demonstrate the earlier assertion (Section 4.1) that, for low spatial frequencies, apparent velocities of more than 1 pixel per frame can be accurately detected. Indeed, for spatial frequencies of 1/16 cycles/pixel, reasonable estimates of apparent motion as high 3 pixel/frame are possible if using the Barron or FOCD estimator. As spatial frequency increases, estimates degrade; however, they do so more gracefully than for the forward-difference estimators. At 0.4 cycles/pixel all the central-difference estimators significantly overestimate motions of less than 1 pixel/frame. Higher velocities are rather unreliable.

Figure 5.5 gives a slightly different analysis. Rather than varying the velocity of the sinusoid, it varies the spatial frequency to see how it affects estimated optical flow[93]. This is repeated for a number of velocities and again repeated at: 0° (solid), 90° (dashed) and 45° (dash-dot) phase. The zero-crossing motion estimates generated using the Horn and Schunck estimator (Figure 5.5b) are the most accurate of all, however this accuracy disappears when estimates are made at a different phase. The FOFD performs even worse (Figure 5.5a). The central difference estimators (Figure 5.5c, d, e) perform quite consistently with respect to phase for velocities of less than 1 pixel/frame though above this phase does seem to play a role. All estimators fail when spatial frequency approaches 0.5 cycles/pixel.

An important issue that surfaces when discussing the hardware implementation of the estimation algorithm is its sensitivity to low optical flow. It is acknowledged that noise and image quantisation make low optical flow difficult to measure, nonetheless it would be useful to have *some* motion estimate available for objects with low optical flow (e.g. distant objects). Figure 5.6 illustrates the results of tests for sensitivity. The upper plot shows flow estimates made at zero crossings while the lower plot shows motion estimates at sinusoid peaks. Results for the zero crossings show that optical flows as low as 0.005 pixels per frame can be resolved with reasonable accuracy provided spatial frequency is neither too low (where quantisation limits resolution) nor too high (where spatial aliasing becomes problematic). Optical flow estimates made at sinusoid peaks are less encouraging. Intensity quantisation greatly reduces the accuracy of the optical flow estimates and leads to a zero optical flow for low spatial frequencies.

In summary, provided spatial frequency is limited to below approximately 0.3 cycles/pixel and temporal aliasing is avoided, the central difference estimators give good results. Barron's estimator provides the best motion estimates over all conditions, however, the FOCD estimator is only slightly worse. This similarity is to be expected since Barron's estimator gives a small weighting to the second order terms. Simoncelli's estimator performs well but it is more affected by high spatial frequencies since it weights first and second order terms similarly, exaggerating derivative estimation error. Since it performs nearly as well as Barron's estimator and incurring significantly less computational cost[94], the FOCD derivative estimator is chosen for use in our algorithm. While the accuracy of the FOCD at very low optical flows is somewhat variable, it can certainly be used to give a rough indication of motion, provided that we accept its accuracy is limited.

Note that the FOCD (or any central difference estimator) has one significant drawback. When used to estimate temporal derivatives, these estimators are non-causal (the estimator requires data from the previous frame and the *next* frame) requiring us to introduce a latency of one frame. We accept this as the cost of obtaining accurate derivate estimates.

---

[93] Note that this thesis does not consider spatial frequency less than 0.05 cycles/pixel since, as spatial frequency falls, the accuracy of the motion estimate is more dependent on numerical accuracy than on the accuracy of the derivative estimator. Of course, in the limit where spatial frequency is zero cycles/pixel all methods will give a zero velocity estimate.

[94] This estimator can be implemented using a single subtraction operation. Barron's estimator requires two bit shifts and three addition/subtraction operations. Simoncelli's estimator is the most complex requiring four floating point multiplications and three addition/subtraction operations.

Motion estimation for autonomous navigation 131

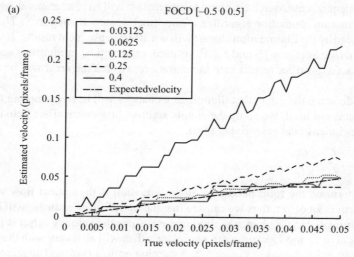

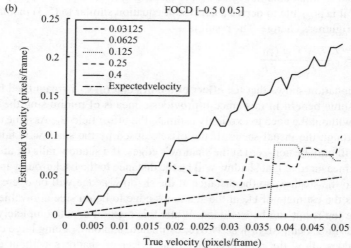

Figure 5.6 *FOCD derivative estimator used to estimate very low optical flows. The top plot was made at zero crossing while the bottom plot was made at peaks in the sinusoid. Note that results are quantised to fit in a 9 bit signed value as per the hardware implementation*

5.2.4 Effect of illumination change

Changes in illumination occur regularly in natural environments – clouds passing over the sun and moving through the dappled shade of a tree all cause the brightness of points in the environment to change over time. Such changes are direct violations of the brightness constancy assumption upon which our motion estimation algorithm is based, and thus it is important to know how changes in illumination will affect motion estimates [253].

In the general case, quick changes in illumination will have a temporally localised effect for motion estimation algorithms using small temporal support[95]. That is, the error induced by the illumination change will not affect subsequent results. If temporal integration (see Sections 3.3 and 5.2.10) is used, error will be minimised (since more faith is placed in earlier results than the current result) but the error will be smeared over time.

To understand the effect that illumination changes will have (assuming temporal integration is not used) we apply the simple additive brightness offset field of (3.37) to the one-dimensional case giving (5.6).

$$u = -\frac{I_t}{I_x} - \frac{C_t}{I_x} \tag{5.6}$$

The first term on the right hand side of this equation is the optical flow while the second term is an offset, thus we can see that the optical flow estimate will be offset by a factor related to the change in illumination. Since the velocity offset is inversely proportional to I_x and I_x varies with phase, the offset will also vary with the phase of the sinusoid as illustrated in Figure 5.7. Assuming optical flow and apparent motion are equal, it is possible to derive a constraint equation similar to (5.3) that takes into account brightness change. The result is

$$U_x = \frac{Z(I_t + C_t)}{I_x} \frac{\zeta_{\text{eff}}}{fF} \tag{5.7}$$

This equation shows that the effect of the offset term C_t is magnified by range. There is some benefit in this since it provides a means of minimising the effect of shadows without the need to explicitly estimate the offset field C_t. Assume a shadow moves through the visual scene. If the offset caused by the shadow is fairly even, then C_t will only be non-zero at the shadow's edges. If a shadow falls on the ground, the range measured to that shadow will be the distance to the background and not the distance to the shadow, so the distance at the shadow's edge will be overestimated. This leads the estimate of U_x at the shadow's edge to be outside allowable limits[96] and hence the result can be excluded. While this method is not completely reliable (it depends on Z being significantly over estimated, and on C_t being large compared to I_t) it does allow the minimisation of the effect of shadows without explicitly estimating the offset field. This is discussed in more detail when considering the numerical accuracy of the hardware implementation.

5.2.5 Robust average

As discussed in Chapter 3, motion estimation must be implemented using robust methods so that results can be smoothed over the largest possible region of support

---

[95] We use three frames to calculate temporal derivatives.
[96] As described earlier, maximum allowable velocity is known to be $V_{\max}$ and from this we can determine an upper-bound for U_x using (4.15).

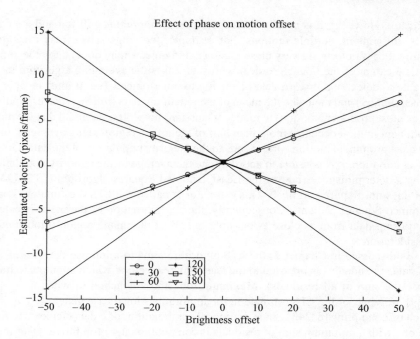

Figure 5.7   Effect of global illumination change and phase. Optical flow is estimated using the FOCD at various phases in the sinusoid. It can be seen that the velocity offset varies with phase as expected. This plot was generated using a sinusoid with a spatial frequency of 0.015625 cycles/pixel, and velocity is of $\frac{1}{2}$ pixel/frame. The X axis indicates the brightness change per frame, assuming a dynamic range of pixel intensity of 0–255 and a sinusoid scaled in this range to avoid saturation

(to minimise noise) without smoothing across motion boundaries. As already seen, sophisticated algorithms able to perform this task using purely visual data do exist, indeed early work focused on such an approach. While this line of investigation was initially promising (e.g. [165, 170, 172]) further investigation showed this approach was not reliable because results depended strongly on parameter tuning[97] and the ideal parameter tuning changed with the nature of the input data. This led to the idea of basing motion segmentation on range data. This allows simple segmentation (only requiring a search for 2nd derivative zero crossings) and smoothing across blobs without the need to determine motion boundaries using motion estimates themselves.

In the simplest case, such smoothing is straightforward – averaging over each blob will suffice (referring to the terminology of Chapter 2, the average is a location

---

[97] At that time we used the Lorentzian as our estimator.

estimate). However, it is possible that the earlier assumptions will fail and a single blob will contain multiple motions. For example, the range sensor may miss discontinuities if objects are very close together, and objects may not occupy the entire vertical extent of the image. To deal with such cases the averaging algorithm must be able to detect and discard data belonging to other objects (i.e. it must be able to detect *outliers*) and then take the mean of the remaining data (*inliers* – corresponding to the most prominent motion) to reduce (Gaussian) noise. It is assumed that the most prominent motion occupies more than half of the image region being averaged, thus the most prominent motion corresponds to the majority population of the data. Given this assumption, it is possible to approximate the average of the majority population using a deterministic variant of the Least Trimmed Squares algorithm (LTS) [239] that does not require sorting[98]. This estimator is called the LTSV estimator (Least Trimmed Squares Variant). Consequently, the approximation of the mean is calculated using data from as close as possible to half of the image region (blob) under consideration.

As discussed in Chapter 2, the LTS algorithm aims to minimise the sum of the h smallest residuals (as opposed to the least squares method which attempts to minimise the sum of all residuals). Maximum robustness is achieved when $h = \frac{1}{2}N$ [239]. Solving for the LTS estimate in the multidimensional case generally requires stochastic sampling [8, 240], since trialing all combinations of h data elements to find the one with a minimum sum of residuals is computationally prohibitive. Deterministic solutions are also possible [8, 232], however, they are still quite computationally expensive. In the case of one-dimensional location estimation (e.g. estimation of the mean) the LTS algorithm can be simplified to calculating average and sum of squares for a number of subsets of the ordered dataset – a task that can be performed recursively [239].

All the above algorithms require either stochastic sampling or ordering of data. We seek a simpler algorithm, friendlier to digital implementation, where sorting[99] of data and random sampling is not required. We also seek an estimator that is scale equivariant (like the LTS/LMedS estimators) so that estimation need not be tuned to the input data's noise variance (i.e. scale [201]). Based on these requirements the LTSV algorithm described below was developed[100]. This algorithm depends on the fact that the average of a dataset will fall closer to the majority data than to the minority data. Because the algorithm explicitly tries to find an estimate based on half the data, it is scale equivariant.

---

[98] To the best of our knowledge this algorithm has not appeared in the robust motion estimation literature or in the general robust estimation literature. Our search included extensive bibliographies and research papers of robust statistical methods prior to 1974 [18, 110, 125], a number of more recent texts on robust statistical methods (including but not limited to [122, 148, 201, 239, 267, 303]) and an extensive literature and internet search.

[99] Fast sort algorithms (e.g. quicksort) tend to be awkward to implement in digital hardware, while simple algorithms such as bubble sort are somewhat slow.

[100] Once again, while we are not aware of any literature documenting this algorithm, we welcome any further information regarding prior publication of this.

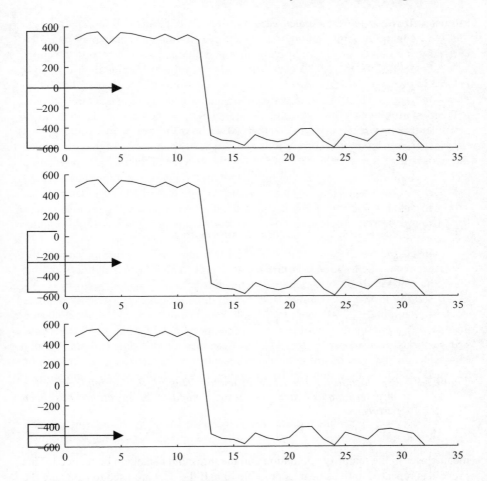

Figure 5.8   Three iterations of the LTSV algorithm

Imagine a one dimensional motion estimate as in Figure 5.8, for which we wish to robustly find the mean. There clearly appears to be two processes at work here, perhaps corresponding to different moving objects. Since the region on the left is the minority population of the data, it should be discarded and the average of the remaining data calculated. The LTSV estimator achieves this by iteratively thresholding data and taking the average of the remaining data. Initially, the mean is assumed to be zero and the threshold is set large enough to encompass the entire dataset (see top Figure 5.8). The mean is recomputed, this time using the data within the threshold. On the first iteration this will encompass all the data, and by its nature, this mean will be closer to the majority data than to the minority data. If the threshold is reduced, recalculation of the mean using only data within the threshold will use more data from the majority population than from the minority population. This will again push the

```
thresh = abs(max possible input value);
current_mean = 0;

for i=1 to number_iterations

        count=0;
        sum=0;

        for j=1:n              // Sum values within threshold
                if abs(current_mean - data(j)) < thresh
                    sum = sum+data(j);
                    count = count+1;
                end if
        end for

        if count >= n/2      // Update Threshold
                thresh = thresh/alpha;
        else
                thresh = thresh*beta;
        end if

        if count ≠ 0         // Calculate mean
                current_mean = sum/count;
        end if
end for
```

Figure 5.9   Pseudocode for the LTSV estimator. This particular implementation only terminates when the desired number of iterations has been performed

mean closer to the majority population and the threshold can again be reduced. This process is repeated until as near as possible to half the data are used to calculate the mean or until the maximum number of iterations have been performed. The algorithm is presented as pseudocode in Figure 5.9.

In practice, reducing the threshold (by a factor of `alpha`) may cause the exclusion of all data. If more than half the data are excluded, the threshold is multiplied by a factor `beta < alpha` that increases the threshold without making it larger than the previous value. The algorithm is designed to successively increase and decrease the threshold until as near as possible to half of the data points are excluded. Potential (though not necessary) variations to this basic LTSV algorithm may include:

- *Dynamic reduction of* `alpha` *and* `beta`. This should be each done each time more than half the data are excluded to reduce overshoot and improve the convergence rate.
- *Recursive implementation.* Data elements included in the average are discarded and a new average calculated using the remaining data. This can continue until

some minimal number of data elements remains or until the difference between successive averages is smaller than the final threshold[101].
- *Improve efficiency through the use of a weighted least squared postprocessing step as described in [239]*. This is simply a least squares estimate where the weights are defined as per (5.8) and t is chosen with respect to the variance of the inliers. Rousseeuw and Leroy suggest using $t = 2.5$ under the assumption that the residuals r_i are normalised using a scale estimate:

$$w_i = \begin{cases} 1 & \text{if } |r_i| < t \\ 0 & \text{if } |r_i| \geq t \end{cases} \tag{5.8}$$

- *Extension to linear regression and multivariate location estimation.* Estimation of multivariate location is quite simple, requiring only a modification of the distance metric as in Section 2.2.7. Extension to linear regression is also straightforward, however, the robustness properties of the resulting estimator are somewhat poor. Our work[102] shows the resulting estimator can only handle a small number of outliers or leverage points. The reason is simple. As we can see in Figure 5.10, once the number of outliers is too high, the initial 'average' line (the one calculated with the threshold set high enough to include all data) is drawn so close to the outliers that the outliers will never be excluded by the threshold operation. In fact more inliers will be rejected by the threshold operation than outliers so the line-of-best-fit may be drawn towards the outliers. A similar thing happens in the case of leverage points as seen in Figure 5.11. A more robust initial fit [122, 201] (something better than the least squares line-of-best-fit that is used currently) would help, however, this is still not enough. Care must be taken to ensure the initial threshold value is not so large that all data points are included in the first iteration thereby removing the benefit of the robust initial fit.

5.2.6 Comparing our robust average to other techniques

While the LTSV estimator is similar to many other robust estimators, it does appear to have a number of features that are unique. In this section we consider a range of other techniques and show how they differ to the LTSV. We do not consider only estimators (such at the M estimator, Trimmed Mean, etc.) but also solution methods for estimators (such as Graduated Non-Convexity and the Downhill Simplex Method), since the LTSV can be thought of as an estimator in its own right, or, as we shall see, it can also be thought of as a solution mechanism for the M estimator. In this section we only show how our LTSV estimator is unique – comparison of

---

[101] This indicates that there is unlikely to be a secondary process in the region of support.

[102] The estimator here was constructed using the same premise as for the location estimator – iteratively remove residuals over a particular threshold from the estimation problem. In this case the residual is the vertical distance from the line-of-best-fit, rather than the simple distance from a point as it was for the location estimator.

138 *Motion vision: design of compact motion sensing solutions*

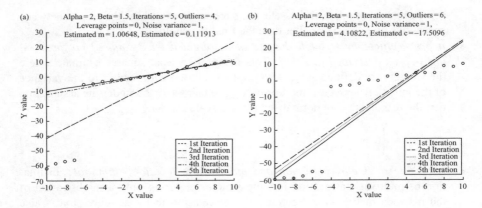

Figure 5.10   *The effect of outliers on linear regression using the LTSV method. We generated test data using parameters $M = 1$, $C = 0$ and added Gaussian noise with a variance of 1. We then made some of the data points outliers by subtracting 50 from the Y values. When the number of outliers is small, the LTSV method is able to reject them, however, as the number of outliers increases, the line-of-best-fit is dragged very close to the outliers and the thresholding operation cannot reject them. Note that some of the lines-of-best-fit overlap hence the result at each iteration is not necessarily visible. In (a) there are 4 outliers which are successfully rejected while in (b) there are 6 outliers which is clearly too many for the LTSV estimator to deal with*

the performance of some more common estimators is presented in the following section.

5.2.6.1 Trimmed and winsorised means

Perhaps the simplest way of making the simple arithmetic mean (i.e. the LS estimate of position in one dimension) more robust is the use of trimming or winsorisation. The trimmed mean is calculated by first ordering the data, removing some fraction of the data from either end of the dataset and calculating the mean using the remaining data. In this way, (some) extreme estimates are removed from the data sample, making the final result (more) robust. The symmetrically[103] trimmed mean has an equal fraction of the data removed from either end of the dataset and can be written as follows

$$\frac{1}{N - 2\alpha N} \sum_{i=\alpha N}^{N-\alpha N} \vec{d}_i \qquad (5.9)$$

---

[103] Symmetrical trimming/winsorising is ideal if the distribution of data is symmetrical (e.g. Gaussian). Non-symmetrical trimming/winsorising is used for non-symmetrically distributed data.

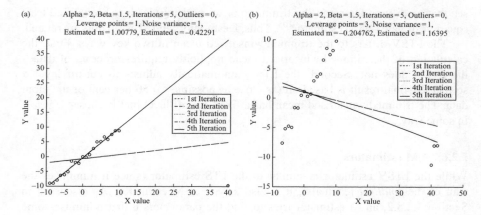

Figure 5.11 The effect of leverage points on linear regression using the LTSV method. We generated test data using parameters $M = 1$, $C = 0$ and added Gaussian noise with a variance of 1. We then made some of the data points leverage points by adding 50 to the X values. When the number of leverage points is small, the LTSV method is able to reject them, however, as the number of leverage points increases, the line-of-best-fit is dragged very close to the leverage points and the thresholding operation cannot reject them. Note that some of the lines-of-best-fit overlap hence the result at each iteration is not necessarily visible. In (a) there is one leverage point which is successfully rejected while in (b) there are three leverage points which is clearly too many for the LTSV to deal with

Here N is the number of elements in the ordered dataset $\vec{d}$ and $0 < \alpha < \frac{1}{2}$ is the fraction of data trimmed from the sample. When $\alpha = 0$ the trimmed mean is equal to the mean, and when $\alpha = \frac{1}{2}$ the median is obtained. A number of values for α have been suggested in the literature, as have adaptive methods for choosing the value of α (e.g. [18]).

The winsorised mean is similar to the trimmed mean, except that rather than completely rejecting data, any data points outside a given range are thresholded to that range. In that symmetric case, the αN smallest samples will be replaced with the $\alpha N + 1$th sample, and the αN largest samples will be replaced with the $(N - (\alpha N + 1))$th sample. In this way, the influence of the outlying data point is reduced without discarding the data completely.

Trimmed and winsorised means are often referred to as L estimators since they are calculated using a linear combination of order statistics[104]. As we saw in Section 2.2.8 the least squares estimate and the arithmetic mean are the same, therefore one can

---

[104] The nth order statistic is simply the nth element of an ordered dataset. For a dataset with N elements (where N is odd), the $(\lfloor N/2 \rfloor + 1)$th order statistic is the median.

see that modifying the mean (by trimming or winsorising) leads to a modified least squares estimate – an M estimator[105]. Thus, L estimators and M estimators are related.

The LTSV differs to the trimmed/winsorised mean in two key ways. First, the calculation of the trimmed/winsorised mean implicitly requires ordering of data – the LTSV does not. Second, the LTSV automatically adjusts to cut-off level α so that the final result is based on (as close as possible to) 50 per cent of the input data. The trimmed/winsorised mean has a fixed cut-off level that is chosen *a priori* in some way.

5.2.6.2 M estimators

While the LTSV estimator is similar to the LTS estimator (since it minimises the sum of the *smallest* residuals), it is also a type of M estimator. As discussed in Section 3.2.8.2, an M estimator tries to find the parameters θ that minimise some function ρ of the residuals r_i (5.10).

$$\min_{\theta} \sum_{i=1}^{N} \rho(r_i) \tag{5.10}$$

The LTSV algorithm as we have described emerges if the estimator function ρ is assumed to be a variant of the truncated quadratic (5.11) that trims over a threshold (though a winsorising cut-off as in (3.58) can also be used):

$$\rho(r) = \begin{cases} r^2 & r^2 < \text{threshold}^2 \\ 0 & \text{otherwise} \end{cases} \tag{5.11}$$

In the case of the LTSV estimator, the resulting minimisation problem is solved using an iterative method where the threshold is updated appropriately at each iteration and convergence occurs when as close to half the data as possible are outside the threshold.

5.2.6.3 Iteratively reweighted least squares

One means of solving robust M estimators is via a process known as Iteratively Reweighted Least Squares (IRLS) [201, 267]. Proceeding naïvely, one can attempt to find the minimum of (5.10) by setting the derivative of (5.10) with respect to the parameters to zero. The derivative of the estimator function ρ is the influence function ψ hence the derivative can be written as:

$$\sum_{i=1}^{N} \psi(r_i) \cdot \frac{dr_i}{d\theta} = 0 \tag{5.12}$$

---

[105] While the trimmed/winsorised mean is just a one-dimensional location estimate L estimators can also be applied in linear regression and multivariate location estimation.

This can be converted into a weighted least squares problem by modifying the influence function as follows:

$$\frac{\psi(r)}{r} = W(r) \tag{5.13}$$

$W(r)$ is usually referred to as a weight function and its most important feature is that it gives smaller weights to larger residuals (above and beyond any weight reduction caused by the influence function), causing them to have less influence on the final result. Substituting the weight function into (5.12) gives us:

$$\sum_{i=1}^{N} W(r_i) \cdot \frac{dr_i}{d\theta} \cdot r_i = 0 \tag{5.14}$$

In the specific case of a location estimate (i.e. $r_i = d_i - \theta$), we can solve for the location parameter θ using (5.15). This is essentially a weighted average where the weights are determined by the residuals.

$$\theta = \frac{\sum_{i=1}^{N} W(r_i) \cdot d_i}{\sum_{i=1}^{N} W(r_i)} \tag{5.15}$$

IRLS can be thought of as a sort of Jacobi iteration [101] with the current estimate of the parameters being used to generate a new estimate. The first step of the IRLS method is to obtain some estimate for the parameters $\boldsymbol{\theta}$ whether by least squares or some other technique. This estimate can then be used to calculate the residuals and consequently to calculate the weight function $W(r)$. Given the weights, it is possible to again solve (5.14) for the parameters $\boldsymbol{\theta}$. This process is iterated until some convergence criterion is met (e.g. change in the residuals is less than some value).

There are many similarities between the LTSV and IRLS algorithms. Both algorithms update an influence function based on the current estimate of the parameters. The critical difference is that, in the LTSV algorithm, the influence function is modified at each iteration (the truncation point of the associated estimator function is reduced), while the influence function is constant in the IRLS approach. Furthermore, because IRLS is simply a solution method for M estimators, all the problems of M estimators (need for an estimate of scale, relatively low breakdown, etc.) remain.

The IRLS solution technique adds another problem. Since estimator functions ρ are generally not convex, (5.12) will not have a unique minimum and the IRLS algorithm may converge to a local minimum rather than the global minimum. To avoid this, the starting point for iteration should be as close to the global minimum as possible. It is not clear how this can be achieved consistently; however, it is common to use the median or least squares estimate as a starting point [122, 201]. The LTSV algorithm avoids local minima by updating the weight function appropriately.

5.2.6.4 Graduated non-convexity

An alternate means of solving for parameters when using an M estimator is known as graduated non-convexity (GNC) [35, 36] (Section 2.3.3.2.6). The key benefit of GNC is that it allows us to avoid local minima and find the true global minimum. GNC is a sort of deterministic annealing that operates by building a sequence of approximations $\rho^{g \to 0}$ to the estimator function ρ with ρ^g being a convex approximation of ρ and $\rho^0 = \rho$.

The GNC process[106] begins by finding a solution for the parameters (via some iterative technique) using the estimator function ρ^g. This solution is used as a starting point for a solution based on estimator function ρ^{g-1} and the process is repeated until we reach a solution based on ρ^0. Originally, a truncated quadratic function was used as the base estimator function [36]; although it is also possible to use other functions. For example, it has been shown that the Lorentzian function provides a far more natural way of constructing the required approximations [35].

Interestingly, Li [191, page 216] suggests that it may be possible to perform GNC using the truncated quadratic by starting with the truncation point α sufficiently large to ensure that no residuals are considered to be outliers, leading to a least squares solution. α is then gradually (monotonically) reduced until the desired cut-off point is reached (based on the variance/scale of the inlying data) and the estimate has converged. This is similar to the use of the Lorentzian function in Black's work [35] and is also similar to the LTSV, which updates the cut-off point of the truncated quadratic estimator function. However, LTSV has no predefined final value for α, hence it does require an estimate of scale. Furthermore, in the LTSV method, the value of α does not strictly decrease. Rather, it can increase or decrease until as close as possible to half the data are rejected.

5.2.6.5 Annealing M estimator

Li [191, 192] introduced the annealing M estimators (AM estimators) which extend IRLS to include the deterministic annealing process of GNC. AM estimators have the additional benefit that they do not require an estimate of scale[107] and unlike IRLS do not need a 'good' initialisation. In an approach analogous to GNC, AM estimators replace the weight function W in (5.15) with a series of approximations to the weight function $W^{k \to 0}$. Equation (5.15) is then repeatedly solved starting with W^k and working towards W^0. The AM estimator is quite similar to the LTSV estimator if we assume that the weight function is based on the truncated quadratic and that updates of the weight function lower the cut-off point of the truncated quadratic as discussed in Section 5.2.6.4. The key difference is that the annealing processes of the AM estimator is strictly monotonic proceeding from a convex approximation towards the target, whereas the annealing process in the LTSV estimator (the reduction in the cut-off point) specifically attempts to find an average based on half the data.

---

[106] See Section 2.2.6 for more detail.

[107] This claim is dubious – while the scale parameter might not need to be used to normalise the residuals it should be considered in choosing ρ^0 which will define which residuals are considered to be inliers and which are considered to be outliers.

5.2.6.6 Adaptive cut-off

You might think of the LTSV estimator as iteratively changing the cut-off value for which data are included in an estimate until as close as possible to half the data have been excluded. Other adaptive methods are also possible. For example in References 103 and 102 a post processing step (similar to that of Rousseeuw and Leroy [239]) is presented where the efficiency of an existing robust estimate is improved by performing a weighted least squares step where the weights are determined adaptively. Such techniques differ to LTSV in that they require an existing robust estimate.

5.2.6.7 Anscombe's method

Anscombe [19] presented a robust technique for estimating the mean of a dataset assuming the variance σ^2 (i.e. the scale – see Section 3.2.8) of the inliers is known. A number of similar methods have also appeared [314]. The thrust of these algorithms is to retain only those data points whose residuals are less than some constant multiple of σ (i.e. $C\sigma$). Anscombe's algorithm proceeds as follows:

1. Calculate the average using all data points. Let m be the index of the residual with the largest absolute value, such that $r_m = \max |r_i|$. If there are many such residuals, m can be assigned arbitrarily to any of these.

$$\bar{y} = \frac{1}{N} \sum_{i=1}^{N} y_i \qquad r_i = y_i - \bar{y} \qquad (5.16)$$

2. For a given value of C, reject the data point corresponding to r_m^k if $|r_m^k| > C\sigma$ then recalculate the mean using the remaining values (k is the iteration number). In other words, the mean is updated (5.17). Here n is the number of data points used to calculate the current mean.

$$\bar{y}^{k+1} = \begin{cases} \bar{y}^k & \text{if } |r_m^k| < C\sigma \\ \bar{y}^k - \dfrac{r_m^k}{n-1} & \text{if } |r_m^k| > C\sigma \end{cases} \qquad (5.17)$$

3. Repeat step 2 until no further data points are rejects.

In his paper Anscombe suggests that there is no advantage in varying C and makes no suggestion as to a reasonable value for C. This technique is reminiscent of the LTSV in that it starts from the least squares estimate (the mean) and iteratively rejects data points with large residuals. It differs from the LTSV in that (a) only a single residual (r_m) can be eliminated at each iteration and (b) the cut-off point ($C\sigma$) is fixed rather than being updated at each iteration.

5.2.6.8 Minimum volume ellipsoid

The Minimum Volume Ellipsoid (MVE) estimator proposed by Rousseeuw and Leroy [239] is a multidimensional location estimator that works by finding the smallest ellipsoid that contains some fraction (usually half) of the data points. Therefore, you can think of the MVE method as searching for a point of maximum density (i.e. the mode) in the distribution of data points. A random sampling approach is used to find the ellipsoid and hence provide the desired location estimate.

The MVE and LTSV are based on similar concepts, however, they are implemented very differently. Recall that a location estimate has a residual of the form

$$r_i = d_i - \theta \qquad (5.18)$$

where d_i is the ith data point and θ is the location estimate (the estimate of the central tendency of the set of data points **d**). In the multidimensional case, the data points and location become vectors, however, the residuals remain scalar values that indicate the distance between a data point and the location estimate:

$$r_i = \mathbf{d_i} - \boldsymbol{\theta} \qquad (5.19)$$

The definition of the subtraction operations (the distance measure) is now open to interpretation. We consider only the Euclidian distance (5.20) here, though other distance measures such as the Mahalanobis distance [2, 4] can also be used.

$$r_i = \sqrt{\sum_{\text{dim}} (d_{\text{dim},i} - \theta_{\text{dim}})^2} \qquad (5.20)$$

Here the summation is over each dimension of the location (i.e. in the two-dimensional case, summation is over the X and Y dimensions). Now we can see that while the MVE tried to find the smallest ellipsoid containing half of the data points, the LTSV iteratively eliminates data points to effectively find the smallest hypersphere (n-dimensional sphere) containing half of the data points. The radius of the hypersphere is equal to the final threshold and the centre of the hypersphere is the location estimate $\boldsymbol{\theta}$.

5.2.6.9 Convex hull peeling

Aloupis [11, 12] presents a range of geometric methods for finding the 'median' of a multidimensional dataset. The method most similar to the LTSV is Convex Hull Peeling (CHP), credited to Tukey [11]. A convex hull is the smallest convex polygon that encloses a set of data points and convex hull stripping uses this idea to consecutively strip away the furthest points (i.e. the outliers) until only one or two points remain as illustrated in Figure 5.12. Both the LTSV and CHP attempt to iteratively remove outliers, however, the LTSV does this by means of iteratively moving and resizing a hypersphere while CHP performs the same function using convex hulls. As with other methods we have examined, CHP is also strictly monotonic in its operation (it always 'moves towards the middle') while LTSV can both increase and decrease the size of the hypersphere.

5.2.6.10 Downhill simplex method

The downhill simplex method (DSM) is a means of finding the minimum of a function (e.g. the M estimator of (5.10)) without the need for derivatives [156, 231]. A simplex is an N-dimensional geometrical figure made up of N + 1 vertices and the resulting line, faces, etc. For example, the two-dimensional simplex is a triangle and the

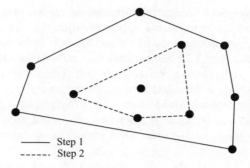

Figure 5.12 Convex hull stripping in the two-dimensional case (adapted from Reference 11)

three-dimensional simplex is a tetrahedron. The basic approach behind the DSM is that the value of the function at each vertex of the simplex is calculated and the maximum and minimum found. The simplex is then reflected about the face opposite the maximum vertex essentially moving the simplex 'down hill'. A combination of such reflections together with contraction and expansion operations will gradually move the simplex towards the minimum and hence a solution for the unknown parameters.

While the exact objectives of the LTSV and DSM are different (the LTSV aims to find the smallest hypersphere enclosing half of the data points, while DSM is searching for a minimum), both methods move a geometrical figure around space to find some optimal location. The LTSV begins by assuming that there is a hypersphere centred at the origin large enough to encompass all data points and calculates the average of all points within the hypersphere[108]. After this step, the hypersphere is successively relocated (to a location based on the current average) and resized (so that at least half the data points are enclosed by the hypersphere) until a solution is found.

5.2.6.11 Mean shift method

The Mean Shift Method[109] (MSM) [62, 201] seeks the point of maximum density in the distribution of data points (i.e. the mode). At its simplest the MSM begins by randomly choosing a data point as an initial local mean. This local mean is updated by computing the weighted mean of points in some local region[110]. This mean will naturally be closer to the highest density point within the local region, hence moving the mean towards the point of maximum density. This process is repeated until the mean shift vector (the distance between the successive means) falls below some threshold.

---

[108] This is the least squares estimate of location.
[109] Also known as the Parzen window method.
[110] This local region is defined as the set of points less than some predefined distance (the Mahalanobis distance [2, 4] is often used) from the current average.

The LTSV works on a similar principle, using the assumption that the mean of a dataset will fall closer to the majority population (i.e. the point of maximum density). The difference between the LTSV and MSM is in the iteration scheme and convergence criteria. The LTSV begins using the entire dataset as the 'local region' then adapts the size of the local region until only half of the data are included. Because the entire dataset is used as the initial local region, no special (random) starting point is required. Traditionally the MSM starts with a random data point (although the sample mean could also be used as a starting point) and searches for a point of maximum density (based on the mean shift vector) using either fixed or adaptive local regions.

If the dataset is multi-modal (i.e. has more than one peak in the density function), a random starting point will lead the MSM to the nearest peak, while using the global mean would lead to the highest peak. Regardless of the starting point, a valid peak will always be found. On the other hand, the LTSV assumes there is a single majority population. If there is no clear majority population the LTSV will not give a useful result. LTSV has an advantage if recursive operation is required since it explicitly divides the dataset into two subsets – inliers and outliers. Once a solution for the inliers (i.e. the majority population) is found, the LTSV algorithm can be applied to the remaining data to search for the majority population in that subset. Finding multiple peaks using the MSM method is more difficult since the data are not explicitly partitioned. To overcome this some method utilising a number of random starting points, or a means of excluding data associated with each peak, must be devised.

5.2.6.12 Maximum trimmed likelihood

Hadi and Luceno [119] proposed the Maximum Trimmed Likelihood (MTL) estimator which attempts to maximise the sum of the trimmed likelihoods λ, rather than minimise the sum of trimmed residuals as is the case with the LTS estimator:

$$\max \sum_{i=a}^{b} \lambda(\theta; d_i) \qquad (5.21)$$

It is shown that this estimator reduces to the Least Median of Squares, Least Trimmed Squared, Minimum Volume Ellipsoid or Least Squares estimators depending on the choice of a, b and λ and the characteristics of the unknowns θ. The interested reader is referred to the source article for further information regarding this estimator – here we simply note that while the MTL estimator is quite general, solving for the parameters is very quite involved.

5.2.7 Monte Carlo study of the LTSV estimator

In this section, we compare the performance of the LTSV algorithm in the case of one-dimensional location estimation with four estimators commonly used in the motion estimation literature:

- the sample mean (mean of all data)[111];
- the sample median[112];
- the Lorentzian M estimator solved using Graduated Non-Convexity (GNC) [35]; and
- the one-dimensional LTS algorithm as described in Reference 239.

In this Monte Carlo study, we use a simple variant of the LTSV algorithm where the alpha and beta values are fixed at 2 and 1.25 respectively and the initial threshold is set according to the maximum and possible absolute value of the data (as observed from Figures 5.13 and 5.14). This threshold is used directly as the starting threshold in the LTVS estimator, and it is used to determine appropriate start and end values for the continuation parameter in the GNC solution[113]. Two steps of the GNC algorithm are used as suggested by Black [35]. Both the LTSV estimator and the Lorentzian estimator are tested at 7 and 100 iterations to ascertain the rate of convergence.

Given this configuration, the performance of each algorithm is compared using four different Signal to Noise Ratios[114] (SNRs) with 100 signal realisations at each SNR. Zero mean additive Gaussian[115] noise is used in each case. Each signal realisation contains 32 elements (corresponding to the height of our input images) of which either 12 (Figure 5.13) or 6 (Figure 5.14) are outliers. In each case bias is measured as the percentage difference between the mean of the true inliers and the estimated mean of the inliers. Note that for signals with an SNR of zero, there are no outliers thus we take the mean of the entire dataset. Results are shown in Figures 5.14 and 5.15 and summarised in Tables 5.1 and 5.2.

The case where 12 of the 32 data elements are outliers is considered first in Figure 5.15 and Table 5.1. Remember that in the one-dimensional case, M estimators can tolerate 50 per cent of the data being outliers [239], as can the 1D LTS algorithm and the LTSV algorithm, hence all algorithms should be able to tolerate this number of outliers. From these results we concluded:

- The sample mean is a poor estimator when there are outliers in the data while the median performs relatively well in a wide range of conditions.

---

[111] The sample mean is calculated to provide a baseline performance indicator.

[112] The median provides a good estimate of majority population in a dataset provided the data distribution has small tails [230]. It can be computed in log N passes through the data with no reordering [230].

[113] We did not attempt to estimate the scale parameter from the data nor did we use the fact that, in this study, we could determine the scale parameter in advance. Rather, we fixed the final value of the continuation parameter at one tenth of the starting value T. Thus, in the final phase of the GNC algorithm, outliers are assume to lie at a distance greater than $(1/10)T$ from the current estimate. Using this fixed scale estimate is valid in our study since the purpose of our work is real-time operation. Solving the M estimator is complex enough without the added complexity of computing the scale parameter.

[114] The signal to noise ratios are expressed as simple ratios – not in decibels.

[115] Whether or not this is a reasonable noise model for motion estimation remains unclear. For comparison we repeated the experiments listed here using additive uniformly distributed noise. The results showed higher mean error and higher error variance than in the Gaussian case.

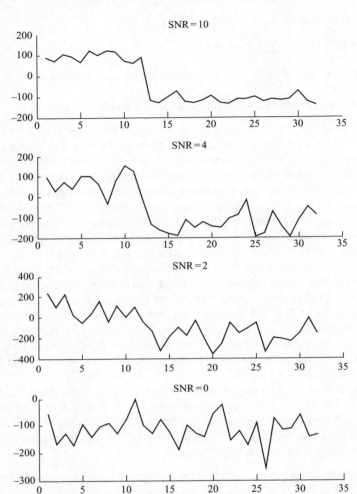

Figure 5.13 Example data used for comparison of average estimators – 12 outliers. Each diagram corresponds to a different signal to noise ratio (SNR). Zero SNR corresponds to a situation with no outliers

- The number of iterations has a strong influence on the performance of the Lorentzian estimator with error being significantly reduced when iteration is increased (although there is also an increase in standard deviation for SNR = 2). The LTSV estimator, however, converges quite well in just seven iterations – additional iterations do not improve results.
- For seven iterations, the median, 1D LTS and LTSV estimators all outperform the Lorentzian estimator. However, at 100 iterations the Lorentzian performs as well (if not better) than the other estimators.

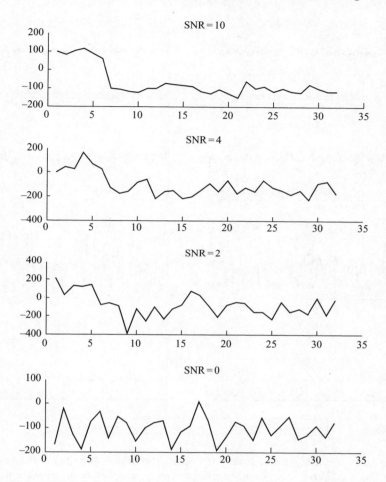

Figure 5.14 Example data used for comparison of average estimation – 6 outliners. Each diagram corresponds to a different signal to noise ratio (SNR). Zero SNR corresponds to a situation with no outliers

- For high noise levels (SNR = 2), the sample mean, median, Lorentzian and the LTSV estimators perform similarly, although the LTSV has higher standard deviation of error. This is as expected – when there is not a clear majority population, the estimators give an approximation to the sample mean of the data. On the other hand, the 1D LTS estimator outperforms other estimators significantly in this situation due its superior ability to reject outliers.
- In the case where there are no outliers (SNR = 0), the median, Lorentzian and LTSV estimators again perform similarly, with the LTSV again having higher standard deviation of error. Interestingly, the 1D LTS estimator performs somewhat poorly when estimating the mean of the entire dataset.

150 *Motion vision: design of compact motion sensing solutions*

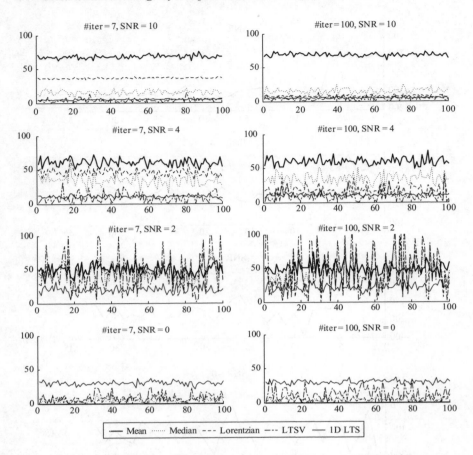

Figure 5.15   Results for Gaussian noise with 12 outliers. Each trace shows the % difference between the estimated mean and the true mean of the majority population.

- The difference in performance between the 1D LTS and the LTSV estimators suggests that, as implemented, the LTSV estimator is not equivalent to the LTS estimator.
- The LTSV has consistently higher than the LTS and Lorentzian estimators. This is because, as defined, the LTSV *always* rejects half the data thus the estimate of the mean of the inliers is calculated using what is essentially a random subset of the majority population.

Robustly averaging a dataset containing 6 outliers is less demanding than finding the robust average of a dataset with 12 outliers, and this is reflected in the results in Figure 5.16 and Table 5.2.

Table 5.1 Summary of the mean error and error standard deviation for each estimator in the case of Gaussian noise and 12 outliers. Note that the grey regions are equivalent since the number of iterations has no effect on the sample mean, median and 1D LTS estimator. Results vary slightly since different datasets were used on each occasion

	Sample mean		Median		Lorentzian		LTSV		1D LTS	
	Mean	Std	Mean	Std	Mean	Std	Mean	Std	Mean	Std
SNR = 10, I = 7	67.95	2.43	15.24	3.26	36.21	0.95	3.62	2.7	4.7	0.78
SNR = 4, I = 7	60.53	5.27	33.88	7.48	46.08	5.44	8.83	6.83	10.54	1.89
SNR = 2, I = 7	49.61	7.72	45.85	13.17	48.64	9.62	43.9	26.23	20.36	4.96
SNR = 0, I = 7	0.0	0.0	4.14	3.1	6.14	3.08	8.14	6.49	29.46	2.9
SNR = 10, I = 100	68.62	2.44	15.19	3.51	7.11	1.36	4.31	3.1	4.76	0.81
SNR = 4, I = 100	59.74	5.45	35.02	7.39	17.32	7.75	9.45	7.72	10.77	1.89
SNR = 2, I = 100	49.42	7.55	44.93	14.7	42.27	22.85	46.8	33.09	22.03	5.29
SNR = 0, I = 100	0.0	0.0	4.0	3.15	3.96	2.87	12.95	8.65	29.65	2.59

Table 5.2 Summary of the mean error and error standard deviation for each estimator in the case of Gaussian noise and six outliers. Note that the two grey regions are equivalent since the number of iterations has no effect on the sample mean and the median. Results vary slightly since different datasets were used on each occasion

	Sample mean		Median		Lorentzian		LTSV		1D LTS	
	Mean	Std	Mean	Std	Mean	Std	Mean	Std	Mean	Std
SNR = 10, I = 7	34.15	1.65	5.13	2.24	2.96	0.74	3.0	2.52	10.0	0.84
SNR = 4, I = 7	30.25	3.07	11.86	4.56	16.44	3.6	8.08	5.88	22.36	2.87
SNR = 2, I = 7	25.32	5.73	18.94	8.53	23.1	6.59	19.63	14.26	37.16	5.73
SNR = 0, I = 7	0.0	0.0	3.89	3.2	5.62	3.16	8.11	5.76	29.52	2.51
SNR = 10, I = 100	34.02	1.47	5.7	2.45	2.71	1.0	5.41	3.73	10.09	0.84
SNR = 4, I = 100	30.28	3.17	12.51	6.0	6.84	4.83	12.82	8.83	21.53	2.37
SNR = 2, I = 100	25.6	6.25	19.65	8.03	18.54	11.74	27.56	20.06	37.67	6.53
SNR = 0, I = 100	0.0	0.0	3.52	3.05	3.37	2.7	12.76	8.94	29.3	2.57

- With fewer outliers, the Lorentzian estimator converges even in 7 iterations.
- The median estimator gives excellent results and the LTSV estimator shows performance similar to the previous example, again with relatively high deviation of error.

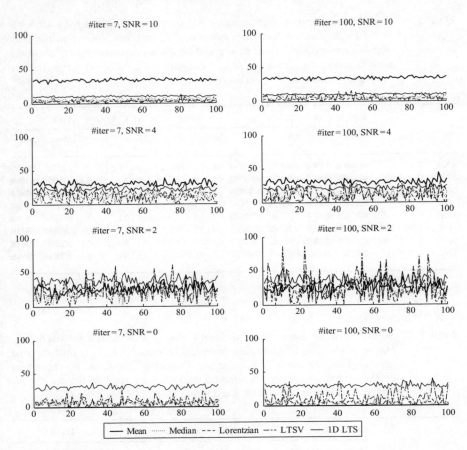

Figure 5.16   Results for Gaussian noise with 6 outliers. Each trace shows the % difference between the estimated mean and the true mean of the majority population. Thick solid line = sample mean, dotted line = median, dashed line = Lorentzian, dash-dotted line = LTSV, thin, solid line = 1D LTS

- Interestingly the 1D LTS estimator gives poorer results when there are fewer outliers.
- Comparing results in the case of 6 and 12 iterations, it can seen that the LTSV gives the most consistent estimates – that is, it is least sensitive to the number of outliers.

This simple Monte Carlo study has shown that the LTSV estimator gives good estimates for the mean of the majority population under a wide range of conditions while being computationally very simple. The primary drawback of this algorithm is its low efficiency; however, it remains an ideal candidate in the context of a hardware implementation where simplicity of implementation is important. It should be noted that all robust average estimators considered have a common weak point – they rely on at least half the data corresponding to the process (e.g. the motion) of interest. If an object is very far away and occupies less than half of the image region under consideration, robust estimators will not estimate the motion of that object, but of the background. While in principle one can eliminate the background data and determine the motion of the 'other object', it is unclear how to determine which subset of the data belongs to an interesting object and which subset corresponds to the background.

5.2.8 Computational complexity

Before moving on, let us briefly consider the computational complexity of the algorithms used in this study in terms of the arithmetic operations required. The Lorentzian estimator requires a multiplication, division and two addition operations at each iteration for each data element and can require a large number of iterations for convergence. The 1D LTS algorithm has a simple implementation but requires a large number of comparisons to order the data, while the median is implemented without the need for reordering, using the algorithm in Reference 230. The LTSV algorithm is simpler again requiring at each iteration only an update of the threshold[116], one division plus two additions for each data element. Since reordering and stochastic sampling are not required, this algorithm is ideal for implementation in digital hardware. Furthermore, the algorithm converges quickly (in as little as seven iterations) and results are comparable (though perhaps with higher standard deviation – i.e. lower efficiency) to a range of other estimators under a range of conditions.

5.2.9 Dynamic scale space implementation

Earlier theoretical consideration of dynamic scale space neglected to address a number of implementation issues. The first issue is with regard to derivatives. Calculation of temporal derivatives requires three frames of data. If the current frame of data indicates a change of scale is necessary, processing is inhibited, the current frame is discarded (since it may exhibit temporal aliasing) and three frames of data are read *at the new scale*. Only then can motion estimation recommence since it is at this point that sufficient data are available at the proper scale. Of course, if a scale change is triggered by any of these three frames, then the frame count is restarted.

Initialisation of dynamic scale space is another important issue. When the system is booted, there is no way of knowing the correct scale until a frame of data has been read. Thus, on the first frame it is important to assume no subsampling (i.e. subSampling = 1) and to force a scale change. This ensures that the system will settle to the correct scale, and it ensures that the requisite three frames of data will be available for processing. Pseudocode for a dynamic scale space is presented in Figure 5.17. Note that while the image data is subsampled by averaging, range data is subsampled by taking the minimum range within each subsampling window. This assists range segmentation (Section 7.2.6.2.3) by ensuring range discontinuities are not smoothed.

---

[116] This requires only a single addition since alpha and beta have been chosen as powers of two.

154   *Motion vision: design of compact motion sensing solutions*

```
count=0;
minRange=0;
subSampling = 1;

while true                                          //repeat forever

    img(2) = img(1);
    img(1) = img(0);
    img(0) = readImage(subSampling);                //Read & subsample image

    range(1) = range(0);
    range(0) = readRange(subSampling);              //Read & subsample range

    //Get the subsampling rate and flag indicating if scale change necessary
    (subSampling, changeScale) = read_ScaleData;

    prevMinRange = minRange;
    minRange = min(range);
    count = count+1;

    // If a scale change is necessary or if this is the first frame
    if changeScale==true or prevMinRange==0

        count=0;                                    // Restart count
        Ux = reSample(Ux,subSampling,prevSubSampling); // Resample Ux

    end if

    // If 3 frames have been read with no scale change,
        allow processing to resume

    if count>=3
        prevUx = Ux;
        Ux = motionEst(img(n-1),img(n),img(n+1),range(n),prevUx);

    end
end
```

Figure 5.17   Pseudocode for the dynamic scale space algorithm

5.2.10   Temporal integration implementation

In Section 3.3, temporal integration was introduced as a process that can both improve motion estimation results and reduce computational load. In principle, the algorithm presented here can benefit from both.

Computational benefits arise if the previous motion estimate[117] is used as a starting point for the robust average algorithm rather than zero. In most cases the previous

---

[117] Because apparent motion is guaranteed to be less than one pixel per frame by our dynamic scale space technique, there is no need for 'warping' of the motion estimation to correct for changes in object position.

motion estimate will be relatively close to the current estimate, hence fewer iterations are required for convergence. This aspect of temporal integration has not been implemented here – it is left as an exercise for the enthusiastic reader.

Our algorithm does, however, make use of temporal integration to improve motion estimates over time. Rather than including a temporal integration term in the constraint equation [35, 59], temporal integration is implemented as a direct refinement of the motion estimate. This refinement is realised as the weighted running average of the motion estimate for each column[118] expressed as a recursive update equation (5.22) [48]:

$$\overline{U}_x^n = \frac{U_x^n + TC \times \overline{U}_x^{n-1}}{TC + 1} \quad (5.22)$$

Here, the result of temporal integration $\overline{U}_x^n$ is the weighted average (with weight TC) of the previous temporally integrated motion estimate $\overline{U}_x^{n-1}$ and the current motion estimate U_x^n (as calculated using (5.3)). This procedure is easily extended to operate correctly when the scale (SD) changes, one simply interpolates appropriately. Interpolation is performed by mapping the column address at time n ($addr^n$) to a column address at time n − 1 ($addr^{n-1}$) using (5.23):

$$addr^{n-1} = \text{round}(2^{currentSD - prevSD} \cdot addr^n) \quad (5.23)$$

Here, *currentSD* is the current level of the dynamic scale space and *prevSD* is the previous level of the dynamic scale space. We can rewrite (5.22) to explicitly show how the change between the current and previous estimates is propagated:

$$\overline{U}_x^n = \overline{U}_x^{n-1} + \frac{U_x^n - \overline{U}_x^{n-1}}{TC + 1} \quad (5.24)$$

Equation (5.24) shows that increasing TC smooths (or damps) changes in the estimate. Lower values of TC give a motion estimate that is more responsive to short term changes in motion, but leads to an increase in noise. Higher values of TC lead to a motion estimate that is less responsive to real changes in motion (since it takes longer for the motion estimate to converge to its true value), however, noise is better suppressed. Higher values of TC may be justifiable despite the increased settling time since an object's momentum means its motion should, in general, change relatively slowly. To simplify processing, we always choose values of TC so that $TC + 1$ is a power of two thus allowing us to replace the division with a bit shift.

---

[118] Temporal integration is not performed at range discontinuities since the velocity measured at these locations at time n − 1, may correspond to a different blob at time n.

5.2.11 The motion estimation algorithm

Drawing together the threads of the previous sections, it is now possible to state succinctly the motion estimation algorithm. Processing is embedded in a dynamic scale space (Section 5.2.9) and proceeds in five steps

1. Solve (5.3) for U_x at each pixel. Calculate derivatives using FOCD and use range data interpolated onto the camera coordinate system.
2. Segment range data. A simple 2nd derivatives zero crossing method is used in this work.
3. Robustly calculate the average of U_x for each column.
4. Smooth U_x by robustly finding the average value of U_x between boundaries identified in step 2.
5. Employ temporal integration by taking the weighted average of the current and previous estimates of U_x, taking special care at blob edges.

In principle, steps 3 and 4 could be combined into a single average over all pixels within a blob (i.e. between range discontinuities). However, the local buffers in the FPGA are too small to allow this. An implementation of this algorithm in FPGA also requires consideration of bit widths and rounding of each variable. This issue is discussed when considering the FPGA implementation of the algorithm, however, the reader should be aware that the software implementation of the algorithm (discussed below) accurately represents the FPGA implementation.

Note that this algorithm does not contain any arbitrary parameters (such as the regularisation parameters of global motion estimation techniques) nor does it require an initialisation or initial guess for the motion estimates (though it could benefit from such information).

5.2.12 Simulation results

Verifying the correct operation of the algorithm is an important preliminary step to hardware implementation. Verification gives us the opportunity to check that the algorithm operates as expected before going to the time and expense of a hardware implementation. It also gives us the opportunity to test variations of the algorithm to see if performance can be improved, again without the time and expense of a hardware implementation. Finally, algorithm verification can be helpful in the later step of hardware verification – testing to ensure the hardware implementation of the algorithm is correct. In this section, we only consider algorithm verification.

If we are going to test the operation of the algorithm, we must first have a means to measure the quality of the results. Defining such a metric for motion estimation is difficult [25], especially if using natural images since there is no way of knowing the 'ground truth' (i.e. the true result). This is further complicated by the fact that different algorithms may calculate different motion parameters (e.g. pixel displacement, affine velocity, etc.). One possible solution is the use of reconstruction error [193] or prediction error [269] as a quantitative metric. In this technique, motion estimates are used to 'warp' a frame of data generating an approximation of a subsequent frame. The difference between the true and approximated frames is used as a metric. However, this metric only describes how well motion estimates map one frame to the next – it does not necessarily tell us how similar the motion estimates are to the true underlying motion in the image. In the literature[119], quantitative comparisons between motion estimation algorithms are often performed using simple sinusoids or 'moving box' sequences (e.g. Figure 3.5 and Reference 25) and a qualitative analysis is applied for natural sequences. In our case, the problem is complicated by the need for both visual and range data.

To overcome this problem, and allow quantitative analysis using natural visual data, a simulation package was developed using MATLAB and a freeware ray-tracing program called Polyray [97]. This simulation package allows the user to fully specify a simple virtual environment including the intrinsic and extrinsic camera parameters, camera motion (all motion is assumed to occur on a flat ground plane), range scanner parameters and motion, as well as the texture and motion of all objects. Objects can be rectangular prisms or cylinders of arbitrary size (more complex objects can be formed by overlapping these simple object types) and the user can specify light sources[120].

Given these input data, the simulation package generates a sequence of images and range scans and calculates the apparent motion for each frame thus allowing quantitative analysis of motion estimation algorithms using more natural test data. While these data will not match exactly the data from real sensors mounted on real vehicles (i.e. there will be no vibration in the data, there will not be sensor noise, etc.) the data do contain noise (e.g. significant aliasing artefacts generated by the raytracing software used) and are structured to have similar resolution and accuracy as might be expected from a true sensor (e.g. range information is accurate to within 4 cm). Therefore, we believe these simulated data will give a good indication of how the algorithm will perform in the final implementation.

We used the simulation package to generate test data that appear roughly similar to the laboratory environment as illustrated in Figure 5.18[121]. Figure 5.18a gives an example of the generated visual data. This image is from a sequence with a translating camera and two independently moving objects. Apparent motion for this scene is shown in Figure 5.18b. The motion of the left object appears clearly since this object is moving quickly from left to right while the central object generates little apparent motion since it moves directly towards the camera. As expected, the forward motion

---

[119] We exclude the video coding literature here, where prediction error is of more importance than accurate estimates of 'true' motion.
[120] Similar approaches have been used before [98].
[121] All sequences used in this chapter were designed assuming $f = 4.8 \times 10^{-3}$ m, $\zeta = 12.5 \times 10^{-6}$ m, $F = 1/16$ s.

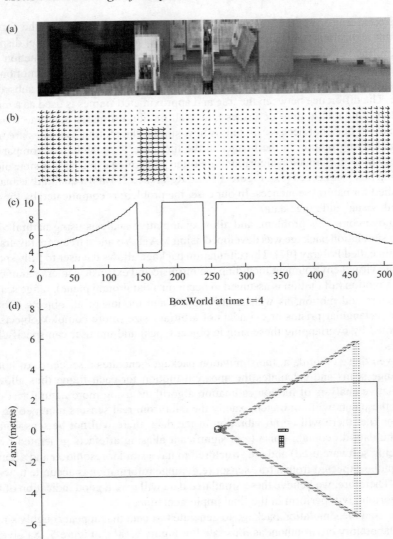

Figure 5.18 Example data from our simulation environment. (a) A single frame from the simulated visual environment, (b) apparent motion is produced allowing quantitative performance evaluation of motion estimation algorithms, (c) simulated range scan (measured in metres) corresponding to the current image frame, (d) 'top view' of the simulated environment showing the change in position of each object, and the changing position of the camera's field of view

Table 5.3 Parameters used when generating simulated data and simulating our motion estimation algorithm

Parameter	Value
f	4.8×10^{-3} m
ζ	12.5×10^{-6} m
F	$1/16$ s
SU_1	256
SU_2	256
D_{min}	0.25 m
V_{max}	0.1 m/s
Safety margin	2.5 s
TC	0, 3 and 7

of the camera has generated an expanding motion pattern. The simulation generates range data with properties similar to data generated by commercial sensors. These data are measured at regular angular increments and must be interpolated onto the image coordinate system. Figure 5.18c illustrates the *interpolated* range data. Finally, Figure 5.18d shows a top view of the virtual environment. It also indicates how object positions and the camera's field of view change.

Given this test data, we can implement our algorithm (we chose to implement it using MATLAB) and see whether the algorithm gives results matching the known 'ground truth'. We can also see how variations of the algorithm affect results. For the purpose of this work, we implemented both the algorithm and the simulated data sequences using the parameters listed in Table 5.3. These parameters match those of the camera and other hardware we use when testing the hardware implementation in Chapter 7.

The first test sequence was generated with a stationary camera. Figure 5.19 illustrates the performance of each stage of our algorithm. At the top is an example of the visual scene. The coarse nature of this image is due to the high degree of subsampling as a result of the dynamic scale space. Experiments tracked the motion of the object on the left, which is translating towards the right. The trace in the second plot denotes the range for each column, and the vertical lines represents the segmentation of the range data. Second from the bottom is the robust, column-wise motion average that remains somewhat noisy despite the robust averaging procedure (due to estimation noise and our estimators' relatively low efficiency). The lower plot illustrates the smoothed motion estimate with the solid trace corresponding to the current estimate, and the dotted trace to the previous estimate. Notice that the motion estimate for the object on the left is quite accurate.

160 *Motion vision: design of compact motion sensing solutions*

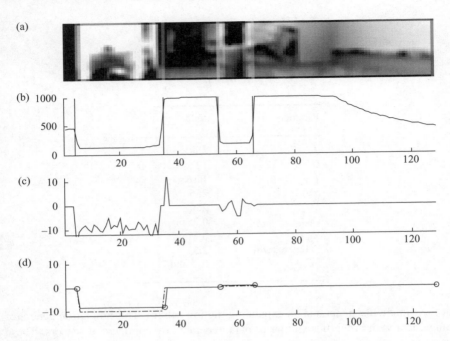

Figure 5.19 *Illustration of the motion estimation algorithm's operation with a stationary camera. In each plot the horizontal axis represents image columns (measured in pixels), (b) shows range (measured in cm) corresponding to the image in (a), (c) shows the unsmoothed motion estimate (cm/s) for (a), (d) shows the smoothed motion estimate (cm/s) for (a)*

To ensure the algorithm performs consistently, it is necessary to extend this analysis over a number of frames and a number of velocities as illustrated in Figure 5.20. Here, solid lines represent velocity estimates and dotted lines indicate the true velocity of the object under consideration. The upper plot in Figure 5.20 illustrates range measurements with the temporal integration coefficient TC set to zero; in the central plot $TC = 3$ and in the lower plot $TC = 7$. The effect of increasing TC is clear: we see a strong reduction in variance. Table 5.4 lists the long-term mean velocity and corresponding variance for each plot. There is marked improvement in variance between $TC = 0$ and $TC = 3$. Results remain relatively constant when TC is increased to 7. Notice that for $U_x = 16$ cm/s, motion is significantly underestimated. This can be attributed to the bit-width limiting (described in the following chapter) which limits motion estimates to an maximum absolute value of 16.6 cm/s (determined using (4.15)). Any results over this value are simply discarded (trimmed) leaving the motion estimate to rely on data less than 16.6 cm/s, hence velocity tends to be underestimated. The accuracy of these motion estimates is exceptional when

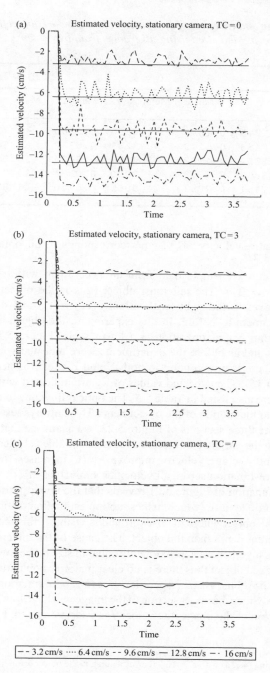

Figure 5.20 Estimated object velocity measured using a stationary camera with different values of TC

Table 5.4 Mean and variance of the velocity of the tracked object for a stationary camera. Measured using all frames from frame ten

U_x cm/s	$TC = 0$		$TC = 3$		$TC = 7$	
	Mean	Variance	Mean	Variance	Mean	Variance
−3.2	−2.7843	0.1648	−3.0850	0.0160	−3.2386	0.0103
−6.4	−6.0458	0.5123	−6.3203	0.0332	−6.5882	0.0437
−9.6	−9.5359	0.4992	−9.7484	0.0426	−10.0686	0.0291
−12.8	−12.3856	0.2928	−12.7026	0.0381	−12.9216	0.0271
−16	−14.2451	0.2093	−14.5065	0.0700	−14.8170	0.0581

one considers that a velocity of 3.2 cm/s corresponds to an optical flow of just 0.14 pixels/frame[122].

Figure 5.21 shows example data from an experiment with a moving camera (camera velocity $= T_z$). The left most object (this is the object whose velocity is tracked[123]) moves to the right (with velocity $T_x = T_z$) and the central object approaches the camera as before. In this experiment, the objects are further away hence the algorithm begins operation at a lower scale. Figure 5.21 illustrates the motion estimates and as before the experiments were repeated with TC set to 0, 3 and 7. Because there is now motion in both X and Z axes it is necessary to determine U_x using (4.15). The expected values of U_x are indicated with dashed lines in Figure 5.21 and are also listed in Table 5.5. The optical flow due to camera motion is of opposite sign to the motion of the object hence U_x is always less than T_x.

Comparing the three sections of Figure 5.22, we again see that TC has a strong effect in reducing the variance of motion estimates and this is reflected in Table 5.5. Both variance and average velocity improve as TC increases, however, there is a degree of underestimation across all velocities and values of TC.

Careful examination of Figure 5.21 reveals that the upper portion of the tracked object has very little texture (which makes motion estimation difficult), and there is a region above the object where the background is visible – this region has different motion and different depth than the object. The cause of the underestimation is that these two regions together are approximately equal in size to the lower, well-textured region of the tracked object thus there is no clear majority population in the column. In this situation, the robust average operation gives an approximation to the average of the entire column. Since the upper section of the image will have low motion compared to the lower section of the image, the velocity is underestimated. Figure 5.23 shows

---

[122] Calculated using (4.13a) with $Z = 1.35$ (corresponding to the object being tracked) and level $= 2$. As shown in Section 5.2.3 optical flow can theoretically be measured at a resolution of 0.0039 pixels/frame though noise will prevent achieving this accuracy.

[123] We actually track a point inside the object marked by the central red line.

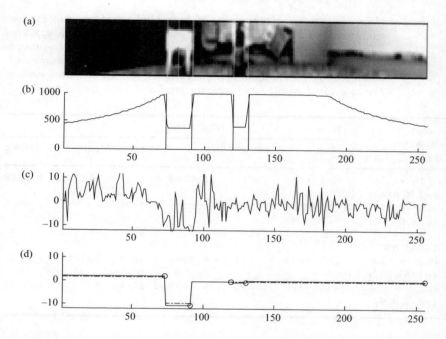

Figure 5.21 *Illustration of the motion estimation algorithm's operation with a moving camera. In each plot the horizontal axis represents image columns (measured in pixels), (b) shows range (measured in cm) corresponding to the image in (a), (c) shows the unsmoothed motion estimate (cm/s) for (a), (d) shows the smoothed motion estimate (cm/s) for (a)*

the result of using just the lower half of the image to estimate motion. The result is a notable reduction both in both noise (through the use of a well-textured region of the image to estimate motion) and underestimation. Note that underestimation for $T_x = -16$ cm/s is again a result of bit width limiting (via trimming) as before.

The final experiment shows the correct operation of dynamic scale space. Figure 5.24 illustrates motion estimates in a situation similar to the second experiment where both the camera and the tracked object are moving. In this case, the camera moves sufficiently that a scale change is triggered. When a scale change is triggered, there is a four frame period during which no processing occurs and the motion estimate does not change. After this time, motion estimation proceeds correctly at the new scale.

In summary, these simulation results show that the motion estimation algorithm will indeed work as expected. Motion estimates are quite accurate although there is a degree of underestimation as motion nears the upper limits of the algorithm's capability and variance increases when the camera moves. Dynamic scale space correctly operates to allow motion estimation for the nearest object and robust averaging allows

164 *Motion vision: design of compact motion sensing solutions*

Table 5.5 *Mean and variance of the velocity of the tracked object for a moving camera. Measured using all frames from frame ten. Note that the expected value of U_x is non-stationary. We expect the mean estimated motion to equal the mean calculated motion. Similarly the variance in the estimated motion should approach the variance of the calculated motion. The calculated average and variance are listed in the first column*

U_x cm/s min → max (avg, var)	$TC = 0$ Mean	Variance	$TC = 3$ Mean	Variance	$TC = 7$ Mean	Variance
−2.43 → −2.49 (−2.46, 0.003)	−1.1438	0.9922	−1.4248	0.1192	−1.6013	0.0668
−4.87 → −5.13 (−5.0, 0.0063)	−3.2908	1.4243	−3.5229	0.2078	−3.9739	0.0715
−7.3 → −7.95 (−7.63, 0.037)	−6.2255	1.8054	−6.5229	0.3989	−6.8758	0.2043
−9.73 → −10.93 (−10.33, 0.12)	−7.9085	4.8981	−8.0980	0.6313	−8.2549	0.3615
−12.17 → −14.17 (−13.2, 0.34)	−10.7320	3.9834	−10.8170	1.1514	−11.0163	0.4892

us to reliably extract the most prominent[124] motion. The simple temporal integration scheme improves and smoothes motion estimates over time; however, one must trade the degree of smoothing against the ability to respond to velocity changes.

5.3 Navigation using the motion estimate

In practice, the motion estimation algorithm gives three outputs for each image column: an estimate of relative projected velocity, a depth value and a binary value indicating whether or not a column corresponds to a blob edge. Given these data, how is it possible to navigate through an environment? This section considers what information is necessary to create a navigation plan, how that information can be retrieved from the data generated by our sensor, and provides a general framework for blob avoidance based on a constant velocity model. This section is only intended to illustrate the use of the sensor – it is not a detailed analysis of navigation methods and implementation of the techniques discussed here is beyond the scope of this work. There are many references for further reading [55, 120, 157, 187, 249].

Ideally, an autonomous vehicle will have sensors able to detect the change in distance to each point in the environment as the vehicle moves. That is, the vehicle will have sensors able to detect relative velocity. In terms of relative velocity, the threat

---

[124] The motion indicated by the majority (i.e. at least 50 per cent) of the data.

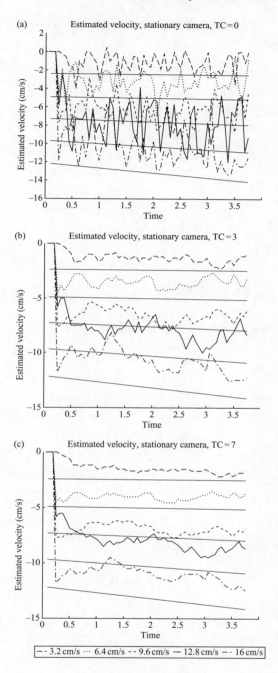

Figure 5.22 *Estimate object velocity measured with a moving camera with different values of TC. Values in the legends correspond to $T_X = T_Z$*

166 *Motion vision: design of compact motion sensing solutions*

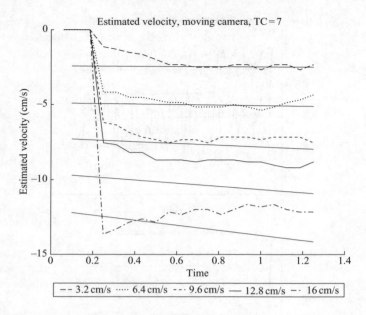

Figure 5.23   *Motion estimates for moving camera when only the lower half of the image is used*

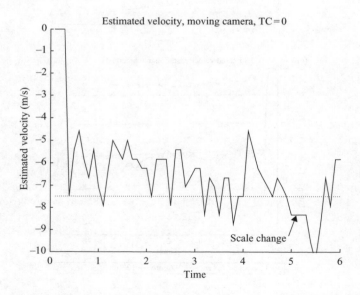

Figure 5.24   *Using a slightly different simulated environment we verify that dynamic scale space operates correctly*

of a collision can be determined by testing whether the current relative velocity of a point will bring that point to the origin[125]. Such an approach allows detection of the threat of collision with both static and dynamic objects. Knowledge of relative motion also allows the vehicle to plan a path that will avoid the obstacle once the threat of a collision is determined.

In practice, sensors have a limited field of view and resolution. Even within this field of view, occlusion limits which parts of the environment a sensor can monitor. Thus, the 'god like' global knowledge assumed above is usually not available – and human experience shows that it is generally not necessary. People can navigate quite well using only local knowledge. There are two parts to this local knowledge: the detection of local relative motion and the assumed motion model that is used to predict the future position of an obstacle. For example, people can easily move around a crowded shopping mall. They do this by monitoring the motion of the people near to them and assuming that they will maintain their velocity. We have all experienced the surprise of someone suddenly changing course! Similarly, while driving, people monitor the nearest vehicles and assume that they will maintain a (generally) constant velocity and will move according to the rules of the road. Of course, global coordination can improve safety; however, the point is that navigation is possible using only local knowledge.

The algorithm we have developed is designed to estimate the relative projected motion[126] (U_x) from the local perspective of a vehicle. From this, it is possible to extract the relative two-dimensional translational velocity (T_x, T_z) under the assumption that any turns have a large radius using (4.15) (repeated here as (5.25)).

$$U_X = \frac{x}{f} T_Z + T_X \qquad (5.25)$$

The focal length (f), physical position on camera (x), and the relative projected motion (U_x) are known. Furthermore, T_z can be determined for each column simply by determining the change in depth divided by the frame interval F. It is important to discard any T_z value calculated near a depth discontinuity; motion may cause these estimates to be invalid.

These data still cannot be used directly to calculate T_x. If U_x is composed of both T_x and T_z components, it should vary across the image due to the x/f term, however, robust smoothing assigns a single (average) velocity to each blob; therefore calculating T_x in a column-wise manner will give incorrect results. Overcoming this is quite simple. Since U_x is the average velocity for the blob, the average (central) value of x for the blob can be used and, since blobs are assumed to move rigidly, the central values of T_z and U_x[127] for each blob can also be used. Thus, calculation of T_x for each image column reduces to calculating T_x once for each blob.

---

[125] The origin of the sensor-centred coordinate system.
[126] Modelled as translation – rotation is assumed to be of large radius.
[127] U_x may vary slightly across a blob due to the temporal integration process, however, the central value of U_x will be representative of the blob's motion.

168 *Motion vision: design of compact motion sensing solutions*

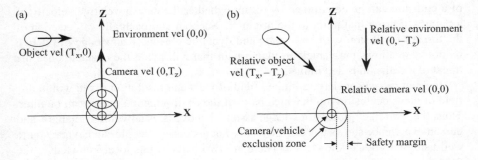

Figure 5.25 *Using relative motion (b) the exclusion zone is fixed relative to the camera coordinate system. If one used absolute motion (a) the position of the exclusion zone would change over time making collision detection more difficult*

Now that the relative motion is available it is necessary to decide upon a motion model. The simplest model is the constant velocity model, and as discussed before, it is widely applicable though other models are available (e.g. constant acceleration). For the purpose of this work, a blob-wise constant velocity model for motion is used where velocity (T_x, T_z) is calculated for each blob. This simple model can be updated directly from (5.25) though it may be preferable to further smooth estimates using a Kalman filter [48]. Such filtering may be necessary if higher order motions models such as a constant acceleration model are used.

Given a motion estimate and a motion model it is possible to predict the future location of points in the environment, hence the likelihood of a collision can be determined and a plan of action to avoid these collisions created. Here, a simple geometric method is presented, though other navigation approaches (e.g. potential fields [183, 296], certainty grids [157] and velocity obstacles [90]) can be adapted for use with the sensor.

Rather than merely searching for points that may intersect the origin (an approach fraught with danger due to noise), an *exclusion zone* is defined around the vehicle which is related to the safety margin of Section 4.3 (see Figure 5.25). A collision is deemed to have occurred if an object enters the exclusion zone. Note that under a constant velocity model, only objects at, or moving towards, the image centre line (i.e. towards the Z axis) can be involved in a collision. Based on the assumption of constant velocity, the future position of a point can be determined using (5.26).

$$P_x^{t+\delta t} = P_x^t + T_x \delta t$$
$$P_z^{t+\delta t} = P_z^t + T_z \delta t$$
(5.26)

Here, (P_x^t, P_z^t) is the current position of the point under consideration, and $(P_x^{t+\delta t}, P_z^{t+\delta t})$ is its position at time $t + \delta t$. To determine if the point will enter the exclusion zone in time δt, (5.27) [43] is used:

$$au^2 + bu + c = 0$$
(5.27)

where
$$a = (T_x \delta t)^2 + (T_y \delta t)^2$$
$$b = 2\{P_x^t(T_x \delta t) + P_y^t(T_y \delta t)\} \quad (5.28)$$
$$c = (P_x^t)^2 + (P_y^t)^2 - (ER)^2$$

Here, ER is the radius of the exclusion zone. To solve (5.27), use the quadratic formula

$$u = \frac{-b \pm \sqrt{b^2 - 4ac}}{2a} \quad (5.29)$$

If $\sqrt{b^2 - 4ac} < 0$ then the point will not enter the exclusion zone. If it equals zero, then the path of the point is a tangent to the exclusion zone and if the value is greater than zero the path will pass through two points in the exclusion zone[128] corresponding to the two roots of (5.27). The smaller value of u corresponds to the first crossing and from this the time at which this crossing occurs can be found using $t = u \delta t$. Note that if $u > 1$, the point will not enter the exclusion zone in the time interval under consideration (δt).

If a collision is likely, a plan of action should be formulated that: (a) is physically realisable by the vehicle (that is, it must take into account the vehicle's dynamic constraints); (b) will avoid the detected collision without causing collision with a different point in the environment; (c) conforms to the 'rules of the road' for the current environment so that there is degree of predictability and to prevent deadlock; and (d) is consistent with any 'global' path plan. This problem can be solved using the concept of 'State-Time Space' [187] where a collision free path is found by finding a continuous curve joining the current location of the robot and its goal location without passing through 'forbidden regions' defined by the vehicle dynamics and 'rules of the road'.

Note that the approach to navigation described here is different to the approach of many authors who use visual data for navigation [33] since it specifically relies on relative motion information. The system does not use a visual stereo to model the environment [46, 49], it is not designed to detect image features such as road lanes [44, 45, 281] and it does not detect obstacles by searching for particular features (e.g. identifying vehicles by searching for symmetry and shape) [34, 45, 112]. This navigation technique also differs from those that explicitly use motion information for navigation. For example, it does not use optical flow as a means of determining egomotion (passive navigation [50, 218]) or the vehicle's future path [73]; also, it does not use peripheral flow or flow divergence to maintain a heading, to detect the likelihood of a collision or to estimate the time to collision [17, 24, 60, 63, 78, 203]. Furthermore, it does not detect obstacles based on optical flow [53, 199, 242, 273, 305][129]. These methods are generally designed for operation

---

[128] Entry and exit points.

[129] There is a fine difference here. This thesis does not segment optical flow – it assigns a relative velocity to each point in the environment based on the assumption that range discontinuities correspond to the edges of coherently moving blobs.

in a static environment, and often they do not directly give a means of planning ahead beyond simple reactive collision avoidance. The approach we use is simple – it aims to assign a relative velocity estimate to each point (column) in the image. These data are calculated using a combination of visual and geometric (depth) data without explicitly modelling the environment. If the relative motions of points in the image are known, those points can be avoided if necessary.

Part 3
Hardware

Chapter 6
Digital design

Finite state machines are critical for realising the control and decision-making logic in digital systems.

Katz [159]

Having established our preferred motion estimation algorithm, the last section of this book considers how this algorithm can be implemented using semi-custom digital hardware. This chapter discusses, at a low level, the techniques we used to implement our prototype motion sensor. We make no claim that our approach is unique, optimal, or appropriate for all design work, but this approach is very straightforward and worked well for us: we hope that you will find it useful too. Assuming the reader is familiar with the basics of digital design, we introduce some concepts from *FPGA* (Field Programmable Gate Array) and *VHDL* (Very-High-Speed-Integrated-Circuit Hardware Description Language) technology. Our descriptions are brief and it would be useful to have a VHDL reference available as you work through this chapter, however, there is sufficient information here for you to understand our design and start on your own projects. In order to keep the discussion generic, we try not discuss specific tools or technologies since these change regularly and there are large difference between tools from different vendors[130]. It is inevitable, however, that there will be some bias towards the tools we are most familiar with – Xilinx Foundation software, and Xilinx Virtex FPGAs. Any links or references to particular vendors in

---

[130] For the reader who wishes to brush up on their digital design skills, the following websites may be of help: http://www.ami.ac.uk/courses/ and http://www.ibiblio.org/obp/electricCircuits/index.htm. Those who would like to know more about FPGA technology should visit the websites of the two major FPGA manufactures: Xilinx (www.xilinx.com) and Altera (www.altera.com). Both manufactures have freely available design tools for you to try, though they do not support the most advanced FPGAs and do not have all the advanced design features of the complete design packages. Finally, those looking for a more thorough introduction to VHDL are referred to Reference 290 and to the FAQ at http://www.vhdl.org/ that maintains a list of VHDL reference books (both free and commercially available books).

this section are for your reference only. They do not imply an endorsement of that vendor's products or services.

6.1 What is an FPGA?

At their simplest, Field Programmable Gate Arrays (FPGAs) are devices containing an array of gates, registers and I/O ports whose interconnections can be determined by the user *in the field* (hence the term 'field programmable'). Modern FPGA devices also include a range of additional resources such as tristate buffers, memories, multipliers and specialised clock distribution networks that minimise skew. Technology is now becoming available that allows an FPGA to be partially reconfigured during execution so that the FPGA can 'redesign' itself to use the most appropriate configuration for the task at hand.

The beauty of FPGA technology is that there is no need to send your design to an IC foundry for manufacture – the function performed by an FPGA is typically determined by a programming pattern that is loaded into the FPGA from a ROM when the FPGA is booted. Typically, Flash ROMs are used so the function of the FPGA can be changed simply by changing the programming pattern in the ROM. This is different to putting a new program into a CPU's, DSP's or micro-controller's memory – in this case, changing the program in the memory causes the processor to execute a different set of instructions (i.e. perform a different function) using the processor's *fixed internal configuration*. Changing the programming pattern for an FPGA will cause the internal configurations and connections of the device to change, thus FPGAs can be thought of as standing at the boundary between hardware and software. FPGAs are rapidly also becoming viable alternatives to the ASIC[131] (Application Specific Integrated Circuit) since:

- The logic density of modern FPGAs is beginning to rival that of some ASICs – that is, you can fit as much logic onto an FPGA as some ASICs.
- FPGAs have a lower up front cost since no manufacturing is required, however, FPGAs are often considered to have a cost advantage only on short production runs due to the higher per unit cost. This distinction is now starting to fade as the per-unit cost of FPGA devices falls.
- FPGAs allow the design to be changed at essentially no cost if an error is found or new features are needed.

6.2 How do I specify what my FPGA does?

There are a number of design entry tools used for FPGA design. Perhaps the most intuitive is *schematic entry* where the designer draws the functionality they require

---

[131] An ASIC is a Integrated Circuit (IC) consisting of an array of components (gates, flip-flops, analogue components, etc.) without any interconnection (i.e. there is no metallisation on the IC). The user determines the interconnections, which are then manufactured onto the IC, usually by the ASIC vendor. More information can be found from ASIC vendors such as http://www.amis.com/.

using a set of library components that includes the primitives components available on the FPGA as well as some higher level functions such as counters. These basic resources can be augmented by *cores* or *Intellectual Property blocks* (*IPs*): third party designs that can be configured to the designer's needs and placed in the schematic as a black box. Many companies now sell cores and IPs for functions ranging from encryption to PCI Interface logic (e.g. [138, 142, 147]). A number of websites also offer free IP (e.g. [139, 144]) and, of course, the designer can also reuse the functional blocks (either schematic or HDL) they created.

HDLs or *Hardware Description Languages* allow the designer to specify the function of a circuit using a type of programming language, the most common of which are VHDL and *Verilog*. These languages are specially designed for describing digital circuits with constructs allowing (among other things) the designer to describe the inherently parallel nature of a digital circuit, to 'wire' different components together, and to *simulate* timing aspects of the circuit. Because state machines are a common construct in HDLs, many vendors provide tools that allow rapid design of these structures. State machine entry tools allow the designer to draw their state machine, inserting actions and conditions using HDL syntax. This state machine is converted to HDL code that can then be included in the overall design

Other common design tools provided by vendors are the floor planner and the low level editing tool. The *floor planner* allows the designer to exert some control over the automated implementation tools by specifying preferred locations for particular design elements, potentially improving the overall performance of the design. The *low level editing* tool gives the user direct access to the resources of the FPGA allowing the user to define functions of the various resources of the FPGA as well as defining the routing. This tool is more often used to improve the results of automated design tools than as a design entry mechanism in its own right.

6.3 The FPGA design process in a nutshell

A typical FPGA design process is illustrated in Figure 6.1. The process begins with the designer entering their design using schematic entry or HDL entry tools (or a combination) together with libraries of primitives available in the target FPGA and logic cores (or intellectual properties). While a schematic design will directly relate to the logic resources in the FPGA (gates, flip-flops, etc.) a HDL design has no such correlation, thus HDL designs must go through what is known as *synthesis* before they can be used further. The synthesis process converts the HDL description into a generic description at the gate and flip-flop level and performs some basic combinatorial and state machine optimisation and perhaps some optimisation based on knowledge of the target FPGA. This optimisation is very simplistic – for example, it will not realise that your large multiplexer might be more efficiently implemented using tri-state buffers. You must specify such design decisions in the HDL. Once a HDL design is synthesised it can be included in a schematic or can be used directly as input to the later stages of implementation.

The schematic design can be *functionally* simulated to ensure the circuit description performs the desired function. However, this can be relatively slow since

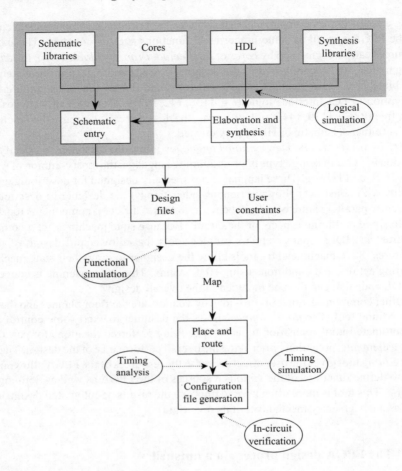

Figure 6.1   FPGA design process (adapted from Reference 74)

the simulation is performed in terms of logic gates and flip-flops. This is where a HDL design can have an advantage. A HDL design can be simulated before synthesis in a process called *logical* simulation. Logical simulation is faster than functional simulation since the simulator only needs to run a 'program' – it does not worry about gates, etc. HDLs also allow for varying degrees of abstraction so you can write simple behavioural models for your design components using high level constructs (e.g. a memory might simply be modelled as an array with particular timing on the inputs and outputs) which can be fleshed out later to include all the design details. This gives a double check mechanism to ensure the detailed design indeed performs the expected function. Note that neither logical nor functional simulation take into account signal propagation delays since at this stage we do not know what paths signals will take.

From this point on, the implementation process becomes more automated. The tools will take the design files as input to a *mapping* process that maps the primitives in the design files to particular components in the FPGA. The *place and route* process assigns these components to particular locations in the FPGA and connects all the components appropriately. These processes can be controlled by the user via a set of *user constraints*. Constraints can appear in the design files, can be generated using special tools, or can come from a text file. There is a range of constraints, the most common of which are:

- Location constraints – these force the design tools to put components in a particular location. For example, it can be useful to force a particular I/O signal to appear on a particular pin of the FPGA or you may wish to position components in a particular configuration to take advantage of advanced features of the FPGA (such as fast carry logic). Floor planning can also help to improve the design's performance by forcing the place and route tool to locate all the logic for a particular function in a given region of the FPGA.
- Timing constraints – these define parameters such as clock speed and time available for a signal to propagate from one point in the FPGA to another. The place and route tools attempt to create a layout that meets the user's timing constraints in the worst case using the known delays of components within the FPGA.

After the Place and Route (PAR) process, two verification options are available. A *static timing analysis* (a point to point analysis of delays in the design [94]) will indicate if timing constraints could not be met. As part of the PAR process, *back annotation* is also performed, directly mapping delays in the final layout to the original network. This allows the designer to perform a *timing simulation* that ensures the design still performs the desired function when all delays are taken into consideration. Of course timing simulation is slower than functional simulation since delays must also be taken into account.

The final step in the design process is to convert the placed and routed design into a configuration file that can be loaded into the FPGA. Once this file is loaded, the designer can perform in-circuit verification to ensure the final design does indeed perform as expected.

6.4 Time

In FPGA design, time must be considered from two perspectives. The first perspective was mentioned above: a signal has only a limited time to propagate from one part of the circuit to the next and this time depends on the clock rate, function of the design, etc. Timing constraints are used to allow the design tools to take these issues into account.

Timing is also important from the algorithmic perspective where we may wish events to occur in a particular order and with particular timing. While HDLs have the ability to model delays precisely (e.g. `X <= 8 after 10ns;`) such descriptions cannot be synthesised (there is no '10 ns' component!). This construct is useful in logical simulation, however, a different way of modelling time is needed if we wish to implement our design.

In our work we use state machines to deal with both aspects of time. State machines provide a straightforward means of ordering operations and ensuring operations occur at the correct moment. State machines also provide a simple way of rectifying any paths in the design that do not meet timing constraints.

6.5 Our design approach

We use a graphical design approach since we believe such designs are quicker to enter, easier to understand, and easier to debug. Rather than using VHDL directly, we use a state machine entry tool to generate most modules[132] then combine the modules using schematic entry. State machines are used to describe the sequential portions of the design while schematics are used for the concurrent portions. Note that our use of VHDL rather than Verilog and other HDLs was more of a personal preference than a technical choice. A key drawback of using a graphical entry tools is that the resulting data files are generally stored in a proprietary format making it difficult to share or reuse designs. Pure HDL designs are less affected by this problem, though use of vendor libraries will complicate sharing of designs.

The beauty of using a state machine entry tool is that it allows us to focus on the logic of the algorithm rather than the syntax of the language and it explicitly shows the structure of the algorithm making debugging easier. State machines also impose a strict order and timing for operations making them similar to programs for normal CPUs. We believe that this makes the leap from software design to hardware design less daunting. The drawback of designing using state machines is that the entry tools often make it difficult to utilise the full power of the VHDL language. For example, the tools may only give access to a limited range of data types. Schematic entry allows quick, direct interpretation of the circuit configuration. Discerning the interconnection of components from HDLs can be quite difficult.

Reflecting our design approach, the following sections give an introduction to the VHDL language so that you can design your own state machines. After this, we give some examples of state machine and schematic entry, however, we purposely limit detail since this chapter is intended to be generic and not focused on any particular design tools.

6.6 Introducing VHDL

VHDL (Very-High-Speed-Integrated-Circuit Hardware Description Language) is a language specially created with digital hardware design in mind and is an IEEE standard. The purpose of this section is to introduce the reader to the VHDL language. While the VHDL language has many constructs that are useful in simulation and testing, only a subset of the language can be synthesised[133]. Since we are focused on

---

[132] The remainder being designed using schematic entry.

[133] For example, only integer arithmetic is supported for synthesis and even then, it is usual that only addition, subtraction, multiplication and division in powers of two can be synthesised. In addition to this any VHDL constructs intended to simulate time are not available for synthesis, nor are other simulation

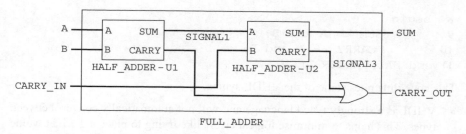

Figure 6.2   Full adder circuit (adapted from Reference 306)

implementation of digital designs, we concern ourselves only with the synthesisable subset of VHDL here. Many of the specific details regarding VHDL in this section are drawn from Xilinx documentation [283, 290, 306].

6.6.1   VHDL entities and architectures

The basic unit of a VHDL circuit description is the `entity`, which has both an external view (i.e. an interface defined by the entity's `ports`), and an internal view (the implementation defined by the `architecture`). The external view (i.e. the `ports`) allows the entity to be incorporated into a larger design without regard to the particular implementation. The designer can then create any number of internal views (or `architectures`) to define the operation of the entity[134]. To illustrate these ideas, let us consider the design of a one bit FULL_ADDER circuit (Figure 6.2).

A FULL_ADDER is composed of two HALF_ADDER modules and the HALF_ADDERs take two inputs (A and B) to generate SUM and CARRY outputs. The VHDL description HALF_ADDER is given below (note that the line numbers are simply for convenience, and are not part of the code).

```
1   -- Description of half adder circuit
2   entity HALF_ADDER is
3   port (A,B           : in bit;
4         SUM,CARRY     : out bit);
5   end HALF_ADDER;
6
7   architecture HALF_ADDER_ARCH of HALF_ADDER is
```

commands such as the `assert` command which causes messages to appear in the simulation tool if a particular condition is satisfied.

[134] If multiple `architectures` are used, VHDL `configurations` can be used to determine which `architecture` to use. Multiple `architectures` can be used to create a model of the design for simulation (e.g. define a memory as an array), and a detailed design for implementation. In all our examples we use a single `architecture` hence we do not discuss `configurations` here. Refer to your VHDL manual for further details.

```
 8  begin
 9       SUM <= A xor B;
10       CARRY <= A and B;
11  end HALF_ADDER_ARCH;
```

This code illustrates some of the VHDL syntax.

- VHDL is a strongly typed language and will not automatically convert between types. This helps to minimise logical errors like trying to place a `10 bit` value onto an `8 bit` bus. While there are many built-in and user definable types, the ones used most often are `STD_LOGIC`, `STD_LOGIC_VECTOR` and `INTEGER`.
- Comments (line 1) are preceded with two dashes and finish at the end of the line.
- A semicolon is placed at the end of each statement.
- VHDL is *not* case sensitive.
- Components and signals must be declared before being used in an entity.
- The symbol `<=` is used to assign a value to a `signal`.
- The names after the `end` statements are optional; however, they do help to clarify the code. Take care with the syntax of `end` statements – the exact syntax can change depending on the statement with which the `end` is associated. For example, `end if` must be used to close an `if` statement.
- White space such as the blank line (lines 6) and indents (lines 9 and 10) can be used to make the code more readable.

In our example, the name of the `entity` is declared in line 2 and the interface to the entity is defined using the `port` statement in lines 3 and 4. A and B are defined as inputs of type `bit` (meaning these ports can take on values of `'0'` or `'1'`) and SUM and CARRY are defined as outputs of type `bit`. In addition to being inputs (`in`) or outputs (`out`), ports can also be

- `buffer` – this is used if a signal is used both as an output to the circuit *and* used within the circuit. The need for this type can be avoided by having one copy of a variable for internal use, and another for output.
- `inout` – this is used where the signal is bi-directional.

For this half adder, we only have one `architecture` (HALF_ADDER_ARCH) to describe the circuit function (lines 7–11). Notice that because A, B, SUM and CARRY were declared in the `port` declaration they are available for use in the architecture.

Given this HALF_ADDER, we can create a FULL_ADDER as shown in Figure 6.2. The corresponding VHDL description is given below.

```
1  -- Description of full adder circuit
2  entity FULL_ADDER is
3  port (A,B,CARRY_IN          : in bit;
4        SUM,CARRY_OUT         : out bit);
5  end FULL_ADDER;
6
7  architecture FULL_ADDER_ARCH of FULL_ADDER is
```

```
 8
 9    signal SIGNAL1, SIGNAL2, SIGNAL3   : bit;
10
11    component HALF_ADDER
12    port (A, B            : in bit;
13         SUM, CARRY       : out bit);
14    end component;
15
16    begin
17         U1:HALF_ADDER port map (SUM=>SIGNAL1, CARRY=>SIGNAL2, A=>A, B=>B);
18         U2:HALF_ADDER port map (SIGNAL1, CARRY_IN, SUM, SIGNAL3);
19         CARRY_OUT <= SIGNAL2 or SIGNAL3;
20    end FULL_ADDER_ARCH;
```

Declarations go here including:
- Type declarations
- Signal declarations
- Constant declarations
- Component declarations

Declarations are local to this entity

Concurrent statements go here.

In the architecture section, declarations are placed immediately after the architecture statement and a series of concurrent statements (explained later) appear between the begin and end reserved words. Once again, lines 2–5 form the entity declaration, this time with three inputs and two outputs. Line 9 declares the signals that join together our components. You can think of signals as wires used to join different components and, as such, a signal should only have one driver. Synthesis tools will not allow you to have more than one driver for a signal unless you follow particular rules. The ports listed in the port declaration are also signals. Lines 11–14 declare the interface to the component HALF_ADDER so that we can instantiate it later.

Finally, lines 16–20 form the architecture body, with lines 17 and 18 creating instances of the HALF_ADDER called U1 and U2 and line 19 performing the OR function for the CARRY_OUT signal. In lines 17 and 18 the port map statement defines how the ports for each instance of the HALF_ADDER component should be connected. In line 17 the ports are connected by name, and in line 18 the ports are connected by position. When connecting by name the following syntax is used:

```
COMPONENT_PORT_NAME => SIGNAL;
```

Since each port of the component (HALF_ADDER) is explicitly named in this syntax, the order of connection is not important. Line 18 maps the ports of the HALF_ADDER U2 on a position-wise basis using the order in which ports were declared. In our HALF_ADDER the ports were declared in the order A, B, SUM, CARRY thus the HALF_ADDER is connected as illustrated in Table 6.1. Regardless of which mapping method is used, all ports must be mapped. If you wish to leave a port unconnected, use the VHDL reserved word open rather than a signal name. For example, the following instantiation will leave the carry port of the HALF_ADDER unconnected:

```
U1: HALF_ADDER port map (A, B, SIGNAL1, open);
```

Table 6.1   Position-wise mapping for instanciated components

HALF_ADDER	FULL_ADDER
A	SIGNAL1
B	CARRY_IN
SUM	SUM
CARRY	SIGNAL3

Note that we did not have to explicitly instantiate an OR gate in order to perform the OR function. We could, if we wished, create an OR_GATE entity and instantiate it as we did with the HALF_ADDER, however, this is unnecessary because there is an OR function defined for the bit type in VHDL. To synthesise this FULL_ADDER, you should enter each entity into a different file[135] and include both files in the design project. You can then specify which file is the top level design (the FULL_ADDER in this case – each design must have one top level entity, much like a C program must have one main function) and begin the synthesis process.

6.6.2 VHDL types and libraries

By default, VHDL provides a range of data types including, among others, the bit type we have previously encountered, the boolean type (taking on values 'false' and 'true') and the integer type (two's complement with the range depending on the design tools used – often 32 bit). VHDL also allows a number of derived types including arrays, enumerated types and subtypes (subsets of other types, for example the natural type is a subset of the integer type including all positive values and zero). Note that because VHDL is a strongly typed language, if you define the same type twice using different names (as shown below) you cannot assign a signal of one type to a signal of the other type without an explicit type conversion.

```
1   type array1 is array (7 downto 0) of bit;
2   type array2 is array (7 downto 0) of bit;
3
4   signal a1 : array1;
5   signal a2 : array2;
6   ...
7   a1 <= a2; -- This is not allowed
```

---

[135] VHDL allows you to create design libraries and packages to structure your project, however, we only consider the standard built in libraries in this chapter. Refer to your VHDL manual for more details on how to create your own libraries.

6.6.2.1 Enumerated types

An enumerated type is a type where all possible values of that type are explicitly listed. For example:

```
1   type bit is ('0', '1');
2   type boolean is (false, true);
3   type sm_states1 is (reset, running, waiting, sleeping);
4   type sm_states2 is (reset, go, pause, rest);
```

The `bit` type we have been using is in fact an enumerated type composed of two *character literals* (character literals must be enclosed in single inverted commas). The `boolean` type is also an enumerated type that makes use of *enumeration literals*. If an enumeration literal appears in more than one type (as `reset` in `sm_states1` and `sm_states2`), it is said to be overloaded. To avoid possible problems, you should qualify any overloaded enumeration literals as shown below.

```
1   current_state <= sm_states1'(reset);
```

The elements of an enumerated type are mapped (or *encoded*) to numbers by the synthesis tools. While there are ways of controlling this encoding in the VHDL code (this can be useful for hand optimisation of a design), it is often easier to allow the design tools to deal with the details of the encoding. Typically, design tools provide a binary encoding option[136] and one-hot encoding option[137]. By default, VHDL uses a binary encoding with the first element being assigned a value of 0. This means that given our earlier enumerated type definitions the following conditions are true if the default encoding is used:

```
1   running = go
2   reset < rest
```

6.6.2.2 Arrays

An *array* is a collection of elements of the same base type. The VHDL language itself allows arrays using any base type and of arbitrary dimension, however, design tools may limit the number of dimensions that an array can have. This can be side-stepped by creating an array using an array *base type*. VHDL also allows for *constrained array types* (whose size is fixed in the declaration) and *unconstrained array types* (whose size is determined in the declaration of a signal of that type). To use an array, you must first declare the array type, and then create a signal of that type.

---

[136] Here elements are assigned successive numerical values based on the order in which they appear.
[137] In one-hot encoding, each element is represented by a binary number with as many bits as there are elements in the enumerated type. The nth element in the enumerated type is represented by a number whose nth bit is set to one with all other bits set to zero. This can be more efficient than binary encoding since less logic is required to manipulate the values (e.g. moving from element to the next in the list only requires toggling of two bits whereas binary mapping would require addition). The cost is that the one-hot representation will require more memory (flip-flops) than a binary representation.

For example

```
1  type UP_BYTE is array (0 to 7) of bit;
2  type DOWN_BYTE is array (7 downto 0) of bit;
3  type arb_vector is array (integer range <>) of bit;
4  signal UP_DATA : UP_BYTE;
5  signal DOWN_DATA : DOWN_BYTE;
6  signal VECTOR1 : arb_vector(-3 to 3);
7  signal VECTOR2 : arb_vector(6 downto 0);
```

All the types declared here are arrays of the base type `bit`. Lines 1 and 2 declare two different 8 `bit` constrained arrays, `UP_BTYE` and `DOWN_BYTE`. In line 1, the `to` directive indicates that the elements of `UP_BYTE` are numbered in ascending order (i.e. the index 0 refers to the left most element of the array). Conversely, in line 2, the `downto` directive indicates that elements of `DOWN_BYTE` are numbered in descending order (i.e. index 7 refers to the left most element of the array). Once an array has been declared using the `to` or `downto` direction, all future accesses to that array must use the same direction. `Signals` of these types are declared in lines 4 and 5.

Line 3 illustrates a type declaration for an unconstrained type called `arb_vector`. In this case the array bounds can take on any `integer` value, however, subtypes of `integer` can also be used to limit the possible array bounds. Lines 5 and 6 illustrate how to declare signals of an unconstrained array type. Note that the array bounds are defined in the `signal` declaration and that negative array bounds are allowed. Assignment of `UP_DATA` to `DOWN_DATA` is illegal since these signal have different types, however, we can assign `VECTOR1` to `VECTOR2` because they are both of the same base type and same size. The difference in array bounds is not important, though the difference in direction means that the value will be reversed. Since `bit` vectors are so common (buses, etc.), VHDL provides a `bit_vector` type. This is similar to our `arb_vector` type except that the array bounds are limited to the natural numbers.

The code fragment below illustrates the range of ways in which arrays can be accessed.

```
1   signal DATA_WORD : bit_vector(15 downto 0);
2   signal DATA_BYTE : bit_vector(7 downto 0);
3   signal DATA_NIBBLE : bit_vector(3 downto 0);
4   signal b1, b2, b3 : bit;
5   signal BACKWARD_WORD : bit_vector(0 to 15);
6   ...
7   b2 <= '0';
8   b3 <= '1';
9   DATA_WORD(1) <= '0';                              -- Element Assignment
10  DATA_WORD(0) <= b2 and b1;                        -- Element Assignment
11  DATA_WORD(7 downto 5) <= B''100'';                -- Assign to array slice
12  DATA_NIBBLE <= (2=>b1 or b2, 1=>'0', others=>'1');-- Aggregate Ass. by name
13  DATA_BYTE <= DATA_NIBBLE & X''B'';                -- Concatenation
14  b1 <= DATA_BYTE(0);
15  DATA_WORD(4 downto 2) <= (b1, b2, b3 and b1);     -- Aggregate Ass. by
                                                                    position
16  DATA_WORD(15 downto 8) <= DATA_BYTE;              -- Assign equal length
                                                                       array
17  BACKWARD_WORD <= DATA_WORD;                       -- Reverse direction
```

Lines 9 and 10 illustrate *element assignment* – the assignment of a value to a particular element of the array by direct indexing. Remember that the bit type is actually an enumerated type composed of two character literals, hence we must enclose the 0 and 1 values in single inverted commas. In line 11 we assign a *bit string literal* (i.e. B'' 100'') to a *slice* of the array (bits 7, 6 and 5). Bits are assigned in order just as ports were, so bit 7 of DATA_WORD will take on the value '1' while bits 6 and 5 will be set to '0'. Bit string literals must be enclosed in double inverted commas and if the string is intended to be interpreted as binary then the *base specifier* B is placed before the string, hex strings use the base specifier X, octal strings use O. In line 15 we see an example of *aggregate assignment* to an array slice. Here again, bits are assigned in order: bit 4 of DATA_WORD is assigned the value of b1, bit 3 is assigned the value of b2 and bit 2 is assigned the result of b3 and b1.

Aggregate assignment can also operate by name rather than position as shown in line 12. The construct 2>=b1 or b2 means that bit 2 of DATA_NIBBLE is assigned a value that is the result of the operation b1 or b2. Similarly, bit 1 is assigned the value '0'. The others=>'1' construct sets all bits that have not previously been set to '1'. In line 13 we see an example of *concatenation* with the & operator[138], where the 4 bit DATA_NIBBLE and the 4 bit hex bit literal X' 'B'' are concatenated and assigned to the 8 bit signal DATA_BYTE. Line 16 shows that we can assign one array to another provided they are of the same length and type. Finally, in line 17, we assign DATA_WORD to a signal with opposite direction. This is a valid assignment (since both signals are of type bit_vector and 16 bits in length), however, it results in the bit string being reversed.

It is important to note that in VHDL, the bit_vector type is considered strictly as a string of bits – this string of bits cannot be interpreted to have a numerical value since no number system (e.g. two's complement) is associated with the type. For this reason, (a) we can only assign bit string literals to bit_vectors, not numeric literals (described later) and (b) arithmetic functions are not defined for bit_vectors[139].

Logical operators[140] (not, and, or, nand, nor, xor) are defined for bit_vector type. These operators require both operands to be of the same length. Relational operators (=, /=, <, <=, >, >=) are also defined for bit_vectors, however, the operands for these operators do not need to be the same length, since comparison is done in dictionary order. That is, the operation begins with the left most element of each array with the ordering being *empty* < 0 < 1. Therefore, ''11011'' is less than ''111'' (due to the 3rd elements from the left), however, ''11'' is less than ''1111'' since the 3rd and 4th positions of the first bit literal are empty.

Before moving on to consider integer types, we must consider how arrays are synthesised. In sequential programming languages it is common to allocate a region of memory as an array when attempting to store a large amount of data. VHDL also

---

[138] Concatenation is defined for one-dimensional arrays and elements of that array.
[139] This is true for standard VHDL, though vendors may provide libraries allowing arithmetic with the bit_vector type.
[140] Logical functions are defined for types bit and boolean and for arrays of these types.

allows such arrays to be created and these arrays can be useful for *logical simulation* purposes (e.g. simulation of a buffer or large memory). However, when you synthesise an array, the design tools will implement the memory using the flip-flops distributed through the FPGA (assuming sufficient flip-flops are available). The resulting memories are called *distributed memories* and while they are quite suitable for small arrays (e.g. small `bit_vectors`), they can become problematic for large arrays because they require large amounts of resources (both flip-flops and wiring) and can take a long time to synthesise.

To overcome these problems, current FPGAs include 'Block Memories', dedicated blocks of RAM available for use in your design. The way these block memories operate varies between FPGA vendors and synthesis tools will not automatically attempt to replace a distributed memory with a block memory. Therefore, to use block memory you must explicitly instantiate it and interface to it. As a result, a design utilising block memory in a particular FPGA device will not be portable to other FPGAs.

6.6.2.3 Integers

As mentioned earlier `integers` are two's complement numbers whose range can depend on the design tools being used, though 32 bit signed integers are common. VHDL also provides the `positive` type (a subtype of `integer` containing all positive integers but not zero), and the `natural` type (a subtype of `integer` containing positive integers and zero). Individual bits of an `integer` type cannot be directly accessed. To access individual bits, the `integer` should be converted to a vector type[141]. To assign a particular value to an integer signal you use *numeric literals*. Any number you enter will be assumed to be in base 10 unless you specifically specify a different base using the hash symbol as follows:

```
100             -- decimal 100 (base 10 -- decimal)
2E6             -- decimal 2000000 (base 10 -- decimal)
2 #100#         -- decimal 4 (base 2 -- binary)
16 #f#          -- decimal 15 (base 16 -- hex)
```

When declaring a signal of type `integer` (or any of its subtypes) you have the option of specifying a `range`. For example

```
1   signal counter_8bit: INTEGER range -100 to 100;
2   signal current_value: INTEGER;
3
4   current_value <= counter_8bit;
5   counter_8bit  <= current_value;
6   counter_8bit  <= current_value + 1;
7   counter_8bit  <= 1000;
```

---

[141] For example, you can use the function `CONV_STD_LOGIC_VECTOR(VALUE, LENGTH)` defined in the `std_logic_arith` package. VALUE is the integer value you wish to convert and LENGTH is the number of bits in the result.

Here, the `counter_8bit` signal is limited to an 8 bit `range` while `current_value` will be a 32 bit integer. The advantage of specifying a `range` is twofold. First, the bit width of an arithmetic operation is determined by the width of the largest operand, hence specifying `ranges` will save logic resources. Second, while assignment between `integers` of different `ranges` is valid (e.g. lines 4 and 5), logical simulation will warn you if you try to assign a value that is out of `range` for the target signal. Trying to assign an out of range numeric literal (as in line 7) will generate an error both in logical simulation and synthesis.

An assignment to a signal with a larger `range` than the source signal (e.g. line 4) operates as expected – the value in `counter_8bit` will be assigned to `current_value`. This assignment will properly take into account any negative values, thus −8 (equal to hex 0xf8 in two's complement) will become hex 0xfffffff8. Line 5 is a little more complicated because `current_value` may be outside the range of `counter_8bit`. This line will synthesise such that `counter_8bit` will be assigned a value using only the lower 8 bits of `current_value`. The examples below explain this in more detail.

```
1    signal counter_8bit: INTEGER range -100 to 100;
2    signal current_value: INTEGER;
3    ...
4    counter_8bit <= -10;                  -- counter_8bit  = -10
5    current_value <= counter_8bit;        -- current_value = -10
6    ...
7    current_value <= 16#00000001;         -- current_value = 1
8    counter_8bit <= current_value;        -- counter_8bit  = 1
9    ...
10   current_value <= 16#80000001;         -- current_value = -2147483647
11   counter_8bit <= current_value;        -- counter_8bit = 1 (16#01#)
12   ...
13   current_value <= 16#7d2456dc;         -- current_value = 2099533532
14   counter_8bit <= current_value;        -- counter_8bit = -36 (16#dc#)
```

The `integer` types have relational operators and some arithmetic operators defined for synthesis. Typically, addition, subtraction and multiplication are available as well as division by powers of two, though this can depend on the design tools used.

6.6.2.4 Other standard libraries

The types and operations we have considered thus far are all built into VHDL, however, extended functionality is available through the use of libraries. Design tool vendors will include libraries to extend the functionality of VHDL, however, to maintain portability it is best to use the standard libraries included in VHDL rather than the vendor libraries. Of course, users can also write their own libraries.

As part of the IEEE VHDL standard, there is an IEEE library defined and this library is included in all VHDL software. The IEEE library includes the `std_logic_1164` package which defines a special type called `std_logic` that we will discuss shortly. The `numeric_std` package defines `signed` and unsigned integer types as well as a range of type conversion functions and operations (arithmetic, comparison, logical and shifting) for these types. The std_logic_arith package is similar though somewhat more expansive, supporting more combinations of data types. To access these libraries in a given VHDL file, add the following lines to the start of the file:

```
1 library IEEE;
2 use IEEE.std_logic_1164.all;
3 use IEEE.std_logic.arith.all;
```

Line 1 indicates which library to use, while lines 2 and 3 indicate which packages to include from the library. The all directive indicates that all the available functionality should be included.

6.6.2.5 STD_LOGIC

The VHDL bit type, while straightforward, does not provide sufficient power to model all circuit functionality. For example, high impedance states cannot be modelled making it strictly illegal for a signal to have more than one driver. To overcome this, new types called std_ulogic and std_logic[142] were created, providing a 9 value logic system. These types are available from the std_logic_1164 library and are defined as follows:

```
type std_ulogic is   (   'U',   -- uninitialised
                         'X',   -- unknown
                         '0',   -- low
                         '1',   -- high
                         'Z',   -- high impedance
                         'W',   -- weak
                         'L',   -- weak low
                         'H',   -- weak high
                         '-',   -- don't care);
```

Of these states, 4 can be used in synthesis ('0', '1', 'Z', '-')[143]. The difference between std_ulogic and std_logic lies in resolution – the ability to arbitrate cases where there is more than one driver for a signal. The std_ulogic type is unresolved – a signal cannot have more than one driver while the std_logic type has defined a resolution function that dictates what the result will be if different values drive the same signal (see Figure 6.3). Table 6.2 illustrates the portion of the signal resolution function (as defined in the std_logic_1164 library) that is relevant for synthesis. The ability to resolve signals means that std_logic is used in design far more often than the bit or std_ulogic types. Note that this resolution function

---

[142] Together with the corresponding vector types std_ulogic_vector and std_logic_vector. The IEEE libraries do not define arithmetic operations for these types; however, vendors often provide libraries containing arithmetic operations for these types.

[143] The don't care state is used by the synthesis software for logic optimisation.

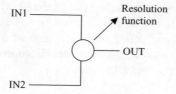

Figure 6.3   Resolved signal (adapted from Reference 306)

Table 6.2   Signal resolution for synthesisable values of `std_logic`

IN2→ IN1↓	'0'	'1'	'Z'	'-'
'0'	'0'	X	'0'	'X'
1	X	'1'	'1'	'X'
'Z'	'0'	'1'	'Z'	'X'
'-'	'X'	'X'	'X'	'X'

is a logical construct used for simulation: if there are two drivers for a signal, the appropriate tristate logic will not necessarily be automatically inferred. Tristate logic must be either explicitly described in your design, or it can be automatically inferred if you make appropriate use of the `std_logic` type as described in Section 6.6.4.

6.6.2.6   Type conversion

VHDL is a strongly typed language and as such does not automatically convert types. There are two mechanisms to explicitly change between types, however. For conversion between different integer types or for conversion between similar array types[144] the syntax below can be used. Note that conversion between enumerated types is not possible using this method.

```
type INT_ARRAY1 is array (1 to 5) of integer range 0 to 10;
type INT_ARRAY2 is array (5 downto 1) of integer range 0 to 10;
signal array1 : INT_ARRAY1;
signal array2 : INT_ARRAY2;
...
--           target_type_name(expression)
array2 <= INT_ARRAY2(array1);
```

---

[144] Array types are similar if they have the same length and the same (or a convertible) base type.

For other types, functions must be written to explicitly convert between types. For example the `std_logic_arith` package contains functions for converting between `integer`, `unsigned`, `signed`, and `std_logic_vector` types and the `std_logic_1164` package contains functions for converting between `std_logic`, `std_ulogic` and `bit` types.

6.6.3 Concurrent and sequential statements

Reflecting the nature of digital hardware, VHDL provides both concurrent and sequential statements. Concurrent statements appear in the body of the `entities architecture` and all concurrent statements execute in parallel. Sequential statements appear within a `process` within the `architecture` body and all statements within the `process` execute in sequence. `Processes` themselves are concurrent statements.

6.6.3.1 Concurrent statements

Concurrent statements are all executed in parallel, the order in which they appear makes no difference to the outcome. For example, the following concurrent signal assignments are identical. In both cases, Z will reach a value of A + B + C.

```
1   X <= A+B;              1   Z <= C+X;
2   Z <= C+X;              2   X <= A+B;
```

We now briefly discuss the most common concurrent statements.

6.6.3.1.1 Concurrent signal assignment
Concurrent signal assignments occur outside `process` statements and have the syntax:

```
SIGNAL_NAME <= EXPRESSION;
```

Each concurrent statement that assigns a value to a signal is considered to be a driver for that signal. As discussed earlier, in such a situation, the `std_logic` type should be used. Appropriate use of the high impedance state will cause the synthesis process to infer the need for tristate buffers and the possibility of two drivers attempting to drive the same signal simultaneously will be removed. The idea of inference is discussed in Section 6.6.4.

6.6.3.1.2 Conditional signal assignment
Concurrent conditional signal assignment is similar to the sequential `if` statement. In this structure, the expression for the first condition found to be true is assigned to the target signal. Conditional signal assignment synthesises to a set of cascaded multiplexers as in Figure 6.4, creating large delays if there are many `else` conditions. To avoid these delays, selected signal assignment should be used in preference to conditional signal assignment where possible. The syntax for a conditional

Digital design   191

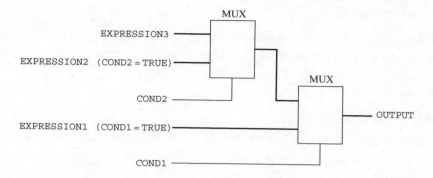

Figure 6.4 *Synthesis of conditional signal assignment. Note that the results of an expression can be a multi-bit value and that conditions are evaluated to a* `boolean` *value (adapted from Reference 290)*

assignment is:

```
OUTPUT <= EXPRESSION1 when COND1 else
          EXPRESSION2 when COND2 else
          EXPRESSION3;
```

6.6.3.1.3 *Selected signal assignment*

Concurrent selected signal assignment is similar to the sequential `case` statement. It assigns a value to an output signal based on the value of a `CHOICE` signal of arbitrary bit width. The `CHOICE_LIST`s in selected signal assignment must obey the following rules:

- they must be static expressions or static ranges (Section 6.6.3.1.6);
- they must be mutually exclusive;
- all possible choices must be enumerated;
- the `when others` clause can be used to implicitly enumerate all choices not explicitly listed.

Selected signal assignment synthesises to parallel structure as illustrated in Figure 6.5. The syntax for a selected signal assignment is shown below.

```
with CHOICE select
OUTPUT <= EXPRESSION_1 when CHOICE_LIST_1,
          EXPRESSION_2 when CHOICE_LIST_2,
                ↓
          EXPRESSION_N when others;
```

6.6.3.1.4 *Component instantiation*

Component instantiation was introduced in Section 6.6.1. In summary, the component must first be declared in the declaration part of the `architecture`, and then the component can be instantiated in the `architecture` body.

192 *Motion vision: design of compact motion sensing solutions*

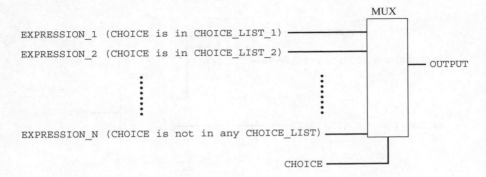

Figure 6.5   Synthesis of selected signal assignment. Notice that all signals can be multi-bit values (adapted from Reference 290)

6.6.3.1.5 Processes
A VHDL `process` is a program region where commands are executed in sequence rather than concurrently. Each process within an `architecture` executes in parallel. We will discuss processes in more detail in Section 6.6.3.2.1.

6.6.3.1.6 A code example
In this section we present a short example entity making use of two concurrent statements – conditional and selected signal assignments.

```
1       -- An example of conditional and selected signal assignment
2       library IEEE;
3       use IEEE.std_logic_1164.all;
4
5       entity concurrent is
6           port (
7               A: in STD_LOGIC_VECTOR (7 downto 0);
8               B: in STD_LOGIC_VECTOR (7 downto 0);
9               C: in STD_LOGIC_VECTOR (7 downto 0);
10              Y: out STD_LOGIC_VECTOR (7 downto 0);
11              Z: out STD_LOGIC_VECTOR (7 downto 0);
12              VAL: in INTEGER range 0 to 15
13          );
14      end concurrent;
15
16      architecture concurrent_arch of concurrent is
17      begin
18
19          -- Conditional signal assignment
20          Y <= A when VAL=0 else
```

```
21              B when VAL=1 else
22              ''10101010'' when VAL=3 else
23              C;
24
25          -- Selected signal assignment
26          with VAL+1 select
27          Z <= A when 5,
28               C when 2,
29               ''00001111'' when 13,
30               B when 6 to 12,
31               ''00110011'' when others;
32
33      end concurrent_arch;
```

6.6.3.2 Sequential statements

Sequential statements occur within a `process` and are executed in sequence. We begin by explaining the concept of a `process` then examine some of the sequential statements available for use within a `process`. We limit ourselves to those statements that are useful in our design approach.

In addition to the sequential statements we consider here, VHDL provides statements for constructing loops (`loop`, `while-loop`, `for-loop`, `next` and `exit`), statements for constructing subprograms (`procedure`, `function` and `return`) and statements for pausing the execution of a process (`wait`). We refer the reader interested in these, and the many other constructs available in VHDL, to their VHDL manual. In our design approach, these statements are not necessary. Loops can easily be constructed using state machines[145]. Subprograms can be implemented using separate state machine/schematic components in a hierarchical manner. We do not explicitly use `wait` statements – the same effect is achieved through the use of sensitivity lists.

As you read this section, keep in mind that VHDL sequential statements are not sequential in quite the same sense as the instructions in a program run on a normal processor. In a processor, a new instruction is executed at every clock cycle while sequential instructions are essentially executed in cascade. This idea is considered in more detail in Section 6.9.1.

6.6.3.2.1 Processes

A `process` is single concurrent statement that contains a sequence of *sequential statements*. Because a `process` is a concurrent statement, all processes within

---

[145] For-loops can be useful in repeating a process a number of times (e.g. once for each element of an array). However `for-loops` are flattened in synthesis leading to a concurrent structure – not the anticipated iterative structure. This means that `for-loops` cannot be used in circuits where data are not immediately available, such as iterating over data stored in a memory. The bounds of a `for-loop` must be *computable* (able to be determined at synthesis time).

an `architecture` operate in parallel. The basic syntax of the `process` statement is shown below. Note that the `PROCESS_LABEL` is optional.

```
1   PROCESS_LABEL : process (SENSITIVITY_LIST)
2   begin
3           [sequential statements]
4   end process PROCESS_LABEL;
```

A process remains suspended until any signal in the `SENSITIVITY_LIST` list changes value. Based on what we include in the `SENSITIVITY_LIST` we can form two types of process: the *combinatorial process* that specifies only combinatorial logic, and the *sequential process* where actions occur in time with a clock. For a combinatorial `process`, *all inputs* must be listed in the sensitivity list so that outputs can be updated immediately. For example:

```
1   architecture comb_arch of combinatorial is
2           signal X,Y,A,B,C : std_logic;
3   begin
4           process(A,B,C)
5           begin
6                   X<=A or B;
7                   Y<=A and C;
8           end process;
9   end comb_arch;
```

Our design approach uses state machines that are inherently sequential structures – a clock dictates progress from one state to the next. At a minimum, sequential `processes` will have a clock signal in the sensitivity list. If an asynchronous reset is required, then the reset signal will also appear in the sensitivity list. The code fragment below illustrates sequential processes with synchronous and asynchronous resets.

```
1   -- Asynchronous reset
2   ASYNC_RESET : process(CLK, Reset)
3   begin
4           if Reset='1' then
5                   -- Sequential statements - reset function
6           elsif CLK'event and CLK='1' then
7                   -- sequential statements -- process implementation
8           end if;
9   end process ASYNC_RESET;
10
11  -- Synchronous reset
12  SYNC_RESET : process(CLK)
13  begin
14          if CLK'event and CLK='1' then
15                  if Reset='1' then
16                          - sequential statements -- reset function
17                  else
18                          - sequential statements -- process implementation
```

```
19            end if;
20         end if;
21   end process SYNC_RESET;
```

Notice that for the asynchronous reset we have placed the reset condition before the clock condition. An `if` statement only executes the first true condition and an asynchronous `reset` signal always has priority over the clock so the `reset` condition must be checked first. The `CLK'event` construct used in lines 6 and 14 uses the `event` attribute which is true only if the clock signal is changing state. VHDL defines a range of attributes for different objects, one of which is the `event` attribute defined for signals. The user is referred to their VHDL reference guide for more information about attributes. The `event` attribute does not directly correspond to any digital component – it is used to allow the inference of flip-flops. See Section 6.6.4 for more details.

6.6.3.2.2 Variable and signal assignment

In a `process` there are two types of object that can be assigned a value – signals and variables. *Sequential signal assignment* differs from concurrent signal assignment in that multiple sequential assignments to the same signal do not lead to multiple drivers for that signal. Consider the following code fragment:

```
1    architecture sig_assign_arch of sig_assign is
2          signal W,X,Y,Z,O1,O2,A1,A2 : std_logic;
3    begin
4            -- Concurrent signal assignment
5          A1 <= not W;
6          O1 <= Y or X;
7          O1 <= A1 and Z;            -- Two drivers for O1
8
9          process(W,X,Y,Z)           -- Driver for O2
10         begin
11             -- Sequential signal assignment
12             A1 <= not W;
13             O2 <= Y or X;          -- No effect
14             O2 <= A1 and Z;        -- Assignment of O2
15         end process;
16   end sig_assign_arch;
```

Lines 6 and 7 are concurrent signal assignments for the signal O1, hence the signal O1 has two drivers. Lines 13 and 14 are sequential signal assignments since they lie within a `process`. In sequential signal assignment, a value is only assigned to a signal at the termination of the `process`, hence only the final assignment has any effect. Because the assigned value is determined when the `process` terminates, the order of sequential signal assignments is not important. For example, placing line 12 after line 14 would not change the final outcome. Note that a `process` forms one driver for the signals that are assigned within it.

When you need to assign a value immediately, you can use a variable. VHDL variables can only be used within a `process` and can only be accessed in the `process` in which they are defined. These variables act like variables in other languages – they are assigned a value immediately. To make this difference obvious, variable assignment uses a different syntax to signal assignment.

```
VARIABLE_NAME := EXPRESSION;
```

Variables are declared between the `process` and `begin` reserved words. To help clarify the difference between signals and variables, consider the follow code fragments which are identical except for A, which is a variable on the left, and a signal on the right.

```
1  architecture arch of test1 is          1  architecture arch of test2 is
2    signal X,Y,Z : std_logic;            2    signal X,Y,Z : std_logic;
3  begin                                  3    signal A : std_logic;
4    process(X,Y,Z) {                     4  begin
5      variable A : std_logic;            5    process(A,X,Y,Z) {
6    begin                                6    begin
7      A := X;                            7      A <= X;
8      Z <= A xor '1';                    8      Z <= A xor '1';
9      A := '1';                          9      A <= '1';
10     Y <= A or '1';                     10     Y <= A or '1';
11   end process;                         11   end process;
12 end arch;                              12 end arch;
```

Because variables are assigned a value immediately, executing the code on the left will lead to the assignments:

```
Z <= X xor '1';
Y <= '1' or '1';
```

In the code fragment on the right only line 9 has an effect – line 7 will be ignored since A is a signal. Thus the net result of executing the code on the right will be the following assignments:

```
Z <= '1' xor '1';
Y <= '1' or '1';
```

It is important to initialise a variable before using it. If the variable is initialised after it is used, a latch will be inferred to store the value of the variable.

6.6.3.2.3 The `if` statement

The sequential `if` statement is similar to concurrent conditional signal assignment so all comments in Section 6.6.3.1.2 hold here. The primary difference between these constructs is that conditional signal assignment assigns a value to a signal based on some condition. The `if` statement is more powerful, allowing any operation to be performed based on a condition. It is left as an exercise for the reader to write a process that performs the same function as the conditional signal assignment in Section 6.6.3.1.6. The syntax for a sequential if statement is

```
if COND1 then
     [sequential statements]
elsif COND2 then
     [sequential statements]
else
     [sequential statements]
end if;
```

Notice the missing letter e in the elsif branch. Be careful to avoid situations where a variable is not assigned before it is read as may occur in the following code fragment:

```
1   process (in1)
2         variable var1 : std_logic;
3   begin
4         if in1 ='1' then
5             var1 := '1';
6         end if;
7         out1 <= var1;
8   end process;
```

In this case, if in1 does not equal '1' var1 is not initialised. This situation is not illegal; however, it may lead to inconsistent results.

6.6.3.2.4 *The* case *statement*

The case statement is similar to concurrent selected signal assignment, hence all comments in Section 6.6.3.1.3 also hold here. The case statement is more powerful than concurrent signal assignment since it allows the execution of arbitrary code rather than just signal assignment. The case statement also allows the use of the sequential statement null meaning that nothing should be executed if the expression takes on a given value. The case statement will only execute those sequential statements in the branch that evaluate to true – there is no need for a break statement as in the C language. The reader should write a process that performs the same function as the selected signal assignment in Section 6.6.3.1.6. The syntax for a sequential case statement is:

```
case EXPRESSION is
     when CHOICE1 =>
           [sequential statements]
     when CHOICE2 =>
           [sequential statements]
     when others =>
           [sequential statements]
  end case;
```

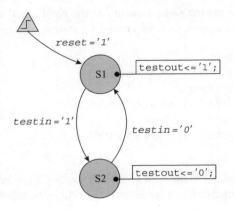

Figure 6.6   A simple state machine

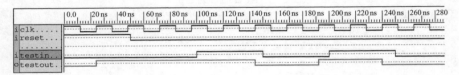

Figure 6.7   Functional simulation of the state machine in Figure 6.6

6.6.3.2.5   Example code

The example code below implements the simple state machine illustrated in Figure 6.6 (for more details about state machines, see Section 6.9.1). The state machine is implemented with synchronous reset and has two states S1 and S2. At reset, the state machine is put into the initial state (S1) and the output (testout) is set high. If the testin input goes high, the state machine transitions to state S2 where testout is set low. If the testin input returns low, the state machine returns to state S1 and testout is set high. This state machine is an (inefficient) implementation of a sort of synchronous inverter. The code included here was generated automatically from a state machine entry tool. You should compare the code and its functional simulation (Figure 6.7) to confirm your understanding of sequential statements.

```
1    library IEEE;
2    use IEEE.std_logic_1164.all;
3
4    entity test is
5    port (clk: in STD_LOGIC;
6          reset: in STD_LOGIC;
7          testin: in STD_LOGIC;
8          testout: out STD_LOGIC);
9    end;
10
```

```
11    architecture test_arch of test is
12
13         type Sreg0_type is (S1, S2);
14         signal Sreg0: Sreg0_type;
15
16    begin
17
18         StateMachine : process (clk)
19         begin
20
21              if clk'event and clk = '1' then
22                   if reset='1' then
23                        Sreg0 <= S1;
24                        testout<='1';
25                   else
26                        case Sreg0 is
27                        when S1 =>
28                             testout<='1';
29                             if testin='1' then
30                                  Sreg0 <= S2;
31                             end if;
32                        when S2 =>
33                             testout<='0';
34                             if testin='0' then
35                                  Sreg0 <= S1;
36                             end if;
37                        when others =>
38                             null;
39                        end case;
40                   end if;
41              end if;
42
43         end process StateMachine;
44    end test_arch;
```

6.6.4 Inference

While it is possible to include latches, flip-flops and tristate buffers in your design by explicitly instantiating them, it is common to allow the design tools to infer the presence of these components by using special structures in your VHDL code. In this section we cover the inference of simple D-latches, D-flip-flops and tristate buffers. Those readers interested in the inference of other latches and flip-flops are referred to their VHDL reference guide.

6.6.4.1 D-latches

D-latches (single input level sensitive memory elements) are inferred when an output is not defined for all conditions as occurs when an `if` statement has no `else` clause. Listed below is a typical code fragment that leads to the inference of a D-latch.

```
1  process(Gate, D)
2  begin
3        if (Gate='1') then
4              Q<=D;
5        end if;
6  end process;
```

A latch will also be inferred if a signal is assigned in one but not all branches of an `if` or `case` statement. Further, variables will be latched if it is assigned after it is read as in the following code fragment:

```
1  process(in1)
2        variable var1:std_logic;
3  begin
4        out1 <= var1 or '1';    -- var1 is read
5        var1 := not in1;         -- var1 is assigned
6  end process;
```

6.6.4.2 D-flip-flops

Inference of D-flip-flops (single input edge triggered memory elements) is similar to the inference of D-latches. As with the latch, an `if` statement without an `else` clause is required, however, one of the conditions of the `if` statement must test for a signal transition. In VHDL, the following construct is used to detect a rising edge:

```
SIGNAL_NAME'event and SIGNAL_NAME='1'
```

Here, the `event` attribute for the signal is true only when the signal changes state. A typical structure that will lead to inference of D-flip-flops is shown below – this is the same sequential process with an asynchronous reset that we saw earlier.

```
1  process(CLK, Reset)
2  begin
3        if Reset='1' then
4              -- reset function
5        elsif CLK'event and CLK='1' then
6              -- process implementation
7        end if;
8  end process;
```

Any signals or variables assigned *anywhere* within the `if` statement will have flip-flops inferred. For inference of D-flip-flops to operate correctly, the test for a signal edge must be the only condition in that branch of the `if` statement, and there must be only one test for an edge within the `if` statement.

6.6.4.3 Tristate buffers

A common requirement when designing systems with buses is to allow multiple processes (or concurrent statements) to access the same signal (bus) under tristate control. This can be achieved through appropriate use of the tristate state defined in the std_logic type. The following code shows how tristate buffers can be inferred so that two processes can share the same output signal. This case assumes that a higher level process will control the select signals so that only one process can assess the output signal (out1) at any one time.

```
%
1      library IEEE;
2      use IEEE.std_logic_1164.all;
3
4      entity tristate is
5           port (in1, in2, select1, select2 : in std_logic;
6                 out1: out std_logic
7           );
8      end tristate;
9
10     architecture tristate_arch of tristate is
11     begin
12          buffer1 : process (in1, select1)
13          begin
14               if select1='1' then
15                    out1 <= in1;
16               else
17                    out1 <= 'Z';
18               end if;
19          end process buffer1;
20
21          buffer2 : process (in2, select2)
22          begin
23               if select2='1' then
24                    out1 <= in2;
25               else
26                    out1 <= 'Z';
27               end if;
28          end process buffer2;
29     end tristate_arch;
```

6.7 Timing constraints

Timing constraints are used to define maximum allowable delay between components in your design. The implementation tools will take this information and attempt to build a circuit that fulfils these constraints thus ensuring the design will perform the desired function. Because different design tools have different syntax for specifying timing constraints and different means of entering those constraints (e.g. dedicated

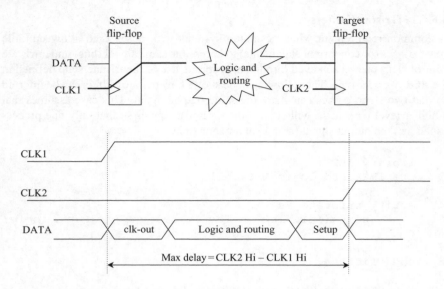

Figure 6.8   Clock-to-setup path timing (adapted from Reference 283)

entry tools, schematic annotation, plain text files), we only consider some basic timing constraint concepts here.

The basic element used to build timing constraints is the timing path. The three most common paths are the *clock-to-setup*, *clock-to-pad* and *pad-to-setup* paths. As might be expected, the maximum delay for each path must be less than the clock period. Let us begin by considering the clock-to-setup path as illustrated in Figure 6.8.

A clock-to-setup path starts at a flip-flop clock pin, propagates through all combinatorial logic and routing (including tristate buffers) and terminates at another flip-flop. The maximum allowable time for a signal to propagate from the source flip-flop to the target flip-flop must be less than the time between rising edges of CLK1 and CLK2 and is equal to the sum of the source flip-flop's clock-to-output time, the logic and interconnection delay[146] and the target flip-flop's setup time. Design tools will also take into account any clock skew (propagation delay). If CLK1 = CLK2 then the maximum allowable delay is equal to the clock period.

Figure 6.9 illustrates a clock-to-pad path. Like clock-to-setup paths, clock-to-pad paths trace through all combinatorial logic (including tristate buffers) but they end at a pad of the FPGA. The time from the clock's rising edge to the signal arriving on the pin is the path delay, however, in determining the clock rate we should also take into account external propagation and setup times (the margin).

---

[146] The logic and interconnect delay takes into account the time it takes for the signal to propagate through all combinatorial logic in the path including through the enable pin of tristate buffers. Worst case delays for each component are specified by the FPGA manufacturer. Design tools will use these delays when performing timing analysis.

Digital design 203

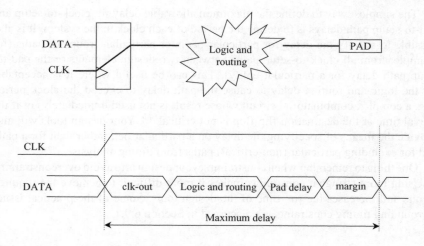

Figure 6.9   Clock-to-pad path (adapted from Reference 283)

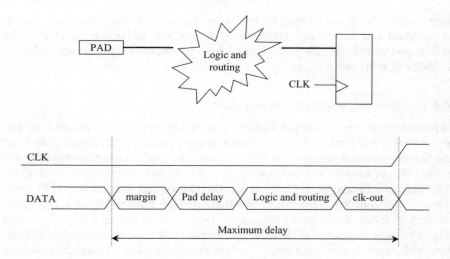

Figure 6.10   Pad-to-setup path (adapted from Reference 283)

Pad-to-setup delays are illustrated in Figure 6.10. As with the clock-to-pad path an external margin is included that takes into account the external clock-to-output delay and propagation delay.

Design tools usually provide a mechanism to specify the external margin for both clock-to-pad and pad-to-setup paths (though this can be difficult to specify). You should refer to your design tool documentation for more information. In practice, it is best to avoid inserting combinatorial logic between a pad and flip-flop to ensure the maximum possible margin.

The simplest way to define the maximum allowable delay for clock-to-setup and pad-to-setup path delays is to define the period of each clock in the system. It is also possible to specify the delay for particular paths or particular groups of paths (for example, from all clock-to-setup paths between two design elements or the pad-to-setup path delay for a particular entity). This can be useful where it is acceptable for the logic and routing delay to cause the path delay to exceed the clock period (e.g. a complex combinatorial circuit whose result is not used immediately) or if the arrival time at the destination flip-flop is not critical[147]. Your design tools will also provide the means of specifying the delay on a particular net, rather than for a path, and for excluding particular (non-critical) paths from timing analysis.

One thing to remember when constraining your design is to avoid overconstraining it. Using more stringent constraints makes it more difficult to achieve the required results and increases the run time of the design tools. Some of the practical issues surrounding timing constraints are discussed in Section 6.9.1.

6.8 General design tips

In this section we list some basic design issues of which you should be aware. Taking these issues into account can improve your designs and reduce debugging times. In this section we only include general principles – state machine specific principles are considered in the next section.

6.8.1 Synchronisation and metastability

Whenever you access external signals, you run the risk of *synchronisation* and *metastability* problems. Any external signal is asynchronous with respect to the local clock (unless special measures are taken to ensure the clocks are synchronised) which means that we have no way of knowing when a transition of that signal could occur. This can directly lead to problems of synchronisation and metastability [159, 266]. To help you understand the synchronisation problem, refer to Figure 6.11a. Here, an asynchronous signal enters the circuit from an input pad and fans out to two flip-flops. The path delay to the flip-flop D inputs will usually be slightly different, thus if the input changes state around the time the flips flops are clocked the flip-flops may actually sample different values. To avoid this, a synchroniser flip-flop is used (Figure 6.11b). This flip-flop synchronises the input signal to the local clock domain ensuring both flip-flops sample the same value.

One problem remains, however. What if the synchroniser samples the asynchronous signal during a transition? In this case, the output of the synchroniser flip-flop may, for a short while, have some intermediate output voltage or even oscillate. Thermal noise and asymmetric delays mean that the output will settle to a valid state eventually, however, the duration is uncertain. This indeterminate state

---

[147] This is sometimes referred to as a slow exception. Naturally, fast exceptions are also possible.

(a)

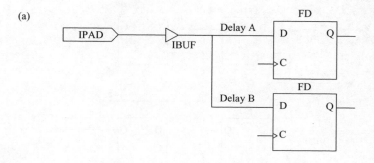

(b)

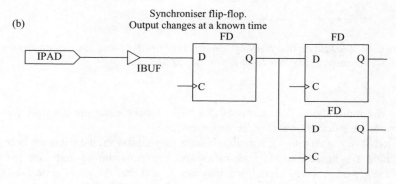

Figure 6.11   (a) If the delays are different the two flip-flops may actually sample different values; (b) synchroniser flip-flops take an external asynchronous signal and ensure it is clocked in the local clock domain so that all internal flip-flops using the external signals receive a consistent input. Note that all flip-flops are assumed to be clocked from the same clock (adapted from Reference 159)

is known as metastability [159] and if a flip-flop remains in the metastable state for a sufficient period of time, this metastability can cascade to other flip-flops. Using special flip-flops with reduced setup- and hold-times reduces the risk of entering a metastable state. For example, special purpose I/O flip-flops in Xilinx FPGAs are designed to ensure the risk of metastability is minimised [10][148].

However, this does not completely eliminate the risk of metastability. One can further reduce the risk by using a pair of flip-flops as shown in Figure 6.12. Provided the clock rate is slower than the time for the metastable state to settle, and clock skew is low, the second flip-flop will never go metastable. Of course, using two flip-flops

---

[148] The risk of a flip-flop going metastable is measured as a statistical measure called the Mean Time Between Failures (MTBF). This measure relates the time period during which a flip-flop can be made to go metastable, and the frequencies of the asynchronous signal and the local clock to indicate how often metastability is likely to occur. Xilinx claims an extremely low MTBF for their more modern devices [9].

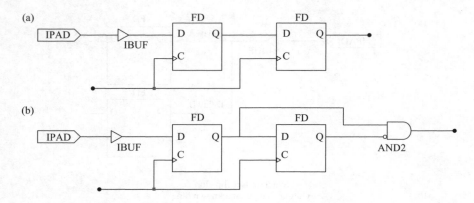

Figure 6.12 (a) Basic two flip-flop synchroniser and (b) an edge detecting synchroniser (adapted from Reference 266)

delays the input signal. Variations of this basic synchroniser are possible for level detection, edge detection and pulse detection [266].

If a `data_ready` signal is available for an asynchronous data, it is not necessary to include a synchroniser on each bit of the input. Assuming that data are valid when the `data_ready` signal becomes active and that the propagation delay for the data is less than the delay for the `data_ready` signal, it is sufficient to use a synchroniser only for the `data_ready` signal. The data can then be safely read when the `data_ready` signal becomes active.

6.8.2 Limit nesting of `if` statements

Using nested `if` statements or `if` statements with many branches leads to a cascaded set of multiplexers which introduce large path delays and can make it difficult to achieve timing constraints. To avoid this problem limit the nesting/branches of your `if` statements and use the `case` statement where possible.

6.8.3 Tristate buffers for large multiplexers

Multiplexers are a common design element. You can save logic resources by implementing large multiplexers using tristate buffers and a decoder to select which buffer should drive the output, as shown in Figure 6.13.

6.8.4 Tristate buffers

Tristate buffers can cause some problems that might only be detected using careful timing simulation. While these problems may not be serious, they are easily avoided. First, it is best to use tristate buffers that are in a high impedance state when the

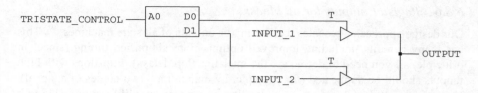

Figure 6.13   2 : 1 Multiplexer implemented using tristate buffers

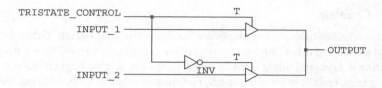

Figure 6.14   Dangerous use of tristate buffers. Net delays may combine to allow both buffers to drive the output

control line (the T input) is low. Using tristate buffers that drive the output when their control line is low could cause problems at power up: at power up all tristate control lines are (presumably) low meaning that all buffers could be driving the output. Second, you should avoid using the tristate buffer control scheme illustrated in Figure 6.14.

Let us assume we are using tristate buffers that are in a high impedance state when the T input is low. Let us further assume that the TRISTATE_CONTROL signal is low so that the lower buffer is driving the output. If the TRISTATE_CONTROL signal now goes high, problems *might* occur, depending on the delay along the TRISTATE_CONTROL net to T inputs of the buffers. If the delay to the T input of the upper buffer is shorter than the delay to the T input of the lower buffer, the upper buffer will start driving the output before the lower buffer has a chance to enter the high impedance state. This might be avoided using a scheme such as that in Figure 6.13 or some other method that ensures all buffers are in high impedance mode before a buffer is allowed to drive the output.

6.8.5   Don't gate clocks

While FPGA design tools allow for multiple clocks (indeed, FPGA can have dedicated resources for routing multiple clocks), the fewer clocks you have in your design, the easier it will be to debug. A particularly important aspect of this general principle is the gating of clocks to generate new clock signals. This should be avoided since glitches may cause unwanted data to be captured on flip-flops and may cause the setup and hold times of the flip-flops to be violated leading to metastability problems. A more appropriate approach is to use a clock enable signal.

6.8.6 Register outputs for all blocks

Our design approach advocates registering the outputs of all state machines[149]. This has many benefits, including improved optimisation, simplified timing (since, in principle, all you need to define are the inter-flip-flop delays). Flip-flops with high fanouts should be duplicated (for example, by maintaining two copies of a signal). Not only does this reduce capacitive loading, it can also simplify routing making it easier to meet timing constraints.

6.8.7 Counters

Consider whether natural counting order is critical for your design. Often the exact order of a counting is not important provided the order is consistent. This is true when generating delays, and might even be true for memory access. In such situations you can save logic resources by using small, fast counters such as Gray code counters and Linear Feedback Shift Registers (LFSRs) rather than full binary counters.

6.8.8 Special features

Use the special features available in your FPGA where appropriate. For example, FPGAs can include dedicated clock routing networks, dedicate reset networks, fast carry logic, memories and so on. Using these features will free up routing resources for your design and improve performance. Schematic library components such as counters will often take direct advantage of such features, however, synthesised VHDL may not unless the special features are explicitly instantiated.

6.8.9 Sequential pipelining

If your design includes complicated combinatorial logic, the design tools may not be able to sufficiently optimise the logic to meet timing constraints. One way to overcome this is by using sequential pipelining where the combinatorial logic is split into simpler stages and a register placed between the stages at the cost of increased processing time (Figure 6.15). Practical use of sequential pipelining is discussed in Section 6.9.1.

6.8.10 Use of hierarchy

Both VHDL and schematic design encourage a hierarchical design approach. This can be very useful, allowing work to be divided between design teams and providing a means of planning the layout of the design. One potential drawback of using a hierarchical design is the inability of some design tools to perform optimisation across hierarchical boundaries. This is a particular problem for combinatorial paths

---

[149] You may have to explicitly choose to register output ports.

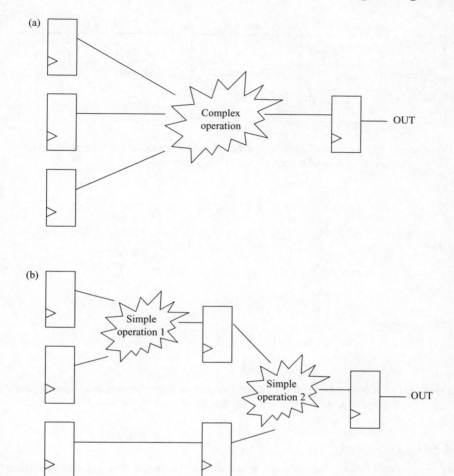

Figure 6.15 *An example of sequential pipelining. In (a) there is a combinatorial operation that is too complex to meet timing constraints. (b) Sequential pipelining involves splitting a complex operation into two (or more) simpler operations with flip-flops between the operations. Sequential pipelining will increase processing time (it will take two clock cycles for a result to appear at the output), however, the simpler operations will allow timing constraints to be met. Note that the extra flip-flop included in the lower path ensures the delays in the upper and lower paths are matched (adapted from Reference 306)*

(e.g. Block B in Figure 6.16). Where practical, you should either include combinatorial paths into a registered path (i.e. incorporate operation 2 into operation 3) or register the output of the combinatorial path (i.e. include a flip-flop after operation 2).

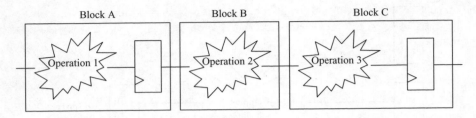

Figure 6.16 *Hierarchical design. Some design tools cannot optimise across hierarchical boundaries (adapted from Reference 306)*

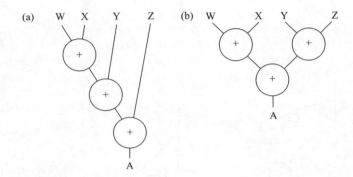

Figure 6.17 *Order of operations. Line 1 results in the structure in (a), while line 2 leads to the structure in (b) (adapted from Reference 290)*

6.8.11 Parentheses

In your VHDL code, consider the way you use parentheses since this can affect synthesis by changing the order of operations. For example, consider the following statements:

```
1    A <= (((W + X) + Y) + Z);
2    B <= (W + X) + (Y + Z);
```

The net result of both statements will be identical; however, these statements will synthesise to different structures. Line 1 will lead to a structure with three levels of addition (Figure 6.17a) while line 2 will result in only two levels of addition (Figure 6.17b), hence the circuit synthesised from line 2 will lead to a smaller path delay making it easier to meet timing constraints.

6.8.12 Bit width

When using integer types always define a range. Arithmetic operations are synthesised at the bit width of the largest argument, thus if you do not specify a range, arithmetic operations will be synthesised with the default bit width of 32 bits. This can be very wasteful of logic resources and can make it more difficult to achieve the desired timing.

6.8.13   Initialisation

Make sure you initialise all variables and outputs. When using state machine design entry, this is most easily done in the reset state. Initialising variables and signals ensures consistent operation (since the state is known) and will prevent unwanted inference of latches.

6.8.14   Propagation delay

Never design your circuit to depend on a particular propagation delay. Let us say you want a signal to arrive at its destination a particular amount of time after some event. For example, you might want an acknowledge signal to arrive at its destination at least 5 ns after a flag is set. To delay the signal you might look through the data sheet of the device you are using and find that an inverter causes a delay of 1 ns. Stringing these inverters together as shown below would create a delay line with a delay of 4 ns plus some routing delay (perhaps totalling 1 ns):

```
Ack <= not(not(not(not(flag))));
```

While this may work in some cases (in some contexts this is a valid delay mechanism) it will not work consistently. First, it is likely that your FPGA design tools will optimise this statement to:

```
Ack <= flag;
```

Thus you will not get the expected sequence of inverters. Second, even if your tools do not perform this optimisation the delay you get will be very sensitive to the circuit layout, which may change as your design is updated. Furthermore, such designs might not be portable between devices and propagation delays are often dependent on external factors such as temperature, thus your design could fail unexpectedly if external conditions change. In short, you will find it far simpler to debug your design if you design a synchronous circuit that does not depend on particular delays.

6.9   Graphical design entry

Having described the basics of FPGA design and VHDL, we now illustrate the use of state machines and schematics in the design process. You will see how we use state machines to order operations, to provide specific timing, to control complexity so that timing constraints can be met and to form constructs such as loops. Schematics are used to interconnect the various design components with each component operating in parallel.

6.9.1 State machines

In this section we do not introduce the traditional state machine theory (e.g. Moore machines, Mealy machines, optimisation, encoding[150], etc. [159]) nor do we consider specifically how state machines are implemented using VHDL (e.g. as one process or two processes[151]). We use state machines simply as a way of entering a series of sequential operations (with no asynchronous logic – all outputs are assumed to be registered) and leave the details to the design entry tools. Hence, we can create components in a manner that is similar to the way we write normal sequential software.

6.9.1.1 Sequential VHDL versus state machines

We have seen that VHDL provides both concurrent and sequential constructs, however, it is important to remember that the sequential constructs are not quite the same as what occurs when a program is executed using a standard processor. At the simplest level, you might imagine that a processor will execute a new instruction every clock cycle. Sequential statements in VHDL are not executed one clock at a time, but in cascade. Consider the following code fragments – on the left we have a simple program (in C syntax) for a processor, and on the right is a similar VHDL program

```
...                      ...
int a,b,c,d;             variable a,b,c,d : integer;
...                      ...
a=1;                     a:=1;
b=2;                     b:=2;
c=3;                     c:=3;
...                      ...
d=a+b;                   d:=a+b;
d=d+c;                   d:=d+c;
...                      ...
```

In principle, the C program on the left will take five clock cycles to execute (ignoring potential complications such as fetching data from memory, etc.). In the first clock cycle, a will be assigned the value one, in the second cycle b will be assigned, in the third c is assigned, in the fourth d is assigned the value of a+b and in the fifth cycle d is assigned d+c. The VHDL program will be implemented somewhat differently. While the operations will 'effectively' occur in order, the actual implementation might look something more like Figure 6.18.

As you can see, the operations certainly happen in order, however, they are not controlled by a clock and there is a danger of glitches occurring at the output d.

---

[150] The current state of a state machine is described by an enumerated type, thus the encoding of this variable refers to how this enumerated type is converted to numerical values, as discussed in Section 6.6.2.1.

[151] A state machine can be implemented in a single process as seen in Section 6.6.3.2.5, or as two processes with the calculation of the next state in one process and the update of the 'current state' register in another. Both forms are equivalent and your state machine entry tools will probably make the decision for you.

Digital design 213

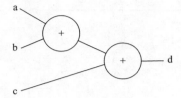

Figure 6.18 Sequential VHDL operations are implemented in cascade and glitches can occur

In fact, the exact timing of this circuit will depend on how the addition operations are implemented and optimised. Furthermore, when this operation is executed will depend on the sensitivity list of the process containing our code. If the sensitivity list contains all the inputs, then the circuit will be asynchronous – a new value for d will be calculated each time one of the inputs changes. This could be problematic if we want d to be updated only at specific times. We would like our design to operate more like our example C program with one instruction being executed every clock cycle. Our first step might be to assume the inputs are all registered and change the code so that a register is inferred for d as shown below

```
1    library IEEE;
2    use IEEE.std_logic_1164.all;
3    use IEEE.std_logic_arith.all;
4
5    entity ADD_ABC is
6    port (a,b,c   : in integer range 0 to 100;
7          result  : out integer range 0 to 300);
8    end ADD_ABC;
9
10   architecture ADD_ABC_ARCH of ADD_ABC is
11   process (clk)
12       variable d : integer range 0 to 300;
13   begin
14       if clk'event and clk='1' then
15           if reset='1' then    -- synchronous reset
16               d:=0;            -- initialise variables
17               result <= d;     -- initialise outputs
18           else
19               d:=a+b;
20               d:=d+c;
21               result <= d;
22           end if;
23       end if;
24   end process;
25   end ADD_ABC_ARCH;
```

214 *Motion vision: design of compact motion sensing solutions*

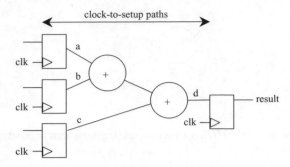

Figure 6.19   Sequential VHDL with registered inputs and outputs. All registers use the same clock

In the code above, a register will be inferred for the `result` signal since we are checking for a clock edge on line 14 (see Section 6.6.4.2). If we assume that a, b and c are registered, at every clock tick this entity will add a, b and c to produce a new `result`. The structure of the resulting circuit will remain the same as before (if we exclude the reset circuit) except that the inputs and outputs are registered as in Figure 6.19. Provided the delay on the clock-to-setup paths (i.e. the paths from the input registers to the output registers) is less than the clock period this circuit will operate correctly (see Section 6.7). If the logic along the clock-to-setup path is relatively simple (as it is in this case) the design tools should be able to ensure proper timing.

6.9.1.2   State machine basics

Before considering what to do if the complicated logic causes clock-to-setup delays to exceed the clock period, let us convert this design into a state machine. This is straightforward: on reset, the circuit should clear the d variable and the `result` signal and on each subsequent clock cycle, the a, b and c inputs should be added. The resulting state machine is illustrated using a state diagram in Figure 6.20.

In Figure 6.20[152] the large circles represent *states* (the triangle is a special state – the reset state) and *transitions* between states are indicated with arrows. If a transition should only occur under particular circumstances, a *condition* (written in VHDL syntax) can be attached to that transition (e.g. `reset='1'`). A state can have multiple conditional transitions and each can be assigned a priority to define the order in which the conditions are evaluated. If no condition is attached then the state machine will automatically transition to the next state on the next clock cycle.

Actions (enclosed in boxes) are a series of sequential VHDL statements that synthesise to cascaded combinatorial logic. Actions can be performed on entry to a state (executed once), within the state (executed once each clock cycle that the state machine remains in that state), on exit from a state (executed once) and can also be

---

[152]   This state machine representation is specific to our design tool (Xilinx Foundation 2.1i), however, all design tools should provide equivalent constructs.

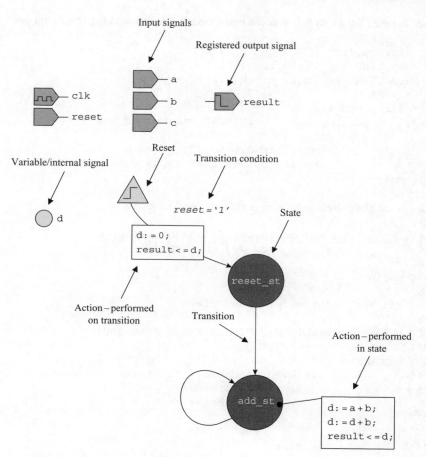

Figure 6.20   State machine for the `ADD_ABC` *entity*

performed on transitions (i.e. when the transition condition is evaluated to be true – executed once). Some tools will treat reset actions specially. The Xilinx tools will place the reset action only in the `reset='1'` branch of the VHDL code if the action is placed on the reset transition (as in Figure 6.20). If the action is placed on the 'reset' state (i.e. the first state entered after a reset), the action will be performed both in the `reset='1'` branch and in the reset state (as in Figure 6.22).

Finally, port signals are indicated with wedge symbols while variables and internal signals are represented with small hollow circles. The properties of each object[153] can be defined through your state machine entry tool, however, these properties may not be shown on the state diagram. The corresponding (automatically generated) VHDL

---

[153] E.g. signals/variables types, whether an output signal is registered, choice of clock signal, synchronous/asynchronous reset.

code is listed below so that you can compare the diagram and the resulting code.

```
1   library IEEE;
2   use IEEE.std_logic_1164.all;
3   use IEEE.std_logic_arith.all;
4
5   entity add_abc is
6           port (a,b,c: in INTEGER range 0 to 100;
7                   clk: in STD_LOGIC;
8                   reset: in STD_LOGIC;
9                   result: out INTEGER range 0 to 300);
10  end;
11
12  architecture add_abc_arch of add_abc is
13
14          type add_abc_type is (add_st, reset_st);
15          signal add_abc: add_abc_type;
16
17  begin
18
19  add_abc_machine: process (clk)
20
21          variable d: INTEGER range 0 to 300;
22
23  begin
24          if clk'event and clk = '1' then
25                  if reset='1' then
26                          add_abc <= reset_st;
27                          d:=0;
28                          result<=d;
29                  else
30                          case add_abc is
31                                  when add_st =>
32                                          d:=a+b;
33                                          d:=d+c;
34                                          result<=d;
35                                          add_abc <= add_st;
36                                  when reset_st =>
37                                          add_abc <= add_st;
38                                  when others =>
39                                          null;
40                          end case;
41                  end if;
42          end if;
43  end process;
44  end add_abc_arch;
```

You should compare this code with our earlier example to convince yourself that both implementations perform the same function – the a, b and c signals will be added at each clock cycle (after `reset`). Notice that states are represented by a *state variable* (called `add_abc` in this case) that is an enumerated type; therefore, the mapping you choose for enumerated types (see Section 6.6.2.1) will affect your design. One-hot encoding will require more memory (flip-flops) to represent the state variable, but will have simpler logic to determine the next state. Conversely, binary encoding requires fewer flip-flops, but has more complex logic. FPGAs can have a large number of flip-flops, hence one-hot encoding is normally preferred.

You may also need to consider what happens if the state variable becomes corrupted. Corruption can leave the state variable with either a valid[154] or invalid[155] value. Both situations are likely to cause the state machine to act improperly at best (e.g. to use corrupted data) and fail outright (e.g. being unable to leave an invalid state and 'hanging') at worst. In many situations (especially if human lives depend on the correct operation of a device) such failures are intolerable, thus it is important that the state machine fails safely – perhaps resetting the system or automatically returning to a stable (safe) state. Your design tools may help you in this regard by offering the option of the 'smallest' state machine[156] (i.e. requiring the smallest amount of logic) or the 'safest' state machine[157] (i.e. returning to the reset state if the state variable is invalid); however, these options may not be suitable in all situations. For example, resetting an aircraft's control systems may be inappropriate. In such situations it is best to explicitly code mechanisms to deal with the corruption of the state variable.

As before, the use of condition checking for clock edges causes all signals and variables assigned within the state machine process to be registered. Your state machine entry tools may require you to explicitly declare that output signals are registered. If you do not do this, your tools might assume that the output is not registered and use asynchronous concurrent assignment statements rather than assigning the outputs within the state machine `process`. While this may well be desirable in some designs, in our design approach we use only sequential, clock-controlled operation.

6.9.1.3 Sequential pipelines in state machines

For the sake of argument, let us now imagine that the cascaded addition operation in state `add_st` is too complex to be completed within the clock period. What options are available to overcome this? While it is possible to use a *slow exception constraint*,

---

[154] This is more likely with binary encoding where most of the possible values for the state variable will be valid states. For example, a state machine with 6 states will have a 3 bit binary encoding. Therefore 6 of the 8 possible values for the state variable will represent valid states.

[155] This is more likely with one-hot encoding where valid values for the state variable have only one non-zero bit. It is relatively unlikely for corruption to put the state variable into a valid state.

[156] Optimisation treats unused values of the state variable as 'don't cares' during logic optimisation thereby giving the smallest possible state machine. There is no guarantee, however, that the state machine will be able to return to a valid state if the state variable becomes corrupted.

[157] All possible values for the state variable are taken into account – if the state machine enters an invalid state it will be returned to the reset state at the next clock cycle. This will require more logic since there will be fewer opportunities for logic minimisation.

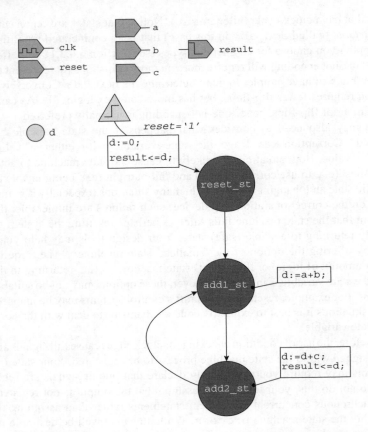

Figure 6.21 An example of sequential pipelining to simplify a state

it can be difficult to specify the desired clock-to-setup path, especially in complex state machines where the signals/variables in question are used in different combinations. The simplest solution is to pipeline the operations as discussed in Section 6.8.9. This is ideal for operations that can easily be decomposed into a number of simpler operations (such as our cascaded additions) although it does increase the time it takes for a result to appear at the output. As illustrated in Figure 6.21, sequential pipelining simply involves splitting the complex state into a sequence of simpler states.

The sequential pipeline is a useful tool for solving delay problems arising from complex actions, however, it will not necessarily help to solve timing problems that arise from the next state logic[158] since adding states can further complicate the next

---

[158] Next state logic is the logic needed to calculate which state the state machine should be in at the next clock cycle. The complexity of this logic depends not only on the complexity of the conditions used to determine the next state, but also on the number of states and the state encoding. This is because states

state logic. Often, in this case, it is necessary to reduce the number of states (by combining states) and reduce the nesting of if statements where possible (e.g. fewer conditional transitions for each state), using asynchronous rather than synchronous reset.

We should point out that sequential pipelines are different to the parallel pipelines we consider in Chapter 7. In that chapter we assume parallel execution of each operation in the pipeline as a means of accelerating processing. In sequential pipelines, operations in the pipeline are executed strictly sequentially.

6.9.1.4 Operation sequencing and loops

The sequential pipeline has uses beyond solving path delay problems. It also gives us a natural way of executing operations[159] in sequence, much as instructions are executed sequentially in a processor. State machines also allow us to create loops that can be used to iterate over data or to create delays[160] that are stable with respect to circuit layout and external factors (see Section 6.8.14).

To begin with, consider the design of a selectable delay circuit with these properties:

- a 3 bit integer input allowing selection of the delay duration;
- the delay should not begin until a StartDelay flag is asserted;
- a DelayExpired flag will be set high for one clock cycle to indicate when the delay has completed.

A state machine fulfilling these requirements is depicted in Figure 6.22 with the corresponding VHDL code. The state machine will perform a delay of 5000 clock cycles if the Delay input is zero, 10 000 clock cycles if it is one and so on.

After reset, this state machine will wait in state S1 until the StartDelay input is set high. In state S2 the state machine assigns a value to the variable Count based on the Delay input, thus determining the length of the delay. The state machine then moves onto state S3 where the Count variable is decremented once each clock cycle until it reaches zero. At this point the DelayExpired flag will be set high and the state machine returns to state S1 where the DelayExpired flag is cleared one clock cycle later.

This state machine illustrates both the sequential execution of operations (operations in each state are executed once each clock cycle) and the design of loops using state machines. There are two implied loops in this state machine. The fact that the state machine forms a loop implies an infinite loop and state S3 and its exit condition

---

are represented using an enumerated type, thus the encoding or mapping (Section 6.6.2.1) can affect the complexity of the logic necessary to calculate the next state.

[159] Of course, these operations can be of arbitrary complexity providing timing constraints can be met.
[160] Such delays will always be a multiple of the clock period.

220 *Motion vision: design of compact motion sensing solutions*

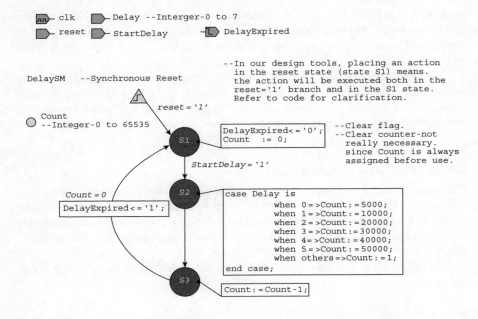

```
1   library IEEE;
2   use IEEE.std_logic_1164.all;
3   use IEEE.std_logic_arith.all;
4
5   entity progdelay is
6   port (clk: in STD_LOGIC;
7         Delay: in INTEGER range 0 to 7;
8         reset: in STD_LOGIC;
9         StartDelay: in STD_LOGIC;
10        DelayExpired: out STD_LOGIC);
11  end;
12
13  architecture progdelay_arch of progdelay is
14
15      type DelaySM_type is (S1, S2, S3);
16      signal DelaySM: DelaySM_type;
17
18  begin
19
20  DelaySM_machine: process (clk)
21      variable Count: INTEGER range 0 to 65535;
22
23  begin
24
25      if clk'event and clk = '1' then
26         if reset='1' then
27            DelaySM <= S1;
28            DelayExpired <= '0';
29            Count := 0;                                       -- Reset Action
```

```vhdl
30          else
31            case DelaySM is
32              when S1 =>
33                DelayExpired <= '0';
34                Count := 0;                 -- Reset action
35                if StartDelay='1' then
36                   DelaySM <= S2;
37                end if;
38              when S2 =>
39                case Delay is
40                   when 0 => Count := 5000;
41                   when 1 => Count := 10000;
42                   when 2 => Count := 20000;
43                   when 3 => Count := 30000;
44                   when 4 => Count := 40000;
45                   when 5 => Count := 50000;
46                   when others => Count := 1;
47                end case;
48                DelaySM <= S3;
49              when S3 =>
50                Count := Count-1;
51                if Count=0 then
52                   DelaySM <= S1;
53                   DelayExpired<='1';
54                end if;
55              when others =>
56                null;
57            end case;
58          end if;
59     end if;
60  end process;
61
62  end progdelay_arch;
```

Figure 6.22 State machine implementing a selectable delay. Note that all comments (starting with two dashes) seen here were placed on the state diagram using our entry tools to clarify operation. The tools allow the placement of arbitrary text on the diagram. This text is ignored in the generation of the VHDL code

effectively form an empty `for loop` — a common means of creating a known delay. In C, the equivalent `for loop` would be expressed as follows (assuming `Delay = 0`):

```
for(Count=5000; c!=0; c--) { }
```

6.9.2 A more complex design

In this section we draw together all the design principles we have discussed to create a relatively complex entity. The entity we consider here is the camera data path buffer discussed in the next chapter (see Section 7.2.6). The purpose of this buffer is to mediate the camera data path's access to primary memory. Pixels will stream into the buffer (a new pixel is available on the input each time the `PixelReady` flag goes high) and will be placed in a FIFO (First-In-First-Out) awaiting transfer to primary memory (RAM). The primary memory subsystem (called the `RAMIC`) polls each buffer in turn, giving each buffer an opportunity to access RAM. If there are any data in the FIFO when the camera buffer is `polled`, the camera buffer will:

- take control of the data and address buses;
- place the data, and an appropriate address (based on the current pixel's address assuming an image with `Width` columns and `Height` rows), on the buses;
- set a `ready` flag to request RAM access;
- wait for the `RAMIC_Busy` flag to go high indicating that the access request has been accepted by the RAMIC;
- wait for the `RAMIC_Busy` flag to return low indicating that the RAM access is complete.

A pair of linked state machines (`InputSide_SM` and `OutputSide_SM`) that perform these operations is illustrated in Figure 6.23. Each state machine is a separate `process` within the same `entity` and these `processes` communicate via the `BufRead` and `BufWrite` signals which are read and write pointers to the FIFO. The FIFO is implemented using one of the small dual ported RAMs available in our FPGA. Port A of the RAM (i.e. the `ENA`, `DIA` and `AddrA=BufWrite` signals) is configured as the FIFO's input (it is used to write new data into the FIFO), while port B of the RAM (i.e. the `ENB`, `DOB` and `AddrB=BufRead` signals) is configured as the FIFO's output (it is used to read data from the FIFO).

Notice that `BufRead` and `BufWrite` are defined to have a `range` of 0 to 511, which is the full range that an unsigned 9 bit number can describe. When these signals reach 511, adding one more will cause the synthesised binary counters to overflow and roll back to 0. This is a simple means of creating circular buffer pointers that will work provided the `range` used starts from zero and goes up to $2^n - 1$.

The purpose of the `InputSide_SM` state machine is to place incoming pixel data into the FIFO[161]. It will write the 8 bit value of the `Pixel` signal to the FIFO when the `PixelReady` flag goes high, if the FIFO is not `Full`. In our design we actually leave one free space in the FIFO (the full condition is `BufRead=BufWrite+1`) to ensure that the `OutputSide_SM` is not tricked into thinking that the buffer is `Empty` (whose empty condition is `BufRead=BufWrite`).

The `OutputSide_SM` state machine is designed to send the data in the FIFO to primary memory. If the FIFO is not `Empty` (i.e. `BufRead /= BufWrite`), the state machine will read the next available element, increment the `BufRead` pointer

---

[161] Reading and writing the FIFO is straightforward and takes only one clock cycle. In a read operation the address is placed on the address bus and the appropriate enable signal is strobed. Data will be available on the following clock cycle. Similarly, a write operation, an address and data are placed on the buses and the enable signal strobed.

Digital design 223

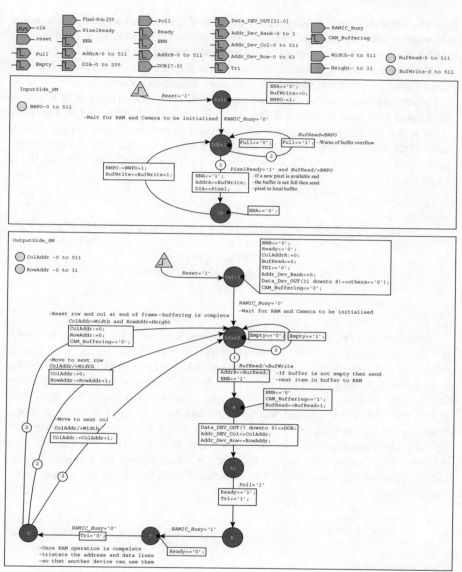

Figure 6.23   Camera buffer implementation. The circled numbers attached to transitions indicate the priority of the corresponding transition condition. Priorities are used to determine the order of evaluation of the conditions hence allowing control in situations where multiple conditions can be true. Conditions with lower numbers are evaluated first

then place the data and address on the buses to the primary memory. These buses are under tristate control and, at this point, are in high impedance mode so the `OutputSide_SM` state machine is not currently driving the buses. When `polled`, the state machine will take control (i.e. drive) the buses, set its `Ready` flag high and proceed through the remainder of the procedure described above.

Notice that there are two types of loop in this state machine, each of which is constructed somewhat differently to the loops considered in Section 6.9.1.4. The states A1, B and C in `OutputSide_SM` and their associated transitions are effectively while loops that wait until a particular condition is true. The overall structure of `OutputSide_SM` is that of two for loops nested within an infinite loop as in the C syntax example below.

```
while (1) {
       for (colAddr=0; colAddr<=Width; colAddr++) {
              for (rowAddr=0; rowAddr<=Height; rowAddr++) {
                     -- perform operations
              }
       }
}
```

While opinions vary, we believe that state machine based design entry is clearer than the equivalent (and often very lengthy) VHDL code. A state diagram shows the operations and order of operations in a clear and concise way making it easier for someone to understand the function of your design. Naturally, large state machines can be very difficult to follow so, wherever possible, state machines should be broken into smaller functional blocks. This can be done in a number of ways, including the use of hierarchical state diagrams where a state can essentially contain an entire state machine, or (as in this book) a combination of state machines and schematic design. Our approach has the advantage that the number of states within any given state machine will be minimal, thus reducing the complexity of the next state logic.

```
1     library IEEE;
2     use IEEE.std_logic_1164.all;
3     use IEEE.std_logic_arith.all;
4
5     entity cambufb is
6     port (clk: in STD_LOGIC;
7           DOB: in STD_LOGIC_VECTOR (7 downto 0);
8           Height: in INTEGER range 0 to 31;
9           Pixel: in INTEGER range 0 to 255;
10          PixelReady: in STD_LOGIC;
11          Poll: in STD_LOGIC;
12          RAMIC_Busy: in STD_LOGIC;
13          reset: in STD_LOGIC;
14          Width: in INTEGER range 0 to 511;
15          Addr_Dev_Bank: out INTEGER range 0 to 1;
16          Addr_Dev_Col: out INTEGER range 0 to 511;
17          Addr_Dev_Row: out INTEGER range 0 to 63;
```

Digital design 225

```vhdl
             AddrA: out INTEGER range 0 to 511;
             AddrB: out INTEGER range 0 to 511;
             CAM_Buffering: out STD_LOGIC;
             Data_DEV_OUT: out STD_LOGIC_VECTOR (31 downto 0);
             DIA: out INTEGER range 0 to 255;
             Empty: out STD_LOGIC;
             ENA: out STD_LOGIC;
             ENB: out STD_LOGIC;
             Full: out STD_LOGIC;
             Ready: out STD_LOGIC;
             Tri: out STD_LOGIC);
    end;

    architecture cambufb_arch of cambufb is

             signal BufRead: INTEGER range 0 to 511;
             signal BufWrite: INTEGER range 0 to 511;

             type InputSide_SM_type is (CR, Idle1, Init);
             signal InputSide_SM: InputSide_SM_type;

             type OutputSide_SM_type is (A, A1, B, C, D, Idle2, Init2);
             signal OutputSide_SM: OutputSide_SM_type;

    begin

    InputSide_SM_machine: process (clk)

             variable BWPO: INTEGER range 0 to 511;

    begin

             if clk'event and clk = '1' then
                  if Reset='1' then
                          InputSide_SM <= Init;
                          ENA<='0';
                          BufWrite<=0;
                          BWPO:=1;
                  else
                          case InputSide_SM is
                              when CR =>
                                      ENA<='0';
                                      InputSide_SM <= Idle1;
                                      BWPO:=BWPO+1;
                                      BufWrite<=BufWrite+1;
                              when Idle1 =>
                                      Full<='0';
                                      if PixelReady='1' and BufRead/=BWPO then
                                              InputSide_SM <= CR;
                                              ENA<='1';
                                              AddrA<=BufWrite;
                                              DIA<=Pixel;
                                      elsif BufRead=BWPO then
                                              InputSide_SM <= Idle1;
                                              Full<='1';
                                      end if;
                              when Init =>
                                      ENA<='0';
                                      BufWrite<=0;
```

```vhdl
77                              BWPO:=1;
78                              if RAMIC_Busy='0' then
79                                      InputSide_SM <= Idle1;
80                              end if;
81                      when others =>
82                              null;
83              end case;
84         end if;
85    end if;
86 end process;
87
88 OutputSide_SM_machine: process (clk)
89
90      variable ColAddr: INTEGER range 0 to 511;
91      variable RowAddr: INTEGER range 0 to 31;
92
93 begin
94
95      if clk'event and clk = '1' then
96              if Reset='1' then
97                      OutputSide_SM <= Init2;
98                      ENB<='0';
99                      Ready<='0';
100                     ColAddr:=0;
101                     RowAddr:=0;
102                     BufRead<=0;
103                     TRI<='0';
104                     Addr_Dev_Bank<=0;
105                     Data_DEV_OUT(31 downto 8)<=(others=>'0');
106                     CAM_Buffering<='0';
107             else
108                     case OutputSide_SM is
109                         when A =>
110                                 ENB<='0';
111                                 CAM_Buffering<='1';
112                                 OutputSide_SM <= A1;
113                                 BufRead<=BufRead+1;
114                                 Data_DEV_OUT(7 downto 0)<=DOB;
115                                 Addr_Dev_Col<=ColAddr;
116                                 Addr_Dev_Row<=RowAddr;
117                         when A1 =>
118                                 if Poll='1' then
119                                         OutputSide_SM <= B;
120                                         Ready<='1';
121                                         Tri<='1';
122                                 end if;
123                         when B =>
124                                 if RAMIC_Busy='1' then
125                                         OutputSide_SM <= C;
126                                 end if;
127                         when C =>
128                                 Ready<='0';
129                                 if RAMIC_Busy='0' then
130                                         OutputSide_SM <= D;
131                                         Tri<='0';
132                                 end if;
133                         when D =>
134                                 if ColAddr=Width and RowAddr=Height then
135                                         OutputSide_SM <= Idle2;
```

```
136                                ColAddr:=0;
137                                RowAddr:=0;
138                                CAM_Buffering<='0';
139                           elsif ColAddr=Width then
140                                OutputSide_SM <= Idle2;
141                                ColAddr:=0;
142                                RowAddr:=RowAddr+1;
143                           elsif ColAddr/=Width then
144                                OutputSide_SM <= Idle2;
145                                ColAddr:=ColAddr+1;
146                           end if;
147                      when Idle2 =>
148                           Empty<='0';
149                           if BufRead/=BufWrite then
150                                OutputSide_SM <= A;
151                                AddrB<=BufRead;
152                                ENB<='1';
153                           else
154                                OutputSide_SM <= Idle2;
155                                Empty<='1';
156                           end if;
157                      when Init2 =>
158                           ENB<='0';
159                           Ready<='0';
160                           ColAddr:=0;
161                           RowAddr:=0;
162                           BufRead<=0;
163                           TRI<='0';
164                           Addr_Dev_Bank<=0;
165                           Data_DEV_OUT(31 downto 8)<=(others=>'0');
166                           CAM_Buffering<='0';
167                           if RAMIC_Busy='0' then
168                                OutputSide_SM <= Idle2;
169                           end if;
170                      when others =>
171                           null;
172                 end case;
173            end if;
174       end if;
175  end process;
176
177  end cambufb_arch;
```

Our next step is to convert this state machine design to a schematic macro and connect the external devices that help our buffer operate. In this case we require a RAM and tristate buffers. The result is shown in Figure 6.24. The macro for our state machine design is in the lower left (named MAC_SM_CAMBUF), above this is the RAM, and in the lower right corner, there are two additional schematic macros[162] that contain the tristate buffers we require.

Components in a schematic design are connected using *nets*. A net can be implemented by drawing a wire (or bus) that connects two or more pins, or the net can connect pins by name without the need to draw a connection between the pins. The

---

[162] These are schematic macros – they were created from another schematic rather than from a VHDL entity or state machine.

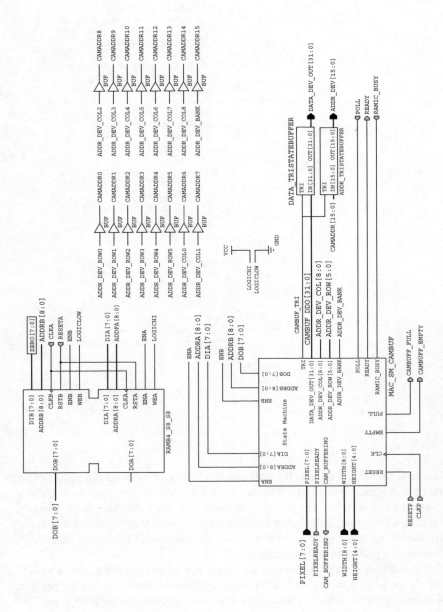

Figure 6.24 *Complete schematic for the camera data path buffer*

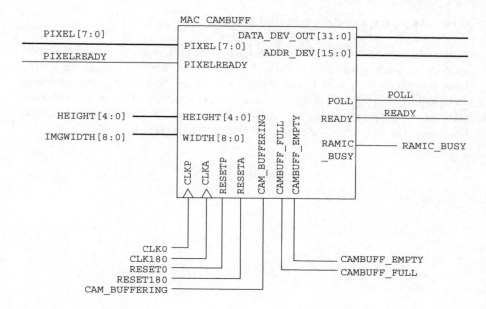

Figure 6.25   Schematic macro created from our camera buffer schematic

tristate control line (CAMBUF_TRI) and the data bus leading to the tristate buffers (CAMBUF_DDO) are examples of wires (buses) connecting pins while the connections from our macro to the RAM are all done by name. Because a net can only have one name, it can be awkward to concatenate a number of nets into a single bus. In our design we wish to concatenate the three partial address buses ADDR_DEV_BANK, ADDR_DEV_COL and ADDR_DEV_ROW into a single address bus called CAMADDR. This is achieved using the buffers in the upper right corner of the schematic. These buffers are optimised away in synthesis, leaving the connection we desire. The same result could be achieved by concatenating the buses in the state machine and your schematic entry tools may provide another mechanism to achieve the same result.

While nets can be used to connect pins, they can also be used to indicate hierarchy. The wires/buses in the schematic terminated with wedge shaped objects are hierarchy connectors. If we convert this schematic into a schematic macro, nets terminated with hierarchy connectors will appear as pins in the macro (see Figure 6.25) and thus the macro can be used in a higher-level schematic design. Each element of the overall schematic design at each level of the hierarchy operates in parallel.

As a final note, we should explain why there are two clock signals in the camera buffer schematic – CLKP (primary clock) and CLKA (auxiliary clock). These clocks have the same frequency but are 180° out of phase. Running our state machine macro using one clock, and the local RAM on the other clock, we are able to access the RAM effectively at twice the clock rate, without the need for a faster clock or stricter timing constraints in the remainder of the circuit.

6.10 Applying our design method

In this chapter we introduced the techniques used to design our prototype intelligent sensor, starting with an overview of FPGA design, working through the VHDL language and showing how FPGA design can be simplified using graphical design entry tools. Given this background the discussion now moves on to consider the overall design of our prototype intelligent sensor by focusing on how data flows through the design; the implementation details are left to the appendices.

Chapter 7
Sensor implementation

> Efforts must be directed towards matching the system architecture both to the image data structure and to the image processing algorithm.
>
> *Cantoni and Levialdi [52]*

Developing an appropriate motion estimation algorithm is only the first part of creating a practical intelligent motion sensor for use in autonomous navigation. Once the algorithm is verified and its operational parameters defined, it must be implemented so that those parameters are met. The algorithm developed in this book is relatively simple, so implementation using modern, general purpose processing hardware such as the humble PC is feasible. Use of a PC would also simplify development and testing thanks to the large and mature set of development tools available. However, a PC based solution has a number of drawbacks when building an intelligent sensor. Using a PC certainly does not lead to a compact solution. Indeed, it leads to a higher parts count since one cannot directly interface to the processor: all I/O devices must be connected to the PC via special purpose add-on cards. A PC also consumes a large amount of power. Another drawback is that a standard PC gives no direct way of utilising the parallelism available in the algorithm. Special purpose image processing hardware may allow the use of parallelism, however, the other drawbacks remain.

Building a CPU or DSP based system (or using one of the many available development platforms) allows more control over system design. Such systems are generally more compact than a fully-fledged PC since they can use smaller form factors (e.g. PC104), however, they usually require support circuitry in order to provide the necessary I/O capability. Furthermore, while the processing capabilities of modern CPUs and DSPs are sufficient for our algorithm, the ability to exploit parallelism is still lacking unless a multiprocessor system is used. It is also possible to combine visual sensors and processing elements within the same devices (e.g. [79, 206]), however, this can be an extremely difficult design process.

These factors led to the selection of semi-custom digital logic and FPGA (Field Programmable Gate Array) technology in particular. In essence, an FPGA is simply an array of I/Os, logic gates and memory that can be interconnected in any way the designer sees fit. This allows the ultimate in versatility at the expense of a more complex design cycle. An FPGA implementation of the motion estimation algorithm described in Chapter 5 can be extremely compact since all interface and glue logic can be built into the same chip as the processing logic, thus creating a 'system-on-a-chip'. Little external hardware beyond memories and sensors is required so the entire processing system can be implemented on a single, small board. Since complete control over implementation is available, it is simple to utilise parallelism.

The current proof-of-concept prototype [166, 168, 173] falls slightly short of this ideal; however, it does show that the ideal is attainable. First, and most critically, a commercial prototype system was used to avoid the cost and complexity of developing a prototype platform of our own. The inevitable result is a loss of compactness; however, the ability to follow the system-on-a-chip paradigm was retained. Second, while only a single processing node has been implemented, it is simple to add additional nodes. A high degree of parallelism is still attained through the use of multiple pipelines. Finally, not all processing is performed in the FPGA. To simplify development, preprocessing of range data[163] is performed using the DSP (digital signal processor) available on the prototype system. The prototype is able to generate frames of motion estimates at a rate of 16 Hz with three frame latency (with latency measured as per Chapter 1).

This chapter begins by discussing the characteristics of the individual elements in the system and then gives a 'top-level' description of the system, indicating how processing is divided between elements and how elements are interconnected. The function of each element is then discussed in detail before a range of experimental results is presented showing the operation of the intelligent sensor. Readers interested in the detailed implementation of our system are referred to the appendices where we present our entire design.

7.1 Components

The intelligent sensor is composed of three key components: the processing platform, the image sensor and the range sensor. In addition to this, a PC captures data and allows visualisation of sensor output (Figure 7.1).

7.1.1 Image sensor

For the purposes of this book, we use a Fuga 15D CMOS camera [146]. This camera was chosen for its simple, RAM-like interface that makes connection to

---

[163] Preprocessing includes scale determination, subsampling, conversion from polar to rectangular coordinates, and interpolation onto the camera coordinate system. As discussed later, a range sensor was not available during development so range data were simulated. For simplicity, range data were preprocessed as part of the simulation.

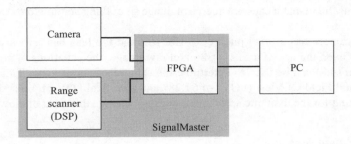

Figure 7.1 Connection of components in our intelligent sensor

other hardware straightforward. This camera is shutterless; the intensity at a pixel is determined at the moment that pixel is addressed. This means that if the image is read too slowly, motion in the environment may lead to a distorted image[164].

In this work we use a 512 × 32 pixel image[165] and a fixed 1 MHz pixel clock. Using (4.16) shows that in the worst case[166], the resulting distortion is approximately 0.13 pixels per image line[167], or a distortion of approximately four pixels from the top to the bottom of the image. At higher scales, distortion falls; for example, at scale $\zeta_{eff} = 16\zeta$ the distortion from the top to the bottom is only 0.25 pixels. At the pixel level, where derivates are calculated, the effect of motion distortion is minuscule and even at the image wide level where distortion is more significant, robust estimation ensures it has little impact. In general, regions of support are far wider than four pixels so, despite motion distortion, the majority of data inside the region of support will still correspond to the target. Since motion distortion will only cause a small amount of the target to be visible outside the region of support, neighbouring targets will be unaffected.

Other features of the Fuga 15D include a logarithmic intensity response making the camera less sensitive to illumination variation, hence reducing the likelihood that the BCA will be violated. The cost of this logarithmic response is relatively low contrast. This camera also has high noise levels in its raw state, however, the noise pattern is fixed and can be corrected using a simple zero or first order calibration procedure. The camera also has a manually adjustable offset control, however, high offsets tend to slow the analogue-to-digital conversion procedure leading to increased motion distortion. To minimise distortion, a large lens aperture was used together with a relatively low gain setting. Larger apertures have a smaller depth of field[168] [22], consequently it is difficult to obtain sharp focus. The defocus that

---

[164] This distortion will appear as shear in the direction of motion.
[165] The camera has a 512 × 512 resolution and it is possible to directly address an arbitrary subset of this.
[166] At the safety margin with $\zeta_{eff} = \zeta$.
[167] This is sufficiently small to prevent error in spatial derivatives.
[168] Range, which appears to be in sharp focus.

234 *Motion vision: design of compact motion sensing solutions*

occurs is useful in that it causes a degree of image smoothing helping to reduce spatial aliasing.

This camera has a pixel pitch $\zeta = 12.5\,\mu\text{m}$ and the lens has a focal length of 4.8 mm. Given the use of a 512×32 pixel raw image, the equations of perspective projection (4.8) can be used to determine the camera's field of view. This gives us a horizontal Field Of View (FOV) of 67.28° and a vertical FOV of 4.62°. Note that subsampling for the dynamic scale space does not change this field of view.

7.1.2 Range sensor

Since the algorithm uses a combination of visual and range data for motion estimation, range data must be supplied. Although a suitable sensor was unavailable during the development of this intelligent sensor, development was able to continue through the use of simulated range data. The Sharc DSP on the prototype board was used to implement what is effectively a ray-tracing program that models the motion of objects through environment and generates range data for that virtual environment[169].

The virtual environment in the simulation is two-dimensional. Objects are described as arbitrary polygons or circles and their motion is defined parametrically, that is, the X and Z positions, and the orientation can be defined as independent functions of time[170]. Egomotion is described in the same way. Although the software allows an arbitrary number of objects, with an arbitrary number of sides to be defined, there is a limited amount of time available for simulation determined by the frame rate and I/O time. I/O and simulation cannot occur concurrently, so only the remaining inter-frame[171] time is available for simulation.

Sending data from the DSP to the FPGA is relatively straightforward since the FPGA is mapped to the DSP memory space. All the DSP needs to do is to write data to a pointer pointing to the appropriate external memory address. For simplicity of implementation, the simulation also performs a range of preprocessing tasks that would otherwise be performed in the FPGA. These tasks are:

- Calculation of all scale parameters. The simulation determines ImgWidth, SS, SD and changeScale (explained below) using the scale cross-over points in Table 4.2. These parameters are calculated from the current scan and are output as an additional burst of data appended to the end of the range data.
- Subsampling. To ensure the true minimum range is used when determining scale parameters, subsampling is performed by taking the minimum range in the given group of SS measurements, not by taking the average.
- Conversion of the range measurements from polar to rectangular coordinates.

---

[169] Interested readers are referred to Appendix D where our code for this program is presented.

[170] Orientation is the rotation of an object about its Y axis. All translations are performed with respect to a fixed world coordinate system, not with respect to the camera centred coordinate system, therefore orientation has no effect on object location.

[171] If a frame of data is passed to the FPGA in 19 ms, and the frame rate is 16 Hz, 43.5 ms are available for simulation.

- Interpolation of the range data onto the camera coordinate system. The simulated range measurements are not made at fixed angular increments, as for a commercial sensor. Instead, the simulation makes range measurements directly on the camera coordinate system so explicit interpolation is not needed. This has the side effect of producing a slightly higher resolution scan than is typically available from commercial sensors[172].

The ImgWidth, SS and SD parameters all convey the same data in different ways. This is done to prevent the need to convert between different forms within the circuit. SS is the degree of subsampling required given the current minimum range. From this, ImgWidth = 512/SS and SS = 2^{SD} are defined. SD is simply the level of the image pyramid corresponding to the given subsampling rate. Functional blocks use ImgWidth to keep track of the current pixel. SS and SD are used in creating an appropriately subsampled image. SS states how many pixels we should average, and SD allows division to be performed as a bit shift operation[173]. The changeScale parameter indicates if the most recent frame of data indicates that a change of scale is required. This is determined using the minimum range and the scale change points listed in Table 4.2

We loosely base the timing properties of the simulated range sensor on the IBEO LD Automotive range scanner [141]. This scanner performs a complete scan every 100 ms, providing range measurements at 0.25° increments with a sample time of approximately 70 μs. Given that only the portion of the scan corresponding to the camera's 67.28° field of view is of interest, only 269 of the available range measurements are used. These data are interpolated onto the camera coordinate system and sent to the FPGA at a rate of SS $\times$ 29 μs per sample[174]. Thus a frame of range data will always take 14.848 ms to be sent to the FPGA. Table 7.1 provides a summary of the simulated range sensor's properties and compares them to a commercially available sensor. Note that the simulated sensor does not experience motion distortion (as does our camera) because the virtual environment can be kept stationary during the range scan. On the other hand, there will be a degree of motion distortion for any practical scanning range sensor since motion continues while a range scan is performed.

Each range measurement is transferred to the FPGA as an 8 bit unsigned value. Since a maximum range of 10 m is sufficient for our application, the 8 bit value encodes distance (in centimetres) divided by four, giving a range resolution of 4 cm. If maximum range increases to 50 m, resolution falls to 20 cm. It is very simple to replace the virtual range scanner with a true range scanner, if that scanner is able to perform the preprocessing tasks listed here. The only change required in the FPGA design is to read data from an external device rather than from the DSP.

---

[172] The IBEO LD Automotive sensor has an angular resolution of 0.25° while pixels in the Fuga camera are spaced at approximately 0.14° increments at the centre of the image down to 0.1° increments at the edges.

[173] If SS is a power of 2, X/SS is equivalent to shifting X to the right by SD bits.

[174] This value was somewhat arbitrary, and is based on early calculations relating to the imaging system. In practice this 29 μs timing can be replaced with anything up to about 37 μs.

236 *Motion vision: design of compact motion sensing solutions*

Table 7.1 Comparison of commercial and simulated range sensors

Parameter	IBEO LD automotive	Simulated sensor
Angular range	270°	67.3° [175]
Angular resolution	0.25°	0.1° min [176]
Maximum range	256 m	10 m
Range resolution	3.9 mm	4 cm
Range accuracy	±5 cm	±2 cm
Scan time	0.1 s fixed	variable [177]
Time per output	69.4 μs (estimated)	29 μs

7.1.3 Processing platform

Rather than building hardware from the ground up, a commercially available prototype development system developed by Lyr Signal Processing [143] was used. The SignalMaster and GatesMaster platforms provide great flexibility with regard to interfacing and allow custom development via FPGA and DSP processing elements.

The key processing element on the SignalMaster platform is the Analog Devices Sharc ADSP-21062 digital signal processor capable of 80 MFlops (peak of 120 MFlops) sustained performance or 40 MIPS. The SignalMaster also provides a wealth of interface options including BITSI mezzanine, PC104, PMC (PCI Mezzanine Card), USB, Ethernet, RS232, and two 16 bit, 48 kHz analogue to digital–digital to analogue. In order to have the flexibility of an FPGA, a GatesMaster Bitsi Mezzanine card was added. This adds a Virtex XCV800 FPGA running at 40 MHz, an additional 16 MB of SDRAM and over 100 digital I/Os providing the required flexibility.

7.1.4 PC

In addition to the processing and sensing elements that make up the prototype, a PC was used to capture and visualise processing results. The PC is not a critical component and can easily be replaced by another data sink, such as the navigation system of an autonomous vehicle. The PC captures data from the sensor using a capture card originally designed for use with the FUGA camera. The sensor mimics the camera interface allowing arbitrary data to be sent to the PC.

---

[175] Based on interpolation to camera coordinate system.
[176] Based on interpolation to camera coordinate system.
[177] User selectable up to a rate determined by simulation complexity and I/O requirements.

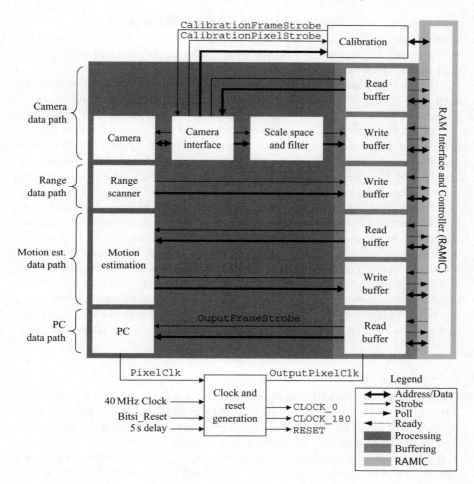

Figure 7.2   Top level FPGA system implementation. See Figure C.1 for more detail

7.2   FPGA system design

As illustrated in Figure 7.2, the FPGA component of the implementation is based around three key sections[178]. The *processing* section contains a number of independent processing modules each designed for a specific task and with different data bandwidths. Each process has a specially designed *buffer* that moderates access to RAM. The buffer allows the processing to proceed at a rate independent of RAM access times. The third section is the RAM Interface and Controller (RAMIC)

---

[178] A more detailed 'top level' schematic view of our design can be found in Appendix C, Figure C.1.

whose primary role is to provide access to off-chip SDRAM. In doing so, it also provides services such as *bus arbitration, address unpacking* and *memory management*. In addition to the three primary sections, there are a number of support components each with a specific task such as clock and reset generation[179].

The use of *bi-phase clocks* in this design allows functional blocks to communicate at 80 MHz while processing proceeds at 40 MHz. Thus, inter-block communication is accelerated without the added design complexity of meeting the demanding flip-flop-to-flip-flop timing requirements of an 80 MHz global clock. Since successive blocks in the various data paths often operate at different data bandwidths (independent of the FPGA clock rate), an array of communication methods are used to maintain synchronisation, including bandwidth matching using FIFOs (First-In, First-Out buffers), two-cycle handshaking and simple strobing.

To simplify design (especially when communicating with devices external to the FPGA), each functional block independently maintains address variables indicating which data element is currently being processed. This allows blocks to determine when a line or a frame of data has been processed and allows data to be written to correct locations in primary memory. Since the address of the current data element is not explicitly passed between functional blocks, the blocks are *implicitly synchronised*. In this scheme, it is vital that data elements are never missed as would occur in a buffer overrun or underrun. If data is missed, the local address will be incorrect hence implicit synchronisation between blocks will be lost and the system will 'hang'.

7.2.1 Boot process

When the sensor is started, a number of initialisation operations must occur before it is ready for operation (Figure 7.3). The FPGA is Reset at power up and remains in the Reset state for five seconds, giving the user time to start the PC for data capture and to prepare for camera calibration. Variables are initialised at this point: most importantly, the dynamic scale space is initialised to assume the first image will be at level 0 of the image pyramid (i.e. it will be 512 pixels wide). When the Reset line is cleared, the SDRAM initialisation process begins (A in Figure 7.3). At the completion of this initialisation process, the low level RAM interface becomes available (indicated by the RCTRL_Busy signal going low) and the camera calibration process can begin (B in Figure 7.3).

When the camera initialisation process is complete (CAM_Init goes low) it relinquishes exclusive use of the low level RAM interface (RAM_Ready goes high) allowing the remainder of the system to begin normal operation. The Calibration_Buffer immediately begins caching calibration data, quickly filling its buffer (C in Figure 7.3). From this point, the timing of the system is entirely dependent on the arrival of range data from the DSP. The frame clock is based on the arrival of the first element of range data for each frame.

---

[179] See Appendix C, Section C.7 for implementation details of these components.

Sensor implementation   239

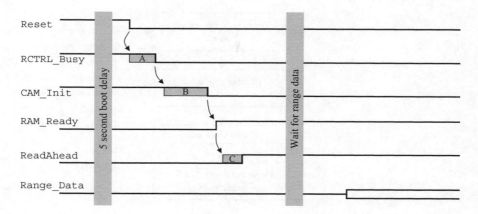

Figure 7.3   Start up sequence for our sensor. This diagram is not to scale. A = Initialisation of SDRAM. B = Camera calibration. C = Calibration data read ahead

7.2.2   Order of operations

Once the FPGA has booted, it enters the operation cycle illustrated in Figure 7.4 (more detailed system timing is provided in Appendix A). The cycle begins when the range scanner (DSP) begins sending range data corresponding to the field of view of the camera. This range data is stored in RAM using the appropriate buffer. Immediately after buffering of range data begins, a 'clobber avoidance delay' occurs. Clobbering refers to the alteration of data before it is read and clobbering could occur in our FPGA because the PC buffer reads data from the same memory locations that the range data buffer and camera buffer store their data. The clobber avoidance delay ensures that the range buffer does not alter memory before the PC buffer has been able to buffer all the previous range data. Once the clobber avoidance delay is complete, the FPGA's frame clock ticks indicating to both the PC (via the OutputFrameStrobe signal in Figure 7.2) and the FPGA that a new frame of data is beginning.

At this point the PC buffer begins reading data and sending it onto the PC. There is a slight delay (about 1.7 ms[180]) between the frame clock ticking and the PC starting to read data. Reading of image data from the camera begins only when the PC starts to read data thereby ensuring the camera buffer does not clobber image data. Upon the completion of image data input and output of data to the PC, scale data for the next frame are made available to the FPGA. These data are actually passed to the FPGA by the range scanner/DSP at the end of the range scan, however, these data are not made available to the FPGA as a whole until all I/O is complete to ensure that the

---

[180] This was determined by watching the OutputFrameStrobe and OutputPixelClock signals on an oscilloscope.

240 *Motion vision: design of compact motion sensing solutions*

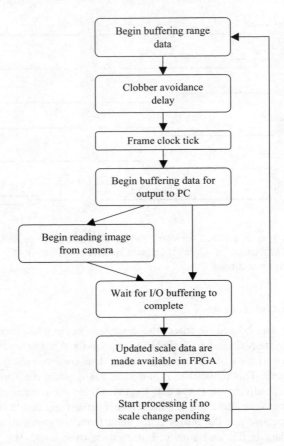

Figure 7.4   *Order of operation for the FPGA. See Appendix A for more detail*

scale parameters (which are used to maintain implicit synchronisation) do not change while they are being used.

At this stage processing will begin if no scale change is pending (i.e. if changeScale = 0 – details of the scale change process are given in Section 5.2.9 and Appendix A). Upon the completion of processing, the system returns to the top of the loop and waits for the next frame of range data to begin. There is an implicit assumption here that processing will be completed before the next frame of range data begins – there is no mechanism to halt processing at the start of the next loop.

7.2.3 Memory management

Memory interface design can have a significant impact on processing performance, so it is important to consider this part of the design carefully. The implementation uses two forms of RAM. Primary storage is provided by 16 MB of off-chip SDRAM clocked at 40 MHz. The Virtex FPGA also provides small blocks of internal

Table 7.2 Summary of primary RAM requirements

Description	Quantity	Bit width
Images	3 frames, up to 512 × 32 pixels/frame	8 bit unsigned
Range	1 value per image column	8 bit unsigned
Velocity	1 value per image column	8 bit signed
Misc.	1 value per image column	32 bit unsigned
Calibration	1 frame 512 × 32 pixels	9 bit signed

RAM known as BlockRAM[181]. These are used for local buffering tasks so that contention on the primary storage is minimised. Design of the memory interface requires consideration of how many data must be stored and how these data will be used. Based on this information, memory access and addressing can be structured so that performance is optimised. The design process begins with a consideration of how much memory is need (Table 7.2).

Clearly the available 16 MB of primary RAM is more than sufficient, thus the discussion can move on to consider data use. Image data is read from the camera and stored one frame at a time; however, three frames of data are required for the processing operation to compute temporal derivatives. To optimise the computation of derivatives we exploit the fact that RAM is organised as 32 bit words while pixel intensity is represented by 8 bit words. The available 32 bits are split into 4 bit planes and when a pixel value is written to RAM, existing pixel data are shifted 8 bits to the left so that the new value can be written to bit plane 0 (see Figure 7.5) effectively implementing a shallow FIFO. This pixel stacking slows down the pixel write operation since a read-shift/append-write cycle is required; however, as will be shown shortly, memory bandwidth is sufficient to support this more complex operation. Furthermore, the slower compound operation required when writing a pixel is compensated for by the gain when calculating derivatives: rather than performing three separate read operations to retrieve three frames of data, the temporal derivatives can be determined using a single read operation.

In addition to image data, three other datasets (referred to collectively as extended data) must be stored in RAM. These are the range measurements and the results of computation (velocity and miscellaneous data). In practice, eight bit values are sufficient for both the range and velocity data in our application. The miscellaneous data section is used primarily to identify the location of range discontinuities and hence a single bit would be sufficient[182]. However, our implementation allows 32 bits per

---

[181] 28 of these dual ported RAM are available in our XCV800 device. Each RAM stores 4096 bits and the user can independently select the address bus width for each port in the range of 8–12 bits.

[182] The first word of miscellaneous data has a special use: it indicates when processing beings, and it is used to convey scale information.

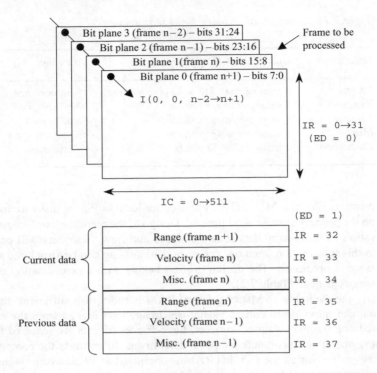

Figure 7.5 *Schematic view of bank 0 RAM allocation. Frame n is the frame currently being processed. Because processing is performed for frame n (which is the previous frame, not the most recently read frame) we must make available the corresponding range data. Our design allows for 32 bit range data (despite the fact that range data are in fact 8 bit) since that is the word size in SDRAM*

column for each of these values simplifying the task of upgrading accuracy if that should be required in the future.

Since processing is performed for frame n of the image data, frame n of the range data must be made available as well. This necessitates a shallow FIFO structure similar to that used for image data, however, a simple bit shift cannot be used in this case because the data are assumed to be 32 bit. Instead, existing data are shifted to a separate location (the 'previous data' section of RAM) resulting in a slightly more complex operation – it is necessary to read the existing value, write it to the 'previous data' section, then write the new value to the 'current data' section. Fortunately, range data have a maximum bandwidth of only 35 kHz so the overhead for writing a value is insignificant. To reduce the number of special cases and simplify design, velocity and miscellaneous data are treated in the same way as range data. Previous velocity and miscellaneous data are not used.

The final use for RAM is storage of camera calibration data. These data are generated via a special process at boot, before the remainder of the circuit begins

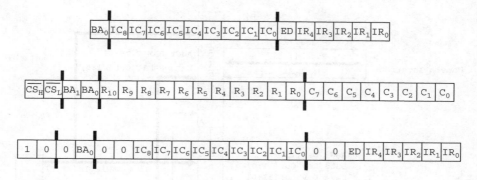

Figure 7.6 Address structure and packing. The top diagram shows the address structure for our SDRAM. Two RAM chips are available and can be selected using the $CS_{H/L}$ lines. The $BA_{1/0}$ lines allow selection of any of the 4 banks of memory in each chip. Each bank is arranged with 2048 rows and 256 columns, which are referenced using the R_x and C_x sections of the address. Our application only requires access to one chip, and two banks of RAM with up to 512 columns and 38 rows access in each bank. The middle portion of the figure shows how this requirement can be mapped onto the address space. Rather than use this entire 23 bit address we use a simpler 16 bit version (bottom of figure). The SDRAM interface block unpacks this 16 bit internal address to a full 23 bit SDRAM address

operation. Since this process has exclusive access to RAM, I/O timing is not critical beyond the need to ensure the unbuffered 1 MHz pixel stream is stored correctly. While all other data are stored in RAM bank 0, the 9 bit signed calibration data are placed in RAM bank 1 in an address space 'parallel' to the raw image.

There remains one further consideration in the low level RAM interface – that is, the address space. As illustrated in Figure 7.6, off-chip SDRAM has an effective 23 bit address space with RAM physically arranged in 256 columns, 2048 rows, 4 banks and 2 separate chips. Each address location stores a 32 bit value. In order to create the simplest possible address space, pixel row and column addresses are mapped to RAM column and row addresses respectively[183] allowing direct use of pixel addresses when accessing RAM. Since only two banks, 512 rows and 38 columns of RAM are used, it is possible to pack the 23 bit address into a 16 bit address (see Figure 7.6). Columns $0 \to 31$ of RAM store image data and columns $32 \to 37$ store extended data, so bit 6 of the RAM column address ($ED = IR5$) can be used to determine when the extended data are being accessed. Note that the mapping from image columns to RAM rows is direct: for example, if $\zeta_{eff} = 4\zeta$ then only rows $0 \to 127$ of RAM are used.

---

[183] Note that image columns are mapped to RAM rows and vice versa. We do this because RAM has only 256 columns, while the image has up to 512 columns.

244 *Motion vision: design of compact motion sensing solutions*

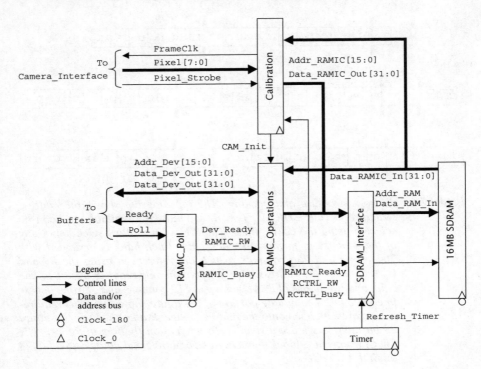

Figure 7.7 Block diagram of the RAMIC subsystem. Timing considerations mean that RAM output is connected directly to the Calibration/ RAMIC_Operations components rather than to the SDRAM_Interface. See Appendix C, Figure C.2 for more detail

7.2.4 RAMIC

The RAMIC[184] provides the system with a simple, transparent interface to primary memory (see Figure 7.7) and is composed of five components: external SDRAM, the SDRAM_Interface RAMIC_Operations, Calibration and RAMIC_Poll.

7.2.4.1 SDRAM_Interface

Low level access to RAM is managed by the SDRAM_Interface[185]. As such, this block has four functions: it initialises RAM at boot, it ensures the RAM is refreshed

---

[184] See Appendix C, Section C.3 for implementation details.
[185] See Figures C.6, C.7 and C.8 in Appendix C for implementation details. Also, see Appendix B for details of timing for SDRAM access.

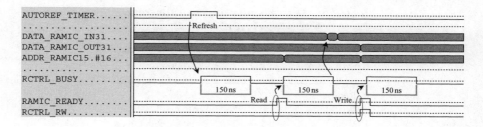

Figure 7.8   Timing for the `SDRAM_Interface`

at 12.8 μs intervals[186], it provides simple read and write access[187] and it unpacks the internal 16 bit address to a full 23 bit address as described in Section 7.2.1. Figure 7.8 illustrates the timing of a RAM access using the `SDRAM_Interface`. This protocol is commonly referred to as two-cycle handshaking [159] and has three phases:

- Wait until the `SDRAM_Interface` is available (i.e. `RCTRL_Busy` low).
- Place address on the `Addr_RAMIC` bus, data on the `Data_RAMIC_Out` bus (if required) and indicate whether a read or write operation is required via the `RCTRL_RW` line (0 = read, 1 = write). The operation begins when the `RAMIC_Ready` line is set high.
- The use of bi-phase clocks allows `RCTRL_Busy` to go high (to indicate an operation is in progress) only half a clock cycle after `RAMIC_Ready` goes high. Upon completion the `RCTRL_Busy` line returns low at which point any requested data will be available on the `Data_RAMIC_In` bus. These data are only available for a short period (minimum of 20 ns) before the SDRAM tristates its outputs.

7.2.4.2 RAM_Operations

The high level RAM operations (pixel stacking/extended data I/O) described in Section 7.2.1 are performed by the `RAMIC_Operations` block[188]. `RAMIC_Operations` determines the type of operation required by monitoring the `RAMIC_RW` bit and the `ED` address bit (see Table 7.3).

Communication between `RAMIC_Poll/Buffers` and the `RAMIC_Operations` blocks uses a handshake protocol similar to that between the `RAMIC_Operations` and `SDRAM_Interface` blocks. Here the `RAMIC_Poll` block polls each buffer in

---

[186] Our SDRAM requires a refresh cycle every 15.625 μs, however, in our design a refresh cannot occur while a RAM operation is being performed. Therefore, we refresh slightly more often than necessary (12.8 μs intervals) in the anticipation that the refresh operation may be delayed. A 12.8 μs interval also requires less logic to measure than 15.625 μs since it is a power of two.

[187] Read and write access is implemented with a RAS to CAS latency of two, and a burst width of one. No byte masking is used.

[188] See Appendix C, Figures C.4 and C.5 for implementation details.

Table 7.3 Determination of memory operation.
X is 'don't care'

RAMIC_RW	ED	Operation
0 (Read)	X	Read data
1 (Write)	0	Write pixel
1 (Write)	1	Write extended data

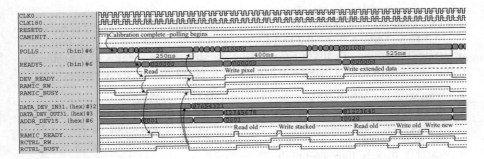

Figure 7.9 Timing for the RAMIC_Operations block. Read, Write pixel and Write extended data operations are shown. The operation sequence has been explicitly indicated for the read cycle. Note the advantage of using bi-phase clocks. The Poll-Ready-Dev_Ready cycle takes only one clock cycle (25 ns) using this scheme since the effective clock rate between the RAMIC_Poll and RAMIC_Operations blocks is 80 MHz. The Poll-Ready-Dev_Ready cycle would require two clock cycles if we used a single clock

turn. If a buffer requires a RAM operation it will, when polled, set its Ready line high and drive the data bus (Data_Dev_Out) and address bus (Addr_Dev) appropriately. In response to the Ready line being high, the RAMIC_Poll block sets the Dev_Ready line high and drives the RAMIC_RW appropriately indicating to the RAMIC_Operations block that an operation is required. The Poll-Ready-Dev_Ready cycle only takes one clock cycle (25 ns) due to the use of bi-phase clocks. At this point RAMIC_Busy goes high[189] and remains so until the operation is complete. If a memory read was requested, data will be available on the Data_Dev_In bus when RAMIC_Busy returns low. This process is summarised in Figure 7.9. Note that the Addr_Dev and Data_Dev_Out buses are tristate between operations.

---

[189] Because the RAMIC_Poll block is the only block using the RAMIC_Operations block, there is no need to check the RAMIC_Busy line before requesting an operation.

Sensor implementation 247

There is no contention for the `SDRAM_Interface`. At boot, the `Calibration` block is given exclusive access to the `SDRAM_Interface` so that it can generate a calibration image. Once this calibration image is complete, the `CAM_Init` line is set low and `Calibration` block relinquishes control of the `SDRAM_Interface` allowing the `RAMIC_Operations` block exclusive access.

7.2.4.3 Calibration

The `Calibration` block[190] generates a zero order calibration image in three stages.

1. Clear allocated section of RAM in bank one.
2. Generate an average image using 16 frames. During this time a translucent sheet of paper must cover the camera so that the camera sees nothing but its noise pattern.
3. The average noise pattern is then shifted to have a zero mean thus calibration values can be directly added to pixel values to remove noise without shifting the mean intensity.

Zero order calibration works well provided illumination does not change significantly. If illumination does change, one may notice an increase in the graininess of the captured images. A first order calibration can correct for changing illumination, however, it was found that zero order calibration works well under most conditions.

7.2.4.4 RAMIC_Poll

The final task performed by the RAMIC is bus arbitration. Bus arbitration is required whenever multiple devices share a common bus: it ensures that only one device is accessing the bus at any given moment. For example, in our design, any of the processing data paths (as shown in Figure 7.2) may require simultaneous access to primary RAM data and address buses.

In our design, the `RAMIC_Poll` component[191] provides the bus arbitration service in a way that ensures each data path has fair access to the memory. By fair access, we mean that no one device should have preferential access to the buses such that other devices are 'starved' of access. If devices were starved of access, buffer overruns would be inevitable leading to the loss of implicit synchronisation and failure of our system. This is achieved using a type of centralised parallel arbitration [1][192]: each device/buffer is polled in a round-robin fashion and is granted access to memory if it requires access to memory.

---

[190] See Appendix C, Figures C.9, C.10, C.11 and C.12 for implementation details.
[191] See Appendix C, Figure C.3 for implementation details.
[192] The two primary forms of bus arbitration are *centralised* where an external arbiter decides which device will be granted access to the bus (i.e. which will be the *bus master*) and *distributed* arbitration where the devices choose who will be the bus master between themselves. Distributed arbitration requires more intelligent devices increasing their complexity while centralised arbitration requires dedicated arbitration logic. Centralised parallel arbitration where each device has a 'request access' and 'access granted' line (such as in our design) is the most common arbitration solution [1].

248   Motion vision: design of compact motion sensing solutions

Table 7.4   *RAM operation timing. These times are measured from the moment a device is polled to the moment polling resumes (see Figure 7.9). RAM refresh cycles occur every 12.8 μs. If a refresh cycle is under way when a device is polled, the times below can be extended by up to 150 ns*

Operation	Timing (ns)
Read arbitrary data	250
Write pixel data	400
Write extended data	525

If the polled buffer requires access to primary RAM, it indicates this by setting its READY line high and driving the data and address buses with appropriate values (see Figure 7.9)[193]. To improve RAM access times, polling is split into two exclusive modes based on the fact that I/O and processing functions occur independently. In the I/O mode, only the four buffers for I/O operations are polled (Camera, Calibration, Range, PC), while in the processing mode, only the two buffers for Motion Estimation are polled.

7.2.4.5   RAM timing

This system is designed with the assumption that buffers will not overflow or underflow[194], and so it is important to consider timing of RAM access to ensure that this is in fact the case. The time taken to perform each of the various RAM operations is listed in Table 7.4. As shown in Figure 7.9, these times are defined as the time from when a device is polled to the moment polling resumes after a memory access is complete. It does not take into account the time a buffer spends waiting to be polled, which is variable since other buffers may access memory. Notice how the use of bi-phase clocks helps to reduce this time. For example, the sequence of operations from a device being polled to RCTRL_Busy going high takes only three clock cycles using bi-phase clocks but would take six clocks cycles without bi-phase clocks. In determining whether under/overflows will occur, the maximum bandwidth of each data path must be considered (Table 7.5).

---

[193] One could easily reverse the order of the Poll and Ready signals so that whenever a device is ready it sets its ready bit. The RAMIC_Poll block then scans the Ready signals and if one is found to be set, the corresponding Poll bit is set indicating to the buffer that its operation has been accepted. In terms of timing and required logic, the two methods are identical.

[194] Underflow and overflow have special meanings defined in Section 7.2.5. Put simply, these terms are defined with respect to the ability of a process to perform a memory operation.

Table 7.5 *Worst case bandwidth of each processing data path*

Buffer	Operation type	Bandwidth	Comment
Calibration	Read	1 MHz	Fixed. Calibration always applied to raw 512×32 pixel image, hence must operate at pixel clock rate.
Camera	Write pixel	1 MHz	Maximum. Based on effective pixel clock which falls with increased subsampling.
Laser	Write extended	~35 kHz	Maximum. Based on timing of range measurements. Falls with increased subsampling.
Processing input	Read	~1 MHz	Limited by the slowest element in the processing pipeline (Robust_Average, ~42 clock per pixel). This bandwidth will increase if parallelism is utilised.
Processing output	Write extended	N/A	Storage of processing results occurs independently of all other operations and thus can occur at the maximum available memory bandwidth.
PC	Read	1 MHz	Fixed. Determined by PC.

In processing mode, under/overflows cannot occur because processing bandwidth (which is controlled by processing rate) is less than the memory bandwidth. In I/O mode the situation is not so clear-cut. Consider peak memory activity (all four I/O buffers accessing RAM) for a period of 1 ms. Table 7.6 shows that RAM will *theoretically* be in use for a maximum of 930 μs or 93 per cent of the available time. Because the round-robin polling scheme polls each device in turn, there is a one clock cycle overhead if a device that is polled does not require an operation. However, this is not sufficient to push *effective* RAM utilisation over 100 per cent so it is guaranteed that the design is over/underrun free.

While long term RAM utilisation remains below 100 per cent, it will peak above 100 per cent for short periods if four or more operations occur within the same polling cycle. Because this occurs rarely, and for short periods, the buffers are able to absorb this extra load. For example, imagine a situation where the calibration buffer[195] is full, when suddenly the RAM utilisation jumps to 160 per cent (the maximum short term theoretical RAM utilisation occurring when all operations and a refresh cycle

---

[195] This is the smallest buffer with only 256 elements and hence is the most vulnerable to underflow. Other buffers have 512 elements.

250 *Motion vision: design of compact motion sensing solutions*

Table 7.6 Peak RAM usage over a 1 ms period. Number of operations is determined using bandwidth from Table 7.5

Operation	Number of operations	RAM Time (ns)	Total (μs)
Calibration read	1000	250	250
Pixel write	1000	400	400
Range write	35	525	18.375
Output read	1000	250	250
Refresh	79	150	11.85
Total operations	3114	Total time	930.225 μs
Average time per operation			298.72 ns

are required within the same poll cycle). The buffer can operate at this high utilisation for 410 μs before an underflow occurs[196] thus the buffer capacity is sufficient for this application.

7.2.5 Buffers

The collection of buffers in Figure 7.2[197] is designed to moderate access to primary memory. These buffers are implemented as FIFOs allowing processes to run at a rate independent of the memory bandwidth. As discussed in Section 7.2.4, the design is assumed to be underflow/overflow free. This is primarily due to the need to maintain a constant pixel clock. Allowing a buffer to halt the camera pipeline would cause the pixel clock rate to vary randomly, resulting in inconsistent and unpredictable motion distortion (see Section 7.1.1). Furthermore, there is no direct way of stopping the PC's pixel clock or the range data clock since they run independently of the system. These issues require construction of a system that can be guaranteed to be under/overrun free, therefore there is no means of halting pipelines[198].

Since the purpose of the buffers is to control contention on primary memory, it is critical that the buffers do not add to that contention. Hence, rather than use primary memory for buffering, the BlockRAM memories available within our FPGA are used. These RAMs are dual ported enabling independent design of the input and output controllers. While the details of each buffer differ significantly, they all have the general structure shown in Figure 7.10.

---

[196] In 1 μs one value will leave the buffer and 0.625 (on average) will enter. Thus the buffer will be empty in 1 μs $\times$ ((1/0.625) $\times$ 256) $\approx$ 410 μs.

[197] See Appendix C, Section C.4 for implementation of the buffers.

[198] The exception to this is the Motion Estimation pipeline, parts of which can be halted depending on the processing load. This is discussed later.

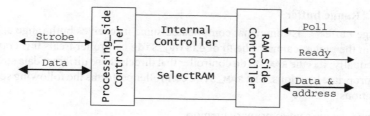

Figure 7.10   Structure of a buffer

Each buffer has two controllers known as `Process_Side` and `RAM_Side`. `Process_Side` provides a process with a simple RAM interface, while `RAM_Side` actually performs requested RAM operations by communicating with the RAMIC. The `Internal_Controller` is a logical device referring to the buffer pointers kept by `Process_Side` and `RAM_Side` to detect buffer empty/full conditions. These buffer pointers are configured so that BlockRAM operates as a circular buffer (or FIFO).

Underflows and overflow are defined with respect to the `Processing_Side` controller being able to perform a valid memory access. Thus, for a read buffer, underflow occurs if the buffer is empty. An empty buffer will continue to output whatever is its final input. For a write buffer, overflow occurs if the process writes data to the buffer more quickly than the buffer can write data to RAM. Once the buffer fills, any new data will be ignored, hence the write operation fails. If either of these conditions occurs, the system as a whole will fail due to the implicit synchronisation of blocks. Buffer empty/full conditions are defined with respect to primary memory access. A full read buffer simply means that the buffer has read ahead as far as possible. An empty write buffer means that the buffer is waiting for a write operation to be requested by the corresponding pipeline.

7.2.5.1 Calibration buffer

The `Calibration_Buffer`[199] provides calibration data to the `Camera_Interface` so that camera noise can be eliminated. This buffer is designed as a read-ahead buffer. It will begin retrieving data immediately after the boot process is complete and will constantly try to maintain a full buffer thus ensuring calibration data are always available. New data are made available on the `Process_Side` data bus when strobed by the `Camera_Interface`.

7.2.5.2 Camera buffer

This simple FIFO buffers the data present at the end of the camera pipeline whenever strobed[200].

---

[199] See Appendix C, Figures C.13 and C.14 for implementation details.
[200] See Appendix C, Figures C.15 and C.16 for implementation details.

7.2.5.3 Range buffer

The range buffer[201] is somewhat complicated since it receives both range and scale data from the range scanner. The Processing_Side controller treats this entire data stream equally; it is the RAM_Side controller that directs these different data streams to their appropriate locations. The RAM_Side controller performs the following sequence of operations:

1. Range data is sent to primary memory.
2. The first element of scale data (ImgWidth) has a 'data valid' flag attached and this combined 16 bit word is written to the first element of miscellaneous data. The PC searches for a flag as it reads incoming data. When this flag is located, the PC knows that the sensor has begun sending valid data so it is safe to begin applying the ImgWidth value. Without this check for valid data, the PC could potentially accept some arbitrary value as the ImgWidth and this would lead to loss of implicit synchronisation.
3. The remaining four pieces of scale data (change_Scale, SD, SS and ImgWidth) again are made available to the remainder of the system so that I/O and processing can occur at the correct scale.

The range buffer reads data from the range scanner directly. As discussed earlier (Section 7.1.2), the DSP simply writes data to an external memory location. The FPGA is able to read these data by monitoring two memory strobe lines. When both strobe lines are low, data are ready on the DSP data bus. Although the DSP runs at the same speed as the FPGA (40 MHz), it is impossible to predict when the strobes from the DSP will change the state with respect to the local clock, thus synchronisers are used for both strobes to avoid synchronisation and metastability problems (Section 6.8.1).

7.2.5.4 Processing input buffer

This buffer[202] retrieves image and range data for the Motion_Estimation block. Rather than simply passing on raw intensity values, the buffer calculates image derivatives using the first-order central-difference estimator described in Section 5.2.3. Data are read from RAM in a column-wise manner matching how the Motion_Estimation block consumes data. Before reading image data for a column, the range values for that column are read from RAM and placed in local buffer allowing rapid access when range data are required. This buffer is designed as a read-ahead buffer that reads ahead until the end of the current frame of data is reached.

7.2.5.5 Processing output buffer

The processing_output_buffer[203] is special since it consists of only the RAM_Side controller. The memory and the Process_Side controller are

---

[201] See Appendix C, Figures C.17, C.18 and C.19 for implementation details. Implementation range data I/O is described in Figures C.35, C.36 and C.37.
[202] See Appendix C, Figures C.22, C.23, C.24 and C.25 for implementation details.
[203] See Appendix C, Figures C.26 and C.27 for implementation details.

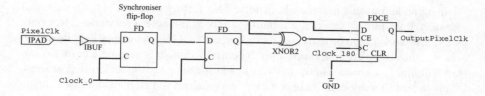

Figure 7.11   Synchroniser used for the PC `PixelClock`

contained in the `Motion_Estimation` block. Once processing is complete, the `Motion_Estimation` block flags the `processing_output_buffer` which then begins to send the completed processing results to primary storage.

7.2.5.6   PC-output buffer

This buffer[204] retrieves image data, range data, motion estimates and miscellaneous data and sends these to the capture card in the PC. Since the capture card is actually designed to capture data directly from the Fuga Camera, it is necessary to mimic the camera's interface, though the address generated by the PC is ignored because the sensor and PC are implicitly synchronised. When the system determines that the PC should start reading, it sends the PC an `OutputFrameStobe`. The PC detects this strobe on its parallel port and begins the reading process.

There is a delay of approximately 1.6 ms before any pixels are read due to an initialisation process that must occur before data are read. Since this buffer is configured in a read-ahead fashion, it is able to completely fill the FIFO before the PC attempts to read a pixel. The PC will read the value on the data bus shortly after each high $\rightarrow$ low transition of the 1 MHz `PixelClock`. To ensure setup and hold times for the capture card are met, new data are placed on the bus at the low $\rightarrow$ high transition. The PC reads a 512×44 psuedo-image where the first 32 lines correspond to visual data, and the range, motion and miscellaneous data follow in groups of 4 lines.

To avoid metastability problems in the `PixelClock` line, a synchroniser flip-flop is used at the input. The XOR gate and the third flip-flop (Figure 7.11) are configured to so that the output signal `OutputPixelClk` does not change unless the input signal `PixelClk` has been stable for two clock cycles. This was done to prevent short glitches observed in the `PixelClk` signal causing a loss of implicit synchronisation.

7.2.6   Data paths

Before discussing the design of the processing data paths (or pipelines), it is necessary to briefly introduce some concepts of parallel processing so that the rationale behind the design is clearer.

---

[204] See Appendix C, Figures C.20 and C.21 for implementation details. Implementation details of I/O with the PC can be found in Figures C.38 and C.39.

Imagine an unusual automobile factory. This factory does not contain the usual assembly line; rather, vehicles are built entirely at a single workstation. Not only would that workstation be extremely complex, but building vehicles would be a slow process since the plant must wait until the current vehicle is complete before it can start building a new one. There are two ways to speed up the manufacturing process. One is to build a number of these complex workstations so that many vehicles can be built at the same time. The rate at which cars are produced would be directly related to the number of workstations. The other possibility is to use the more traditional assembly line method where the task of building a vehicle is split up into a number of simpler tasks performed at separate, simpler workstations. A vehicle will pass from workstation to workstation until a completed automobile appears at the other end of the line. The time it takes for the first complete car to appear to the end of the assembly line is the time it takes to pass through all the workstations, however, once the first vehicle appears at the end of the line, subsequent vehicles will appear at a rate determined by the slowest workstation.

In direct analogy to the automobile factory, there are two ways to speed up computation through the use of parallelism. The most common conception of computational parallelism is the use of an interconnected array of *processing elements* (PEs). This form of parallelism is general decomposed further [182] into SIMD (*Single Instruction stream, Multiple Data stream*) and MIMD (*Multiple Instruction stream, Multiple Data stream*) parallelism[205]. SIMD is used for data-parallel problems; that is, problems where the same set of operations is performed on each element of a large dataset. Parallelism is achieved when each processor in the array performs the operation on different sections of the dataset. SIMD machines have a single control unit to coordinate processing and issue instructions to each PE so that each PE will always execute the same instruction during a given clock cycle. An ideal candidate for SIMD parallelism is global weather forecasting. Here the globe is split up into a number of small regions and the same weather model is applied to each region. Mapping each region to a separate PE provides a significant speed-up of processing. Many image processing tasks (such as motion estimation) can also achieve significant speed-up from SIMD parallelism since they apply the same set of operations to each pixel in the image sequence.

While it may seem that the speed-up achieved will be equal to the number of PEs used, this upper bound is rarely achieved due to memory bandwidth constraints and because processing at each PE is usually dependent on other PEs. The automobile factory analogy shows the ideal case: cars can be built completely independently of one another, thus, if ten complex workstations are built, and they are supplied

---

[205] A recent trend enabled by the Internet has been distributed computing (e.g. the Seti project [145]). Here different datasets are sent over a network and processed by independent computers. This is a weak form of SIMD parallelism called message passing since the memory of each PE is completely independent and data can only be transferred by explicitly sending messages over the network. Distributed computing differs from parallel computing in that the PEs of a parallel computer have hardware-supported direct access to the memory space of other PEs.

```
if (condition==true)}
then
        perform complex operation 1
else
        perform complex operation 2
end if
```

Figure 7.12 A simple example program illustrating when MIMD machines have an advantage over SIMD machines

with materials sufficiently quickly[206], then cars can be built ten times more quickly. Many computational tasks are not entirely independent, however. For example, when forecasting the weather, individual PEs must communicate their partial results to neighbouring elements (regions) so that a consistent global weather forecast is created. The need for such communication is normal when a large dataset is partitioned and it inevitably reduces the achievable speed-up.

MIMD parallelism differs from SIMD in that each PE can perform different instructions on different data. There is no central control in a MIMD computer – PEs share memory but are otherwise independent. This is the most flexible form of parallelism, since it can both mimic SIMD parallelism and can be used to accelerate problems which are not highly data-parallel. A pseudocode example of a program where MIMD processing is beneficial is shown in Figure 7.12. This program can perform one of two complex operations depending on a condition. In a SIMD machine, the program runs serially in each PE. Entering the first branch of the 'if' statement, PEs where the condition is not true will be idle and thus processing time will be wasted. The remainder of the PEs will be idle when the 'else' branch is executed. A MIMD machine would not waste time in this way. If the condition is false it will automatically jump to the 'else' branch.

A variant of MIMD parallelism arises when special purpose PEs are chained so that the output of one forms the input to the next. This is called a parallel pipeline and is analogous to the assembly line of our automobile factory. After some latency (determined by the time it takes data to move completely through the pipeline), results are obtained at a rate determined by the slowest processing element in the pipeline. Parallel pipelines are ideally suited to tasks that can be broken into a number of subtasks of similar complexity. They also provide an ideal way of decomposing a complex problem into a number of simpler subproblems. For this reason, this is the primary form of parallelism used in this design.

The processing section of our design is split into four data paths. Of these, the PC and range data paths are the simplest and have already been described completely in the discussion of the corresponding buffers. Below we describe the camera and motion estimation data paths in more detail.

---

[206] I.e. if we have sufficient 'memory bandwidth'.

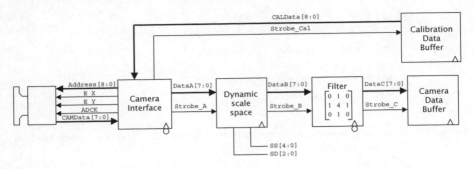

Figure 7.13 Block diagram of the camera data path. See Appendix C, Section C.4.1 (and in particular, Figure C.28) for implementation details

7.2.6.1 Camera data path

The camera data path[207] (Figure 7.13) is a pixel parallel pipeline where a sequence of three operations is performed simultaneously on different pixels. Data are first read from the camera and calibrated in the `Camera_Interface`. They are then subsampled as required in the `Dynamic_Scale_Space` block, and finally, the subsampled data are filtered in the `Filter` block. Each block in sequence uses an alternate clock allowing data to pass between blocks at an effective rate of 80 MHz.

The `Camera_Interface` block[208] reads data from the camera. First, the y and x addresses are strobed into the camera using the `E_Y` and `E_X` signals. The y address is latched by the camera so it need only be set at the start of a new image row. Pixels are always read in a row-wise manner. Shortly after a pixel has been addressed, the `ADCK` line (analogue to digital conversion strobe) is set low to indicate that the camera should perform a conversion. When the `ADCK` is returned high data are read from the `CAMData` bus. The timing for a pixel read is illustrated in Figure 7.14. Once an entire image has been read, the sensor is parked at an address outside the region of interest to prevent sensor artefacts that arise when addressing a given pixel for an extended time.

Each pixel value must be calibrated before being passed to the `Dynamic_Scale_Space` block. A new calibration value is obtained by strobing the `Calibration_Buffer`: one clock cycle after the `Strobe_Cal` line goes high the calibration value for the current pixel is available on the `CALData` bus[209]. This 9 bit signed calibration value is subtracted from the 8 bit unsigned pixel value and the result thresholded to an 8 bit unsigned value before being strobed into the `Dynamic_Scale_Space` block.

The camera interface always reads a 512 × 32 pixel image from the centre of the camera's vertical field of view. The `Dynamic_Scale_Space` block[210] takes the stream

---

[207] See Appendix C.4.1 for implementation details.
[208] See Appendix C, Figures C.29 and C.30 for implementation details.
[209] A simple strobe is sufficient due to the implicit synchronisation between blocks.
[210] See Appendix C, Figure C.31 for implementation details.

Sensor implementation 257

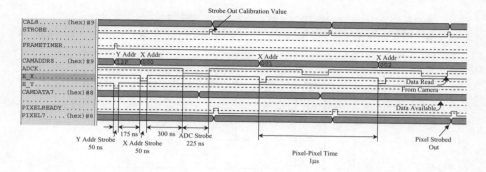

Figure 7.14   Timing for a pixel read operation

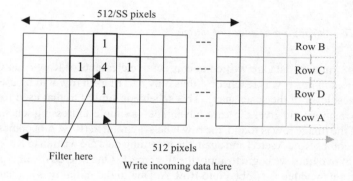

Figure 7.15   Structure of queue used in the Filter *block*

of pixels from the Camera_Interface, and for every SS input provides one output corresponding to the average of these pixels, effectively reducing pixel bandwidth by a factor of SS. Since SS is always a power of two, division is performed by shifting the sum to the right SD bits.

The Filter block[211] applies a 2D convolution mask (7.1) to the incoming pixel stream. This can only be done if three image lines are available, hence a means of storing the incoming data that copes intelligently with images at different scales is required. The solution uses a modified queue or pipeline structure (as in Figure 7.15) where each row of the buffer is implemented using a separate BlockRAM[212].

$$\frac{1}{8} \begin{bmatrix} 0 & 1 & 0 \\ 1 & 4 & 1 \\ 0 & 1 & 0 \end{bmatrix} \tag{7.1}$$

[211] See Appendix C, Figures C.32, C.33 and C.34 for implementation details.
[212] Configured as 512 8 bit words.

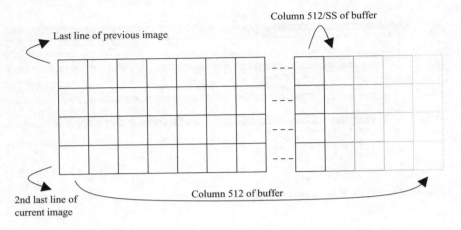

Figure 7.16 Wrapping at edges of image

In this structure, data are written to consecutive elements of Row A until the end of the current *image* row is reached[213] thus only part of each row will be used except at the lowest level of the scale space. This scheme is similar to that used for writing images to memory and gives the same advantage – a simple addressing scheme. When the end of the image row is found, the row labels are rotated (Row A becomes Row D, Row B becomes Row A, etc.) and again the incoming data are written to Row A. After a latency of two lines we begin to apply the filter mask. One filtered result is generated each time a new value is strobed into Row A. Due to the initial two line latency the input strobe (`Strobe_B`) will stop with two lines left in the filter pipeline so another method is required to clock out the final two lines. In principle, it may be possible to do away with pixel strobe for the final two lines since all available data are already in the buffer. However, this would lead to an unnecessarily heavy load on the RAMIC. We avoid this using an external 1 MHz strobe clock for the final two lines.

In this filter, edges are not handled in a special way – the filter simply 'wraps' around at image edges and uses whatever data are available in the queue (Figure 7.16). For example, at the left edge the filter wraps around to the opposite end of the row and at the top edge it 'wraps' around to the last row of the previous image. One simply must remember that, at the image extremes, filtered results are invalid.

7.2.6.2 Motion estimation data path

Despite the relative simplicity of the motion estimation algorithm, the motion estimation parallel pipeline[214] is the most complex section of the design (Figure 7.17).

---

[213] I.e. `ImgWidth` elements have been written.
[214] Implementation details of this block are given in Appendix C, Section C.6. In particular, Figures C.40 and C.41 give a high level overview of this block.

Sensor implementation 259

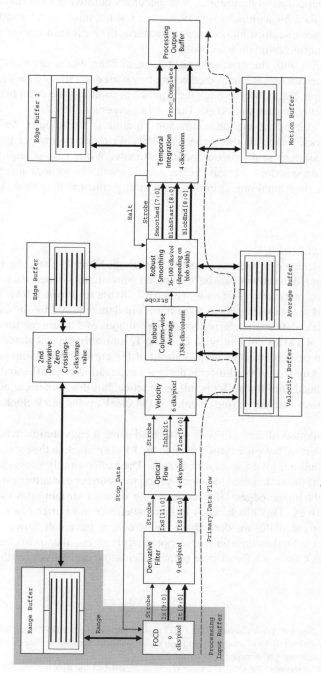

Figure 7.17 *Block diagram of the motion estimation pipeline. For clarity, address, data and strobe lines to the local buffers are indicated with a single bold arrow. The processing times indicated are the minimum calculation time. Note that buffers are split into two sections reflecting the fact that the BlockRAMs used to implement these buffers are dual ported. See Figures C.40 and C.41 in Appendix C for more detail*

Fixed-point arithmetic is used throughout, with accuracy tailored at each functional block to minimise data bit widths. It is important to use the minimal bit width that retains the desired accuracy/resolution since this reduces the logic resources required to implement the motion estimation algorithm.

Processing begins with the Processing_Input_Buffer where derivatives are calculated. These derivatives are smoothed before being used to estimate the velocity at each image column. Data flows through the pipeline in column-wise order to match the way it is consumed by the Robust_Column-wise_Average block.

The slowest block in the motion estimation pipeline is the Robust_Column-wise_Average block which requires approximately 43 clock cycles per pixel[215] (1388 clocks per column) to produce a motion estimate. While care is required to ensure that this bottleneck does not cause buffer overflows, this bottleneck also makes the design of other functional blocks easier since timing constraints in those blocks can be relaxed.

7.2.6.2.1 Derivative calculation
Derivatives are calculated in the Processing_Input_Buffer[216] using the FOCD estimator. Temporal derivatives can be calculated immediately when data corresponding to the Processing_Pointer (Figure 7.18) are read from RAM. This is because each RAM access retrieves the three frames of data required for the calculation, however, to calculate spatial derivatives three columns of data are required. The buffer reads from RAM in column-wise order so I_x can only be calculated when the buffer has read one column (33 pixels) ahead of the current processing pointer. The Processing_Input_Buffer buffer is designed as a read-ahead buffer and could in principle read ahead until the buffer is full. In practice, this does not occur because RAM access is slower than derivative calculation (11 clock cycles and 9 clock cycles respectively).

Since the calculation of derivatives is performed using a convolution mask, two issues occur here that we have previously seen in the Filter block in the camera data path. First, edges are not given special treatment. The buffer simply uses whatever data are available at the anticipated location leading to incorrect derivative estimates at the left and right image edges[217]. Second, is the issue of strobing out the final column of derivatives. This block reads in data whenever the buffer has read far enough ahead, then calculates the derivatives and strobes at the result, however, this technique fails for the final column where it is not possible to read ahead. To overcome this we do not check the read-ahead state before calculating derivatives for the final column.

---

[215] In Appendix A, Section A.5, we show that at 16 Hz (a limit imposed by our DSP based simulation of a range sensor), approximately 89 clock cycles per pixel are available for processing, therefore we have significant scope for improved performance.
[216] See Appendix C, Figures C.22, C.23, C.24 and C.25 for implementation details.
[217] The convolution mask for derivative calculation is one-dimensional.

Sensor implementation 261

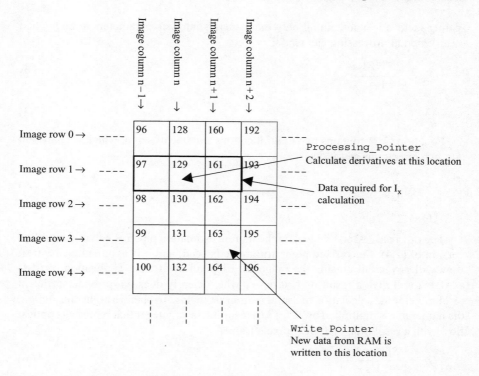

Figure 7.18   Buffer structure for derivative calculation

To reduce noise, derivatives are filtered prior to use. The `Derivative_Filter`[218] block applies the same filter kernel used for image filtering to the derivatives; however, the structure of this block is nearer to the `Processing_Input_Buffer` buffer than the `Filter` block from the camera data path due to the column-wise data ordering in the `Motion_Estimation` pipeline. A pair of buffers is used to store I_x and I_t and each buffer is addressed identically allowing parallel application of the filter mask to both data streams. The filter is applied as in Figure 7.18 with appropriate adjustment for the 2D convolution mask (7.1).

7.2.6.2.2   Solving the constraint equation

Now that derivatives are available, it is possible to solve the constraint equation (5.3). Rather than solving this entire constraint, we solve part of it (7.2), leaving a constant

---

[218] Implementation details for this block are given in Appendix C, Figures C.42 and C.43.

scaling factor (7.3) containing only the camera's intrinsic parameters to be applied by an external processing element[219].

$$\hat{U}_X = \frac{2^{level} Z I_t s}{I_x s} \tag{7.2}$$

$$\frac{\zeta}{fF} \tag{7.3}$$

Equation (7.2) contains two multiplications and one division, which we partition into two operations as follows:

$$OF = \frac{I_t s}{I_x s} \tag{7.4}$$

$$\hat{U}_X = 2^{level} OF \cdot Z \tag{7.5}$$

The `Optical_Flow`[220] block in the motion estimation pipeline solves a modified version of (7.4). Since fixed point arithmetic is used, and it is assumed that optical flow will never fall outside the range of ± 1 pixel per frame, application of (7.4) as stated will give a result of 1, 0 or -1 which clearly does not provide sufficient resolution for the calculation of useful motion estimates. To obtain a higher resolution, the numerator is multiplied by SU_1 (7.6) resulting in an integer that represents optical flow with a resolution of $1/SU_1$ pixels/frame:

$$OF = \frac{SU_1 \times I_t s}{I_x s} \tag{7.6}$$

Choice of SU_1 is important. If it is too small there will be insufficient resolution to make reliable motion estimates, however, making SU_1 excessively large is wasteful in terms of logic resources and does not necessarily give improved motion estimates due to noise and quantisation of image brightness (see Figure 7.19). The choice of SU_1 becomes more important when considering dynamic scale space.

The motion estimation algorithm uses dynamic scale space to 'focus its attention' on the nearest blob so that an accurate assessment of that blob's motion can be made. Of course, despite this focus on the nearest object, it is still important to have some sense of the motion of other visible blobs – even if they are significantly further from the sensor than the nearest blob. Such blobs will have low optical flow, hence detecting their motion requires a relatively high resolution motion estimate. For example, suppose the camera is stationary and the nearest blob is at a depth of 30 cm so that SD = 4. Also assume there is another blob visible in the environment moving at 5 cm/s. If SD remains constant (4.13a) shows that setting SU_1 to 256 gives sufficient resolution to detect the blob's motion at a distance of 19 m. If, on the other hand, $SU_1 = 32$, it would only be possible to detect the motion of the same object at 2.4 m.

---

[219] This removes the dependence of (7.2) on the cameras intrinsic parameters – SD is determined in part by the pixel-pitch ζ and frame rate.

[220] See Appendix C, Figures C.44 and C.45 for implementation details.

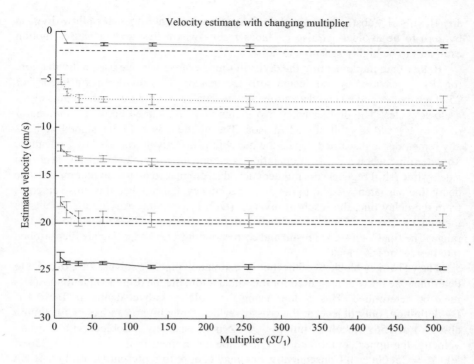

Figure 7.19 The effect of changing SU_1 at a range of velocities. Mean and standard deviation of the velocity estimate are determined as per Section 5.2.12. Dotted lines indicate the true motion of the tracked object and standard deviation is indicated with error bars. Values of SU_1 above 32 make little difference to the motion estimate

Of course, estimation of optical flows as small as $1/256 = 0.0039$ pixel/frame is rather noisy. Furthermore, when solving for $\hat{U}_x$ (7.5), blobs with small OF may well have a large depth Z, thus small errors in OF can lead to large changes in $\hat{U}_x$. This variance can be controlled (to a degree) by using robust averaging and temporal integration, although it is certainly not eliminated. Fortunately, it is relatively straightforward to assign a confidence measure to motion estimates – the further an object is from the current scale cross-over depth, the lower the confidence assigned. Since $D_{max} = 10$ m (the same order of magnitude as the 19 m calculated earlier), $SU_1 = 256$ was selected. This is supported by experiments in Section 5.2.3 showing that optical flow can be determined (with limited accuracy) to a resolution of 1/256 pixels/frame.

The possibility of division by zero in (7.6) is dealt with by assigning an `inhibit` flag to each motion estimate. Any estimate whose `inhibit` flag is set will be excluded from the robust average operation. There are a number of conditions where the `inhibit` flag for a motion estimation may be set. Two of these conditions are: $(I_x s) \leq 1$, $(abs(I_x s) \leq 4$ and $abs(I_t s) \leq 4)$. Not only does this allow division by zero to be avoided, it also excludes results where low texture causes motion estimates to be highly sensitive to noise.

Rather than implementing the division routine ourselves, we used a divider core (or IP)[221] provided by the design software vendor. This divider has a latency of 25 clock cycles and has no data strobe – new data can be placed on the inputs every 4 clock cycles. The divider takes two inputs: a 17 bit signed dividend (numerator of (7.6)[222]) and a 9 bit signed divisor. The quotient is a 17 bit signed value – any remainder is discarded, hence the quotient is effectively rounded by truncation. To determine when a division operation is complete, a one-clock pulse is sent into a delay line (shift register) the moment new data are placed on the divider inputs. The delay line has latency equal to the divider's latency, hence when this pulse emerges from the delay line, the result of division (OF) is available. Any quotients outside the range $-256 \rightarrow 255$([223]) have their `inhibit` flag set and are thresholded to this range. The final 9 bit signed result and corresponding `inhibit` flag are then strobed into the `velocity` block.

The `velocity` block[224] of the motion estimation pipeline solves (7.5). In order to determine the bit width required for this calculation, the maximum value for $abs(\hat{U}_x)$ must be determined. This is done using the scale cross-over depths in Table 4.2. By definition, only objects at the cross-over depth maximum can have a maximum $abs(OF)$ of 256. Since the multiple $OF \cdot Z$ is constant for an object with constant velocity, the upper bound on $abs(\hat{U}_x)$ can be determined from (7.5) as $1 \times 256 \times 400/4 = 25600$([225]). Consequently, one may choose to represent $\hat{U}_x$ using a 16 bit signed value, however, this allows for a far higher resolution than we can reasonably be expected to achieve. Furthermore, such a high resolution would also greatly increase the logic resources required for the robust average operation. The necessary bit width is reduced by dividing by SU_2 as in (7.7):

$$\hat{U}_x = \frac{2^{SD}}{SU_2} OF \cdot Z \qquad (7.7)$$

Since, eventually, the motion estimate will be sent to the PC, it is convenient to set $SU_2 = 256$. In this case, the upper bound on $\hat{U}_x$ falls to 100 and the motion estimate can be represented by an 8 bit signed value which can be directly read by the PC. $\hat{U}_x$ falls in the range $-100 \rightarrow 100$ and from this it is easy to show that motion estimates U_x will fall in the range $-16.66 \rightarrow 16.66$ cm/s[226]. Thus, an 8 bit signed

---

[221] Recall that a logic core is a design that is used as a 'black box'. The user is generally able to define a number of parameters for the core (such as input and output bit widths) from which a drop-in module is created.

[222] 256 is an 8 bit unsigned integer, $I_t s$ is a 9 bit signed (two's complement) integer – multiplying these values gives a 17 bit signed result.

[223] A 9 bit two's complement number representation cannot represent +256.

[224] See Appendix C, Figures C.46 and C.47 for implementation details.

[225] Division by four accounts for the division of range by four (Section 7.1.2). Here we assume we are at a depth of 400, corresponding to $SD = 0$.

[226] This result follows directly from (4.15) with $V_{max} = 10$ cm/s, $x = 256 \times \zeta$ and $f = 4.8e - 3$.

value represents motion estimates with a resolution of 0.1666 cm/s. Note that the upper-bound of abs($\hat{U}_x$) and the scaling factor SU_2 are determined by the scale cross-over points. If the cross-over points change, then the upper-bound and the scaling factor must be corrected. This will in turn change the resolution of velocity estimates.

While considering the issue of resolution, it was observed in Section 4.6 that motion estimates from higher scales will have a lower resolution. Suppose there is an object at a depth equal to a scale cross-over point and with an OF of V_{OF}. If the object moves away from the camera, its OF will fall to $\frac{1}{2}V_{OF}$ as it approaches the next scale cross-over point. Regardless of scale, the resolution of OF is always the same ($= 1/SU_1$). However, the velocity calculation does not use the raw optical flow OF, it uses the projected optical flow $2^{SD}OF$. This projection scales the optical flow from the current level in the dynamic scale space, to the lowest level of dynamic scale space (SD = 0). The result of this is that data from higher levels of scale space (i.e higher SD values) have a lower resolution for the corresponding projected optical flow. Fortunately, this variation in resolution makes no difference to the motion estimates since division by SU_2 swamps the effect of projection.

The constraint $|\hat{U}_x| \leq 100$ *should* never be broken (assuming no noise) since objects further than the cross-over depth will necessarily have a smaller OF than objects at the cross-over depth. The one case where this may not be true is at the edge of a shadow (see Section 5.2.4). The shadow edge may have an optical flow of less than 1 pixel/frame[227], however, its range will be overestimated introducing the possibility of results where $|\hat{U}_x| > 100$. This gives a limited ability to reject the false motion created by moving shadows. The algorithm sets the inhibit flag for any $|\hat{U}_x| > 100$.

A multiplier core (or IP) with a latency of 5 clock cycles implements the $Z \times OF$ operation and a simple bit shift implements the $\times(2^{SD}/512)$ operation. When new optical flow data are strobed into the Velocity block, they are immediately passed to the multiplier[228]. While a multiplication is underway the previous result remains on the multiplier's output (see Figure 7.20). Rather than wasting this time it is used to perform three operations on the previous result: perform $\times(2^{SD}/512)$ operation, check if the previous value of $\hat{U}_x$ was over 100 (and set inhibit flag if necessary), and send the result and the inhibit flag to the Velocity_Buffer.

Although the design performs the division operation of (7.2) first, a design that performs the multiplication operation first is also possible. Such a design would be less logical since the value $2^{level}ZI_t$s does not represent any physical quantity whereas division calculates the optical flow and allows rejection of any obviously erroneous motion estimates (> one pixel/frame). Another design alternative is to take the robust average of OF over each column and then apply (7.7) once for each column. This

---

[227] It may well be more than this since shadow motion depends not only on object velocity but lighting conditions as well. Even if it moves with a velocity less than V_{max}, the shadow may fall closer than the current scale change-over distance leading to the possibility of optical flow greater than 1 pixel/frame. Of course, motion estimation will fail for optical flows greater than 1 pixel/frame.

[228] Range data are read from the Processing_Input_Buffer between columns of optical flow data.

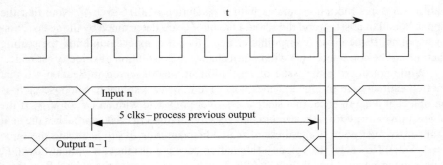

Figure 7.20 *Timing for multiplication operation. While a multiplication operation is underway, the previous result remains on the multiplier's output. Time t between inputs is determined by the bandwidth of upstream blocks*

would greatly reduce the number of multiplication operations required, however, there are two arguments against such a design. First, such a design would not allow the elimination of motion estimates in shadow regions thus it would lose a degree of robustness. Second, while this design requires one multiplication for each pixel (rather than each column), these extra multiplications are essentially free. There is no hardware penalty since the velocity block remains essentially the same regardless of which design is used and there is no time penalty since the velocity block operates in parallel with other blocks in the pipeline when the input data rate is high[229].

7.2.6.2.3 Robust smoothing

Robust smoothing is split across two blocks, the first calculating an average velocity for each column and the second smoothing between range discontinuities. In principle, only a single smoothing stage that robustly averages all pixels between range discontinuities is necessary, however, this would require a buffer able to store an entire frame of $\hat{U}_x$ values, which is not possible using BlockRAMs and excessively slow if primary RAM is used. Note that the initial threshold for the robust average algorithm is currently hard-coded as the current upper-bound of abs($\hat{U}_x$). If any system parameters (Table 5.3) are changed, the initial threshold must be updated.

Calculation of a new column-wise average begins when a column of velocity estimates is available in Velocity_Buffer. Because the Robust_Column-wise_Average[230] calculation is far slower than the upstream blocks, data will be consumed from the buffer more slowly than they are produced. To avoid overflows, the Robust_Column-wise_Average block monitors the state of the buffer and if the buffer is almost full, the Processing_Input_Buffer block is signalled to stop

---

[229] For low data rates, a pipeline is essentially a sequential processing unit.
[230] See Appendix C, Figures C.48, C.49 and C.51 for implementation details.

producing data. This signal occurs with sufficient space in the buffer to store any data still in the pipeline. Note that the 1388 clks/column processing time indicated in Figure 7.17 is the worst case time. This time can be as low as 940 clks/column depending on how many pixels are inhibited.

The `Robust_Column-wise_Average` block implements the robust average algorithm presented in Section 5.2.5 with three minor modifications. First, only data whose `inhibit` flag is clear are used when calculating the robust average. If for some reason all data in a column are inhibited, the velocity estimate for that column will be zero. Second, this block is able to calculate a weighted average rather than just a simple average. This is useful if one has some *a priori* knowledge regarding the scene, however, in this implementation weights for each row are equal.

The final modification pipelines operations much as they were in the `Velocity` block. In this case, the division and summation operations in Figure 5.9 are pipelined. The division operation updates the `Current_mean` value, however, it takes 25 clock cycles for the division to complete[231]. Rather than wasting time waiting for the division result, it proceeds with the next iteration using the *previous* value of `Current_mean` to calculate the new sum. This pipelining makes little difference to the operation of the Robust Averaging beyond a requirement for an extra iteration to compensate for the use of the previous result. The implementation uses seven iterations of the robust averaging algorithm. The average for each column is stored in the `Average_Buffer` as it is calculated. Only when a complete frame of column-wise averages has been calculated does the `Robust_Column-wise_Average` strobe the `Robust_Smoothing` block to begin operation.

The `Zero_Crossings` block[232] monitors data passing between the `Velocity` block and the `Range_Buffer`, capturing column addresses and the corresponding range measurements. These range data are segmented using a zero crossing method where sign changes in the 2nd derivative indicate a step discontinuity (Figure 7.21). This approach works well for our simulated range sensor, however, it may need to be improved in cluttered environments or to account for sensor noise. The 2nd derivative is approximated by applying the FOFD derivative estimator twice giving the mask in (7.8):

$$[1 \quad -2 \quad 1] \tag{7.8}$$

This mask was chosen for its simplicity; accuracy of the second derivative is not important here provided the zero crossing property is maintained. The `Zero_Crossings` block queues three consecutive range values to which it applies this mask. Any result less than $h/w = 4$ is thresholded to zero to avoid spurious discontinuities with the consequence that step discontinuities in depth of less than 16 cm[233] are not detected. Zero crossings are detected as a change of sign between two consecutive 2nd derivatives with the condition that the second derivative passes through zero – a

---

[231] A delay line like that used in the `Optical_Flow` block determines when a division is complete.
[232] Implementation details for this block are available in Appendix C, Figures C.51 and C.53.
[233] Remembering that range is divided by four for use in the FPGA.

268 *Motion vision: design of compact motion sensing solutions*

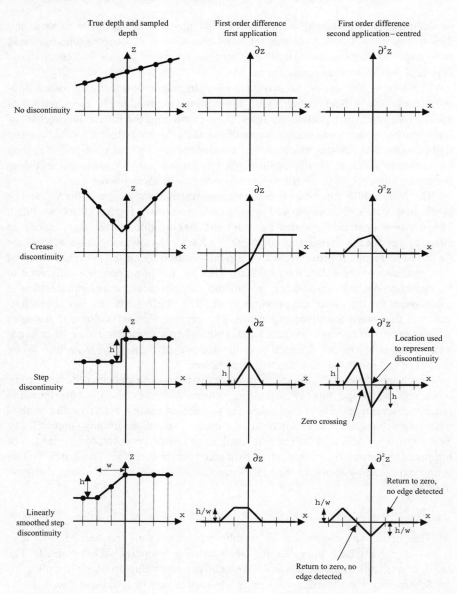

Figure 7.21   Operation of our zero crossing operator. Discontinuities are only detected for step edges. Only steps of h > 16 cm will be detected. No discontinuities are detected for smoothed step changes

zero crossing does not occur if the 2nd derivative returns to zero. The location of a range discontinuity is defined as the column to the right of the zero crossing point. If a zero crossing (i.e. a discontinuity) is detected, its address is stored in the Edge_Buffer and these addresses are subsequently used by the Robust_Smoothing block.

The Robust_Smoothing block[234] calculates the robust average of the data in the Average_Buffer between each pair of range discontinuities. The velocities either side of the discontinuities are not used in the robust averaging procedure since range discontinuities are assumed to correspond to motion discontinuities and velocity estimates at these discontinuities are likely to be incorrect due to failure of the brightness constancy assumption. A side effect of this is that a blob must be a minimum of three pixels wide at the current scale. Any blob less than three pixels wide will be absorbed into the blob immediately to the right. Of course, the motion estimates for such a small blob will be somewhat unreliable since the robust averaging procedure will never use more than half the data to calculate the average.

7.2.6.2.4 Temporal integration

Once the robust average for a blob is calculated, the start[235] and end column addresses for the blob, together with the blob's smoothed motion estimate are strobed into the Temporal_Integration block[236]. To allow for the possibility of temporal integration for blob n taking longer than the smoothing operation for blob n − 1 the Temporal_Integration block has the ability to halt the Robust_Smoothing block. This only occurs if blob n is significantly wider than blob n − 1.

Temporal integration is performed for all frames, except the first where there is no prior data to integrate. As per Section 5.2.10 the temporal integral is calculated as the weighted average of the current and previous motion estimate for each column spanned by the current blob, except the start and end columns. This prevents smoothing across blob boundaries, which may have moved up to one pixel since the previous frame. The matching of the current and previous motion estimates allows for changes of scale. While blob-wise temporal integration may appear a more efficient approach than our column-wise method, difficulties arise if the number of blobs changes between frames.

The previous motion estimate is stored in a local buffer to enable rapid calculation of the temporal integral and once the new temporal integral for a column is calculated it replaces the previous value. The temporal integration block also modifies how range discontinuities are stored. Whereas the address of discontinuities is stored in the Edge_Buffer, Edge_Buffer_2 is mapped to have one location for each image column. A location will contain the value 0xFF if it corresponds to a range discontinuity, otherwise it will contain zero. When all temporal integrals have been calculated, the Processing_Output_Buffer is strobed and begins

---

[234] See Appendix C, Figures C.51 and C.52 for implementation details.
[235] The start address for blob n is equal to the end address for blob n − 1 plus one.
[236] See Appendix C, Figures C.54 and C.55 for implementation details.

270 *Motion vision: design of compact motion sensing solutions*

storing motion estimates and range discontinuity information to primary RAM. While the `Temporal_Integration` block and the `Processing_Output_Buffer` could be pipelined, little would be gained by doing so since the Temporal Integration process is much faster than the writing of data to primary RAM.

7.2.6.2.5 Recovering the motion estimate

The numerical value given by the calculations described above is proportional to the true motion estimate – a number of proportionality constants have been ignored in the calculation to allow implementation using fixed point arithmetic. The true value for the motion estimate U_x can be recovered from $\hat{U}_x$ using (7.9). In this equation, the factor of 4 corrects for scaling of range data. We use the manufacturer's specification for the pixel-pitch ($\varsigma = 12.5\ \mu m$) and focal length ($f = 4.8\ mm$).

$$U_x = \hat{U}_x \frac{4\varsigma}{fF} \frac{SU_2}{SU_1} \tag{7.9}$$

7.3 Experimental results

Now that the intelligent sensor has been fully described we show its operation in an experimental environment.

7.3.1 Experimental setup

We performed our experiments in a manner similar to the simulation-based experiments in Section 5.2.12, and the sensor was configured using the parameters in Table 5.3. Due to high noise levels encountered in simulation when $TC = 0$, only $TC = 3$ and $TC = 7$ are used here.

The prototype intelligent sensor and experimental environment are illustrated in Figure 7.22. The upper diagram shows the primary components of the intelligent sensor with the FUGA camera and the SignalMaster prototype board clearly visible. While the Xilinx FPGA is not visible in this picture, the Bitsi extension connector used to join the camera, PC and FPGA can be seen clearly. The lower picture shows the experimental environment. The camera is mounted on one of the ICSL's Cooperative Autonomous Mobile Robots (CAMRs) allowing the camera to be moved in a predictable way. Because the cable from the camera to the SignalMaster must be kept relatively short, both the PC and SignalMaster platform are placed on a trolley that is moved manually alongside the CAMR. In the experimental environment, CAMRs are used as obstacles since their motion is predictable allowing accurate simulation of the range sensor.

7.3.2 Aligning the camera and range sensors

An approximate alignment of the virtual range sensor and camera can be performed by defining a virtual environment for the range sensor that matches the physical configuration of the test environment. One can then manually position the camera

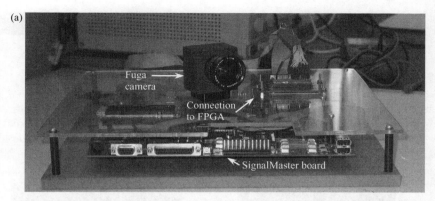

Figure 7.22 (a) Intelligent sensor prototype and (b) experimental environment

so that calculated range discontinuities are aligned with the object edges visible in the camera output. A similar process could be used if a true range scanner were used.

This process shows that the manufacturer's specifications for pixel-pitch and focal length are accurate. Indeed, rough experiments showed a correspondence of ± 2 pixels between expected location (based on the simulated location of blob edges) and actual object location in the image. This work assumes that this error is negligible (certainly small enough to be dealt with by the robust averaging algorithm), however, if more accurate measurements of the true pixel-pitch, focal length and any lens distortion were available, then it would be simple to make the appropriate corrections. This can

272 *Motion vision: design of compact motion sensing solutions*

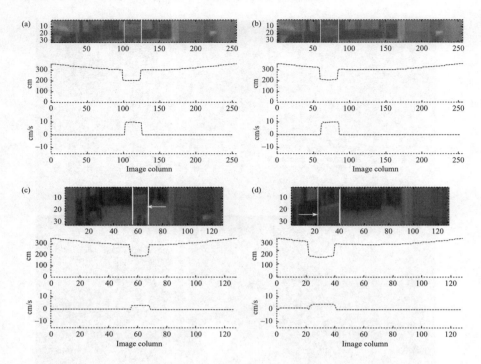

Figure 7.23 *Data captured from our motion sensor prototype. The upper part of each figure shows data captured by the camera, the central plot indicates range and the lower plot indicates estimated velocity*

be done without changing the FPGA component of the design. The corrections must occur at the input, where the mapping of range data to pixels in the range sensor must be corrected and at the output side where the effect of variation in pixel spacing or focal length can be corrected by allowing f or ζ in (7.9) to be space variant.

Figure 7.23 shows data captured from the sensor in two different experiments. Parts (a) and (b) are from different time instants in an experiment with a stationary camera while parts (c) and (d) were captured at different time instants in an experiment with a moving camera. In each diagram, the upper section shows data from the camera: white lines indicate blob edges. Due to difficulties in starting the sensor and vehicles at precisely the same moment the blob edges do not correspond exactly to the true edges of the vehicle (as indicated by the white arrows in parts (c) and (d)). The simulated range data are relatively accurate, however, and this misalignment does not significantly change over time. This is clear from comparison of the diagrams on the left (taken from the start of each experiment) with the diagrams on the right (taken near the end of each experiment). The central section of each diagram shows the interpolated range data while the lower section shows the motion estimate.

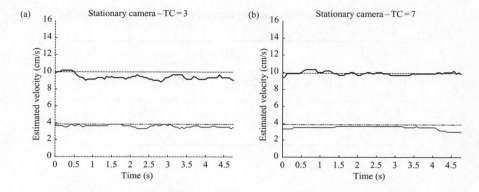

Figure 7.24   Experimental results with a stationary camera. Solid lines correspond to motion estimates and dotted lines represent true velocities

7.3.3 Stationary camera

Figure 7.24 shows motion estimates for the case of the stationary camera with part (a) being performed with $TC = 3$ while part (b) is performed with $TC = 7$. In each case two velocities are tested; $T_x = 10$ cm/s and $T_x = 3.85$ cm/s. Before analysing these data it is important to list some issues regarding how experiments were performed.

- Expected velocity values were approximated experimentally and hence include some error.
- While we were able to repeatedly use identical input data in our simulation based experiments, each trace here is generated using unique input data (i.e. data from a different experimental run). Because the input data are different at each run it becomes difficult to compare results given by different parameter settings (e.g. different values of TC).
- Before each experiment the camera is aligned using the method described in the previous section.

The plots show that motion estimates are excellent for a stationary camera and this is reflected in the mean and variance listed in Table 7.7. As with the simulation based experiments it can be seen that increasing TC from three to seven makes only a marginal improvement. Indeed, allowing for a small error in T_x, a comparison of Table 5.4 and Table 7.7 shows that these results have similar accuracy to those obtained in the simulation experiments.

7.3.4 Moving camera – effect of barrel distortion

Initial results for the moving camera were less encouraging. Figure 7.25a shows motion estimates for low velocities ($T_x = T_z = 3.85$ cm/s). The motion estimate is noisy, underestimated, and does not follow the expected velocity trend. Figure 7.25b

Table 7.7 Long term mean and variance of measured velocity.

TC = 3				TC = 7			
T_x	T_z	Mean	Variance	T_x	T_z	Mean	Variance
3.85	0	3.6	0.0211	3.85	0	3.49	0.0405
10.00	0	9.4	0.0939	10.00	0	9.89	0.0376

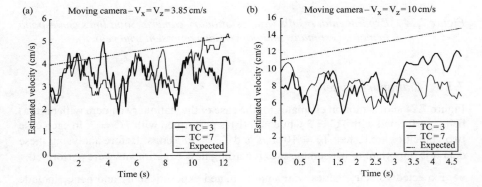

Figure 7.25 Experimental results for a moving camera. Expected value calculated from (4.15)

shows that similar results (with far greater underestimation) are obtained in the case for $T_x = T_z = 10$ cm/s. These results are not consistent with simulation results where good motion estimates were obtained at up to $T_x = T_z = 9.6$ cm/s.

Closer examination of Figure 7.23 shows that there appears to be significant barrel distortion[237] (distortion due to the wide angle nature of the lens) in the image. This is visible in all images though it is clearer in Figure 7.26a where the straight lines in the test pattern appear bent. The effect of barrel distortion is exaggerated in the sequence with a moving camera since this sequence is processed at a higher level of the image pyramid where subsampling compresses the curvature. These experiments used data from the centre of the camera's sensor assuming this was aligned with the camera's optical axis (i.e. centre of the lens) and hence barrel distortion would be minimised. However, looking at Figure 7.26a we see there is significant barrel distortion at the centre of the image (marked with a black arrow). Measuring the height of the lens

---

[237] There is a small amount of anticlockwise rotation about the optical axis, however, barrel distortion is still evident with horizontal lines sloping in the opposite direction on either side of each image.

Sensor implementation 275

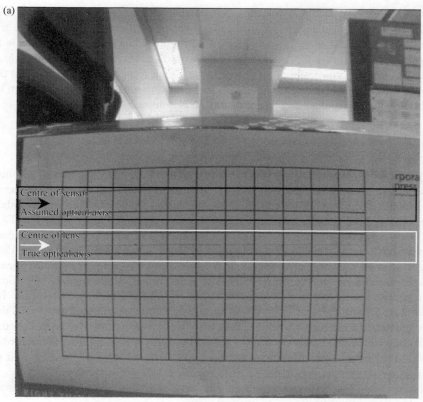

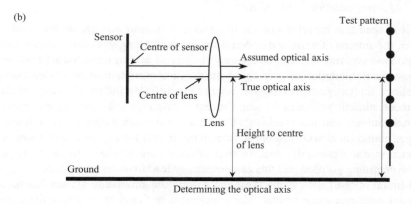

Figure 7.26 *Barrel distortion with our lens. Top – test pattern, bottom – determination of optical axis. As we move further from the centre of the image, the angular spacing of the pixels becomes smaller leading to barrel distortion. This distortion is most visible with a wide angle lens such as ours. Boxes indicate regions used for motion estimation*

276 *Motion vision: design of compact motion sensing solutions*

above the ground and matching this to a point in the image (as per Figure 7.26b) showed that there is a significant misalignment of the lens and sensor – the true optical axis is marked with a white arrow in Figure 7.26a. Notice that there is far less barrel distortion at this point. It is hypothesised that this barrel distortion induces noise into the motion estimation procedure resulting in poor estimates. To test this hypothesis experiments with the moving camera were repeated using data from the white region of Figure 7.26a where distortion is minimal.

7.3.5 Moving camera – elimination of barrel distortion

Figure 7.27 illustrates that using data centred on the optical axis appears to eliminate barrel distortion. While horizontal lines do slope slightly, this is consistent across the image indicating not barrel distortion, but a slight rotation of the camera about the optical axis. Results using these data are illustrated in Figure 7.28. For $T_x = T_z = 3.85$ cm/s, an improvement is shown in results with less underestimation, closer tracking of the expected velocity and lower variance. In the case of $T_x = T_z = 10$ cm/s, a significant reduction in noise is seen and the motion estimate clearly tracks the trend of the expected motion estimate though there remains significant underestimation. Note that for one run at $TC = 7$ (corresponding to the solid line in Figure 7.28b), there was significant misalignment between range and visual data (as illustrated in Figure 7.27c). Despite this, the motion estimate is no poorer than the other estimates indicating that the robust average process is correctly discarding motion estimates from the region between the expected edge of the blob and the true edge.

7.3.6 Moving camera – image noise

Thus, it is seen that barrel distortion has had a significant impact on the results; however, the amount of noise and underestimation is still far higher than indicated by the simulation experiments. Motion estimates using real data are expected to be noisier than those from simulation since images from the camera contain some residual noise after calibration (e.g. see speckling left edge of Figure 7.23a and b). The cause of the gross underestimation remains unclear. Currently, suspicion falls upon the additional noise in the images causing a relatively large number of motion estimates to fall over the upper-bound for abs($\hat{U}_X$) (more so than in the simulation results where a similar phenomenon was observed). Such motion estimates are ignored (trimmed) in the robust averaging algorithm and this can potentially lead to underestimation.

An initial investigation along these lines using the simulation system has been performed with promising results. The experiment in Figure 5.23 where the camera and the object are in motion was repeated this time thresholding (Winsorising) rather than trimming large values of abs($\hat{U}_X$). Figure 5.23 is replicated here as Figure 7.29a for comparison against the new results illustrated in Figure 7.29b. Notice the improved results at higher velocities. Further investigation of this issue is left for future work.

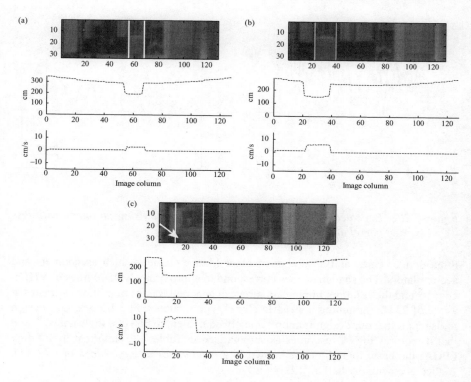

Figure 7.27   *Example data with reduced barrel distortion. Parts (a) and (b) are taken from an experiment with $T_x = T_z = 3.85\,cm/s$ while part (c) is taken from an experiment with $T_x = T_z = 10\,cm/s$. Note the tracking error in part (c)*

7.3.7   Moving camera – noise motion and high velocities

Finally, while we have assumed explicitly that image sequences are stable, it is interesting to consider the response of the sensor to bumps and other high frequency motion noise. A crude experiment was performed where the camera was shaken by hand while observing a moving object. The result was a zero motion estimate. Similarly, the sensor's response to high velocities (velocities where temporal aliasing occurs) was tested by waving our hand in front of the camera. Once again, the sensor gave a zero result.

7.4   Implementation statistics

Our design is made up of 54 blocks (29 state machines and 25 schematics as described in Appendices A, B and C) developed by ourselves together with a multiplier core,

278 *Motion vision: design of compact motion sensing solutions*

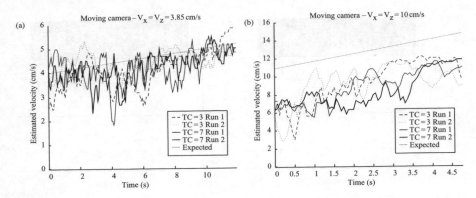

Figure 7.28   *Repeated experiments with moving camera using an image with less barrel distortion*

three divider cores and a number of other library elements such as counters and accumulators. The state machines correspond to approximately 4500 lines of VHDL code[238] and the resulting implementation can generate frames of motion estimates at a rate of 23 Hz (assuming a 40 MHz clock) with a latency of 3 frames. Our design tools report that our design is approximately equivalent to 368 000 logic gates[239] and that it requires FPGA resources as summarised in Table 7.8. Note that in a Xilinx FPGA, the basic logic block is known as a *combinatorial logic block* or *CLB* [5]. A slice is essentially half a CLB.

7.5 Where to from here?

While the previous sections have shown that our intelligent sensor is indeed effective in our application, there may be some research questions to explore with the aim of achieving improved performance if required by certain applications. Some of these have been mentioned elsewhere, and in this section we summarise them.

7.5.1 Dynamic scale space

As formulated, dynamic scale space chooses a scale that guarantees temporal aliasing is avoided *in the worst case* for the nearest object, however, it may be useful to use a scheme where the scale is chosen using a combination of minimum range, and the

---

[238] This is the approximate number of lines in the VHDL code generated by the state machine entry tool, excluding comments automatically generated by that tool. It is difficult to estimate how many lines of code the schematics may translate to and this is not included in our estimate. More details are available in Appendix C, Section C.1.

[239] This should be taken with a grain of salt since the conversion of slices to gates is somewhat complex [3, 87].

Sensor implementation 279

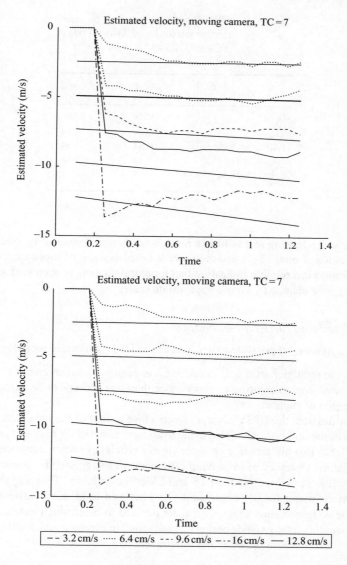

Figure 7.29   Part (a) is replicated from Figure 5.23. Part (b) shows the result of thresholding (Winsorising) rather than trimming unexpectedly large values

minimum resulting optical flow. Thus, if the optical flow at the chosen scale is found to be very small (less than half a pixel/frame), the algorithm could move down the image pyramid (reducing subsampling) until a reasonable optical flow is found, thus reducing the effects of noise. Temporal subsampling could also be applied to 'amplify' the apparent motion. The danger with such schemes is that apparent velocity may

Table 7.8 FPGA resources used by our design assuming a Xilinx Virtex XCV800 BG432 FPGA

	Available	Used
Slices	9408	3286
I/O pins	316	154
Tristate buffers	9632	434
BlockRAMs	28	18
Global clock networks	4	2

suddenly change leading to a failure in motion estimation. Fortunately, such a failure can be avoided if blob edges are monitored. Displacement of more than one pixel between frames is a reliable indication that temporal aliasing is occurring and can be used to trigger a change to a more appropriate scale.

7.5.2 Extending the LTSV estimator

There are a number of possible extensions of the LTSV estimator including:

1. Dynamic reduction of `alpha` and `beta` to improve convergence.
2. Recursive implementation to allow more than one process to be extracted from the region of support.
3. As formulated, the LTSV always rejects close to half of the data elements, thus mean of the inliers is calculated using a random subset of the majority population. This leads to a higher error variance (lower efficiency) when compared to other estimators. One way of improving efficiency was suggested by Rousseeuw and Leroy [239] in relation to the LTS and LMedS estimators. They suggest that the robust estimator (in our case, the LTSV) be used to identify outliers, then the simple weighted least squares estimate be used to determine the final estimate, with weights set to exclude outliers. A reduction in variance may lead to improved motion estimates in the case of a moving camera.

7.5.3 Temporal integration

As implemented, the motion estimation algorithm uses temporal integration to improve motion estimates over time, however, it does not use temporal integration to reduce computation. Since dynamic scale space guarantees apparent motions of less than one pixel per frame, we can reasonably assume the previous motion estimate at a pixel is a good indication of the current motion at that pixel. This can be used as a starting point for the robust averaging algorithm that would allow us to reduce the number of iterations required for convergence.

7.5.4 Trimming versus Winsorising

As indicated by our experiments, trimming data in an attempt to minimise the effect of optical flow induced by shadows, adversely affects estimation of motion near the upper limit of the sensor's capability. While simulation results show that thresholding (Winsorising) the motion estimate $\hat{U}_X$, rather than trimming it, improves results, this has not been confirmed in practice.

7.5.5 Rough ground

The sensor is designed under the assumption of motion on a smooth ground plane. On an uneven road, noise motion will be induced and as crude experiments have shown, this will adversely affect the sensor. Stabilisation of the image and a more thorough evaluation of the sensor's sensitivity to noise motion are open to investigation.

7.5.6 Extending the hardware

The prototype intelligent sensor operates at 16 Hz (limited by our DSP based simulation of the range sensor) with a latency of 3 frames[240]. The frame rate is limited by the simulated range sensor (as it would be if a true range sensor were used). The sensor's ability to respond to velocity changes is also related to the temporal integration parameter TC as described in Section 5.2.10. Lower values of TC lead to a more responsive estimate at the cost of increased noise, while higher values of TC decrease noise while making the motion estimate less responsive.

At present the achievable processing rate for the FPGA component of the design is limited by the `Robust_Column-wise_Average` block, which requires about 43 clock cycles per pixel to produce a result. In the worst case (i.e. 512×32 pixel image) approximately 26 ms are required for I/O and 17.62 ms ($43 \times 512 \times 32 \times$ clock period) are required for processing, so the minimum time in which a frame of data can be read and processed (without changing the FPGA and assuming a sufficiently fast range scanner is available) is $26 + 17.62$ ms which corresponds to a processing rate of approximately 23 Hz (assuming a 40 MHz clock).

There are three steps we may take if a particular application requires a higher processing rate. First, a faster range sensor is required since it is impossible to generate frames of motion estimates faster than frames of range data are obtained. Of course, improvements in the range sensor will only improve performance until the limit of the FPGA's processing ability is reached.

Second, time spent in the I/O and processing modes must be reduced. A faster camera and range sensor would assist in reducing the I/O time, however, as currently designed, the `RAMIC` is nearly at 100 per cent utilisation during I/O. Therefore any increase in I/O device bandwidth must be matched by improvements to RAM access (e.g. by reading and writing in bursts of eight rather than bursts of one, improvements

---

[240] One frame is due to the non-causal nature of the temporal derivative estimate, the other two frames account for data input/output as per Chapter 1.

282 *Motion vision: design of compact motion sensing solutions*

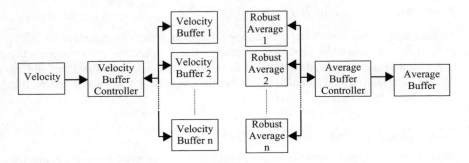

Figure 7.30   Duplication of the Robust_Column-wise_Average operation. For simplicity arrows indicate data, address and control signal flows

to the buffer polling scheme and use of ZBT memory). Modifications to the camera, range and PC data paths would not be necessary since the components in these paths are currently capable of far higher bandwidths than the devices they serve.

Finally, it is possible to increase the processing rate by reducing the time taken for motion estimation. The most obvious way to do this is to increase the clock rate, though this may require significant redesigning of the system to meet the tighter timing constraints and to ensure I/O functions still operate correctly. Perhaps a simpler alternative is to utilise parallelism. The design of the motion estimation data path utilises a pipeline structure, however, the benefit of this structure is largely lost due to the existence of a single block in the pipeline that is significantly more complex than the others. Figure 7.17 shows that the Robust_Column-wise_Average operation is over 100 times slower than other operations in the pipeline and this operation alone limits the achievable processing rate making the Robust_Column-wise_Average block a prime candidate for improvement.

One way to increase the throughput of this block is to fully utilise the temporal integration (Section 5.2.10). Further reductions in average operation time can be realised by duplicating the block, however, this duplication is not entirely straightforward since it is necessary to consider contention on the Velocity_Buffer and the Average_Buffer. One possible method of duplicating the Robust_Column-wise_Average block is illustrated in Figure 7.30. Each column of data is placed into a different buffer by the Velocity_Buffer_Controller. In this way, the individual Robust_Average blocks do not contend on buffer access and can operate at their full rate. The Average_Buffer_Controller moderates access to the Average_Buffer ensuring that each Robust_Average block is able to store its results.

Part 4

Appendices

Appendix A
System timing

This appendix gives details of our sensor's internal timing and control flow. For illustrative purposes we show the timing and control flow for both a 512×32 pixel image (i.e. the case where there is no subsampling due to scale space) and a 32×32 pixel image (where there is maximal subsampling). After the timing diagrams there is a table detailing the meaning of labels in the diagrams. Note that all diagrams assume 10 Hz operation for the sensor. At higher rates only the inter-frame time changes – the timing for individual data elements remains the same.

A.1 Timing for a 512×32 pixel image

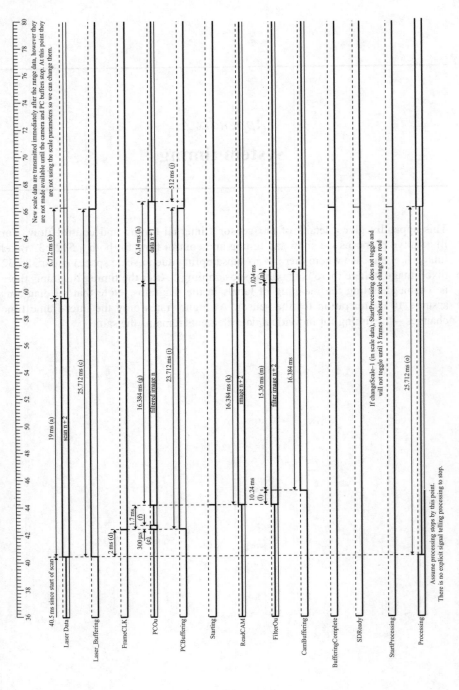

Figure A.1 Timing for a 512×32 pixel image

A.2 Control flow for a 512×32 pixel image

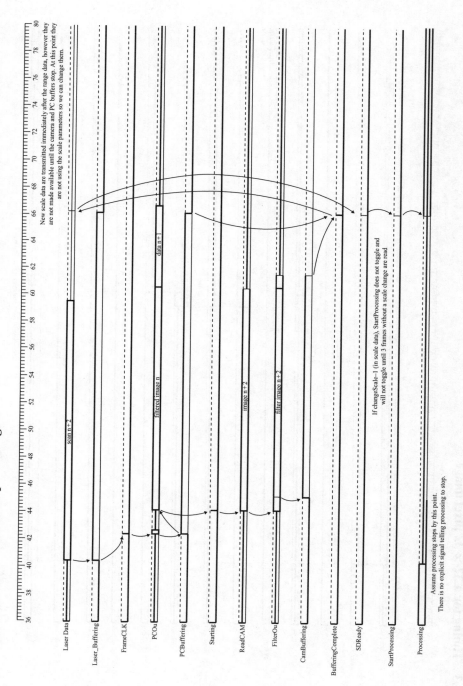

Figure A.2 Control flow for a 512 × 32 pixel image

288 Motion vision: design of compact motion sensing solutions

A.3 Timing for a 32 × 32 pixel image

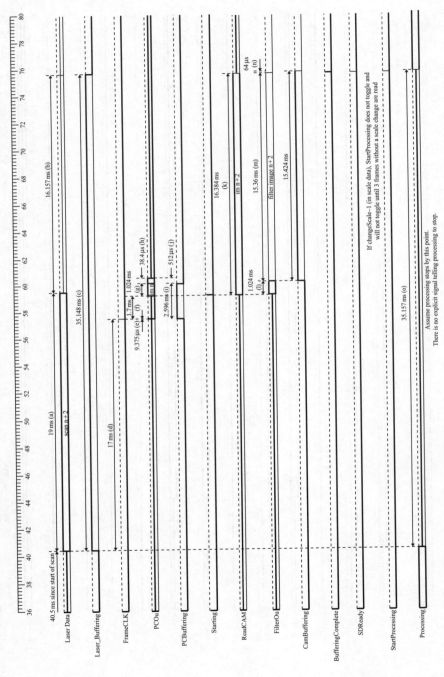

Figure A.3 Timing for a 32 × 32 pixel image

A.4 Control flow for a 32 × 32 pixel image

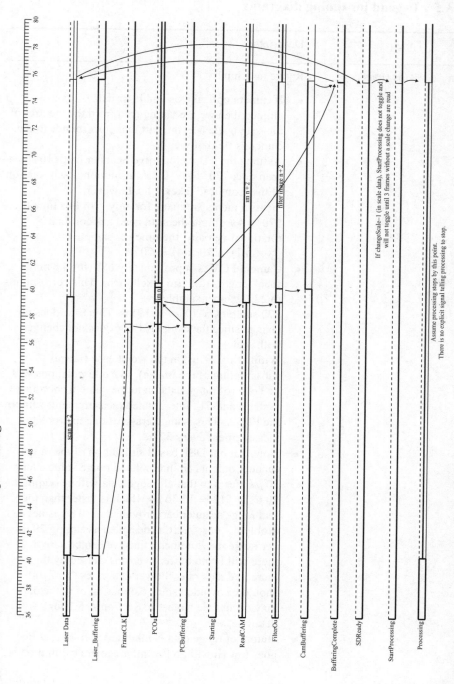

Figure A.4 *Control flow for a 32 × 32 pixel image*

A.5 Legend for timing diagrams

	Time	Description
a	19 ms (max)	Range data input

- If camera optical axis and laser data 0° axis are aligned then we are taking data from the 'centre' of the scan cycle (i.e. the part that corresponds to the camera's field of view).
- Assume the LD Automotive Scanner has a 10 Hz scan rate, 270° FOV, 0.25° resolution. Each rotation of the scan head takes 0.1 s, however, measurements are made for only $\frac{3}{4}$ of this time (i.e. 270°). Assuming the scan head has constant rotational velocity, the time per sample is $(\frac{3}{4} \times 0.1)/(270 \times 4) = 70\,\mu s$.
- Camera FOV is approximately 68° (for 4.8 mm focal length, 12.5 μm pixel pitch). This corresponds to 272 scanner samples. 272 samples $\times$ 70 μs = 19 ms. This time does not change since the cameras FOV does not change with scale.
- Within this 19 ms, in the worst case (i.e. no subsampling of the image), 512 interpolations will be done (of range data on to the camera coordinate system) and 512 interpolated measurements sent to the FPGA. This is an average of one interpolated measurement every 37 μs.
- Note that our DSP based simulation of the range scanner outputs an interpolated range value every 30 μs meaning that all range data will be output in 30 μs $\times$ 512 = 15.36 μs which is faster than the real range scanner can provide data. This is not a problem – our RAMIC is designed assuming 29 μs per range measurement, thus we can be sure that there will be no buffer overruns caused by this increased data rate. Slower data rates (>37 μs) should be avoided since clobbering of range data may occur (see Note 1 in Section A.5.1 for details).
- Output of range data is inhibited until a reset line goes low (it will go low after camera calibration is

Continued

Time	Description
b 16.157 ms (max) 6.712 ms (min) (assuming 19 ms to read all range data)	complete). This means that processing WILL NOT begin before the reset line goes low. Wait between end of Range Data input and output of new scale data New scale data are not made available until all buffers have completed their operations. Changing the scale parameters part way through a buffering operation would cause the system to fail since the buffers will immediately use the new values even though: • the PC is unaware of the change (it receives the scale parameters as part of the 'break data'); • we have not read an image at the new scale yet.
c 35.148 ms (max) 25.712 ms (min)	Laser buffering period This is the period during which the laser buffer is active. During the period between the end of reading the range data and output of scale data, the laser buffer will be idle, but not inactive. This time period is equal to the maximum of $(d + f + g + h - j)$ and $(d + f + l + m + n)$.
d 2 ms + DELAY	Clobber avoidance delay • We must avoid clobbering of range data. That is, we must ensure that the writing of new range data into RAM is always ahead of the reading of old range data from RAM for output. The easiest way to ensure this condition is to ensure reading of old data begins after writing of new data. • This task is slightly complicated by the fact that the time it takes to output data to a PC can vary since data are always clocked out to the PC at 1 MHz. However, it always takes 19 ms to read in all the necessary range data regardless of scale because we still scan over the same angular range. • To account for this variation we make the delay dependant on scale. For SD = 0, DELAY = 0 ms, SD = 1, DELAY = 8 ms, SD = 2, DELAY = 12 ms, SD = 3, DELAY = 14 ms, SD = 4, DELAY = 15 ms. Note that these delays are a little longer than necessary. This is because we are not sure how much f (the time it take the PC to start reading data) can vary. We have designed this scheme to allow f to be as low as 1 ms, though it is typically closer to 1.7 ms.

Continued

Time	Description
e 300 μs (max)	Output read ahead

- The PC output buffer reads ahead until full. How long this takes is difficult to determine since it is an operation where data moves from RAM to a buffer, so this timing will be determined by contention for RAM.
- We saw in Section 7.2.4.5 the average time per RAM operation is approximately 300 ns. The buffer can hold 512 data elements so it will take 154 μs on average to fill the buffer. Under no circumstances will it take any longer than the 300 μs indicated on the timing diagrams and this time does not change with different scales. Clearly the buffer will be full before period f expires.

f 1.7 ms (typ)	PC start delay

This is the time it takes the PC to start reading data (i.e. `OutputPixelClock` to begin toggling) after we send the `OutputFrameStrobe` signal and is typically around 1.7 ms (measured with an oscilloscope). It does not seem to vary much but we have allowed for some variation (down as low as 1 ms, no upper bound) in this design – see Part d of this table for details.

g 16.38 ms (max) 1.024 ms (min)	Output of image data to PC

Time to send data from buffer to PC. Worst case time = $512 \times 32 \times 1\,\mu s = 16.38$ ms. The pixel clock here is always 1 MHz so this time will reduce for higher scales. For example, with SD = 4 (32×32 pixel image), the time is 1.024 ms. Note: this is the time it takes to output data to the PC from the buffer, it is not related to the time it takes to read data from RAM into the buffer. This process assumes sufficient data are available in the buffer to avoid underruns.

h 6.14 ms (max) 384 μs (min)	Output of other data to PC

Time to send Range + Velocity + Break data from buffer to PC. Each of these vectors is as wide as the image and consists of 4 bytes (each byte is treated as a single pixel) so the total worst case time is $512 \times 4 \times 3 \times 1\,\mu s = 6.14$ ms. For higher scales, this will be less e.g. SD = 4 (32×32 pixel image) = 384 μs.

Continued

	Time	Description
i	23.712 ms (max) 2.596 ms (min)	PC output buffering Time to buffer all output data (counting time where the buffer is idle because it is full). This time is equal to: $$f + g + h - j = 23.712 \text{ ms } (512 \times 32 \text{ image})$$ $$2.596 \text{ ms } (32 \times 32 \text{ image}).$$
j	512 μs	Allowance for read ahead During the read ahead phase (e) the buffer is filled. Once the PC starts reading data (g) one element will be removed from the buffer every 1 μs (equal to the PC's pixel clock), hence a new value can be read into the buffer every 1 μs. If we assume it never takes longer (on average) than 1 μs to retrieve a new value from RAM we can assume that buffering of output data will be complete 512×1 μs before the PC reads its final value.
k	16.38 ms	Camera read Time taken to read a 512×32 pixel image from the camera ($= 512 \times 32 \times 1$ μs). We ALWAYS read a 512×32 pixel image from the camera. This image is subsampled to the appropriate scale downstream. This subsampling effectively reduces the pixel clock rate to 1 MHz/(2^scalediv). If we have scaleDiv $= 1$ (i.e. image width halved) then the pixel clock rate after subsampling is 500 kHz. Thus, no matter the scale, the time for 1 frame of data to be captured and subsampled is ALWAYS 16.38 ms (plus small pipelining time for calibration and subsampling).
l	1.024 ms	Filter input delay Filtering introduces a 2 image line latency (i.e. 2 full lines are read before filtering begins). No matter what scale is currently being used, this delay is $512 \times 2 \times 1$ μs $= 1.02$ ms. This value does not change because (see above) if we halve the image width, the pixel clock halves. (In practice, this delay is a bit longer due filter pipelining.)

Continued

	Time	Description
m	15.36 ms	Main filter processing Filter the first ImgRows −2 lines of the image. Pixels emerge from the filter at a pixel rate of 1 MHz/(2^scalediv). Total time (assuming 512 × 32 pixel image − scalediv = 0) 512 × 30 × 1 μs = 15.36 ms. Once again, this value does not change with scale since a change in scale leads to a corresponding change in the pixel clock.
n	1.024 ms (max) 64 μs (min)	Final filter processing Filter the last 2 lines of the image. The pixel clock at this stage is fixed at 1 MHz so the timing will change with changes in scale. The worst case time occurs at the finest scale $\longrightarrow$ 512 × 2 × 1 μs = 1.02 ms (best case = 32 × 2 × 1 μs = 64 μs).
o	35.157 ms (max) 25.712 ms (min)	I/O time • While the FPGA is in I/O mode no processing can occur since the buffers associated with processing are not polled by the RAMIC, hence they cannot access RAM. The total I/O time for a 512 × 32 pixel image is 25.712 ms and 35.157 ms for a 32 × 32 pixel image. • To get an idea of how much time is available for the process let us assume 16 Hz processing rate and a 512 × 32 pixel image. In this case there is 36.5 ms available for processing which is equivalent to 2.23 μs/pixel or 89 clock cycles/pixel (at 40 MHz, 512 × 32 pixel image).

A.5.1 Note 1: image data clobbering

This design ensures that image data are not clobbered before they reach the PC. We have to be careful to ensure that the writing (of filtered data) to RAM is always behind reading of data from RAM (for output to PC). If writing gets ahead it will change (clobber) the data we are expecting to read. This design is clobber free because:

1. the PC buffer reads ahead before new data are read from the camera;
2. the PC and the camera read data at the same rate, so the camera can never 'overtake' the PC.

A.5.2 Note 2: the use of n in the timing diagram

In order to get a result for frame n, data for frames $n-1$, n and $n+1$ must be in memory. Thus, once we have 'processed' frame n we need to read in image/scan $n+2$. Once processing for frame n is complete, we output image n and the result for frame n. We have a dilemma because the current system set up will only have access to scan $n+1$ and $n+2$ by the time we get around to output the range data for frame n. Range data for frame n is clobbered. For simplicity of implementation we output range data for frame $n+2$ (memory organisation makes this more convenient than outputing range data for frame $n+1$) together with image and velocity data for frame n. This is just a visualisation issue – it does not affect processing. In this work we do not worry about it – the difference between range scans is small anyway. If we are really desperate to get the right range data we can do it after capture on the PC. At courser scales our margin increases further because data are written more slowly to RAM.

A.5.3 Note 3: scale space change over process

1. After all buffers have stopped using the scale parameters, update all the scale parameters.
2. If new scale data indicate scale change:
   a. do not process existing data (frames $n-1$, n, $n+1$);
   b. wait for three frames to be read WITHOUT a scale change occurring;
   c. process these new data.
3. If no scale change is indicated then process existing data.

A.5.4 Note 4: the first frame

The first frame is a special case. At the first frame both the Range Scanner and FPGA assume finest scale (i.e. 512 pixel image width). Once this is read a change of scale is forced even if it is not strictly required. This has the advantage of utilising scale change logic to read three frames before processing beings. If there was no scale change after the first frame we would need to implement additional logic to force the system to wait for three frames before processing began (so that derivatives could be calculated). The first frame is not read until calibration is complete and the range scanner begins operating.

Appendix B
SDRAM timing

Our design makes use of the SDRAM[241] available on the Gatesmaster/SignalMaster board used in this work. This appendix details the timing used to access that memory. This appendix is not intended as a tutorial on the use of SDRAM (see Reference 274 for a useful introduction) but rather to clarify the timing used in the design of our FPGA. We access SDRAM using single word bursts and a CAS latency of 2.

B.1 Powerup sequence

When power is first applied, the following sequence of operations must be performed in order to ensure the memory is properly initialised:

1. Apply power and clock signal.
2. Wait a minimum of 200 μs with the DQM and CKE lines high.
3. Precharge all banks.
4. Initialise mode register.
5. Carry out 8 Autorefresh cycles to stabilise the device.

Note that in Figure B.1 we have a fixed 40 MHz (i.e. 25 ns) clock, therefore, we must round timing (e.g. $t_{RP} = 24$ ns) to the nearest clock cycle (e.g. 1 clock). The grey regions in this timing diagram indicate 'don't care' states and MR indicates any valid setting for the mode register.

---

[241] This memory is specifically Toshiba TC59S6432CFT-10 SDRAM – this RAM appears to have been discontinued – the datasheet is no longer available from the Toshiba website. Further details regarding the operation of SDRAM can be found in Reference 274 or in any SDRAM datasheet.

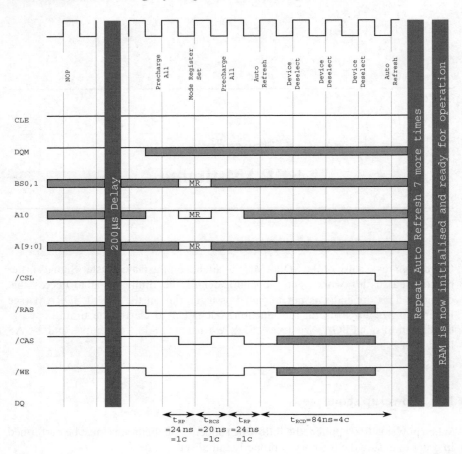

Figure B.1   Powerup sequence for SDRAM

B.2   Read cycle

We use an 'autoprecharge' read operation with a burst length of one. The read cycle begins by placing a row address on the address bus, then the column address is placed on the address bus and a read command issued. The requested data are made available 2 clock cycles later (Figure B.2).

B.3   Write cycle

We use an 'autoprecharge' write operation with a burst length of one. The write cycle begins by placing a row address on the address bus, then the column address is placed on the address bus, data are placed on the data bus and a write command issued (Figure B.3).

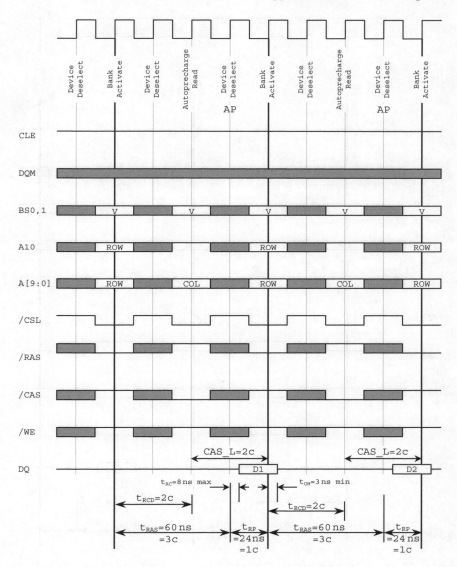

Figure B.2 Two SDRAM read cycles. Note that we have a fixed 40 MHz (i.e. 25 ns) clock therefore we must round timing (e.g. $t_{RAS} = 60$ ns) to the nearest clock cycle (e.g. 3 clocks). The grey regions in this timing diagram indicate 'don't care' states and V indicates any valid state

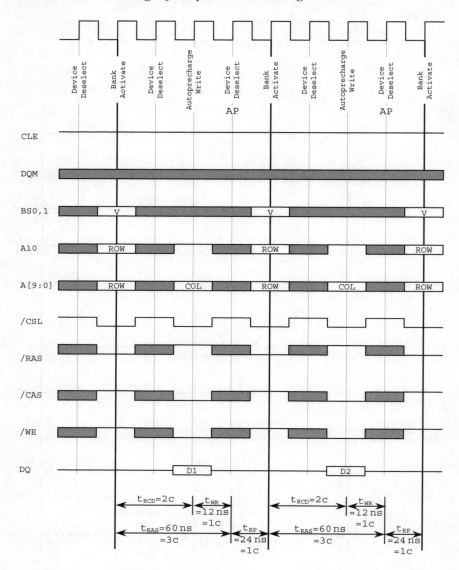

Figure B.3 Two SDRAM write cycles. Note that we have a fixed 40 MHz (i.e. 25 ns) clock therefore we must round timing (e.g. $t_{RAS} = 60$ ns) to the nearest clock cycle (e.g. 3 clocks). The grey regions in this timing diagram indicate 'don't care' states and V indicates any valid state

Appendix C
FPGA design

This appendix includes all design files (schematics, state machines and constraint files, all summarised in Section C.1) for the FPGA based component of our design. Please remember that this is one of the first prototype designs, so while the design works, it should not be considered to be the 'best possible' implementation. No floor planning (beyond specifying which I/O pins to use) was performed in this design – the final FPGA layout is entirely the result of automated tools. This implementation is targeted to the Xilinx Virtex XCV800 BG432 FPGA, and in particular, to the GatesMaster BITSI daughterboard (manufactured by Lyr Signal Processing [143]) using this FPGA. Naturally, this design will reflect the particular hardware we have used, particularly for I/O blocks, however, our design is quite modular so you should easily be able to modify the design to work with your hardware.

Use of this design (or any part thereof) is entirely at your risk. We cannot provide support and will not accept any liability for your use of this design (in part or whole). This design (or any part thereof) must not be used to maintain the safely or security of people or equipment.

We do not provide a detailed description of the function of design elements here – please refer to Chapter 7. The name of a design element is generally prefaced with 'MAC_SM' if the element is a state machine and 'MAC' if the element is a schematic. The 'TMN' labels refer to timing groups used in the user constraint file (included in Section C.8) to define the allowable delay on particular nets.

C.1 Summary of design components

In this section we present a summary of the components in our design. Where one name is listed the name corresponds to the name used in the design. For components with two names listed, the first name corresponds to the component names used in Chapter 7. Often these components are made up of a number of subcomponents in the final design, therefore, we include a second name indicating

302 *Motion vision: design of compact motion sensing solutions*

the name of this subcomponent in the design. There are two types of component – SCH = schematic and SM = state machine. The number of lines of VHDL code is included for state machine components – this is the number of lines in the code automatically generated by the state machine tools and includes some automatically generated comments.

System section	Name	Type	Description	Lines of code	Figure
RAMIC	RAMIC_Poll RAMICPol	SM	Controls polling of buffers.	228	C.3
	RAMIC_Operations RAMICP	SCH	High level RAM interface – instantiates RamicMem.	N/A	C.4
	RAMIC_Operations RamicMem	SM	High level RAM interface.	136	C.5
	SDRAM_Interface SDRAMInt	SM	Low level RAM interface.	244	C.6
	SDRAM_Interface RstTimer	SCH	Timer to control SDRAM refresh cycles and the SDRAM power up process.	N/A	C.7
	SDRAM_Interface RAMIO	SCH	Physical connection from the FPGA to the SDRAM.	N/A	C.8
	Calibration Calibrat	SCH	Interconnection of subcomponents for generation of camera calibration image.	N/A	C.9
	Calibration Zeroram	SM	1st part of calibration image generation – zero RAM.	126	C.10
	Calibration SumPixels	SM	2nd part of calibration image generation – sum 16 frames.	161	C.11
	Calibration CalcCal	SM	3rd part of calibration image generation – average pixel values and remove image bias.	125	C.12
Buffers	Calibration Buffer CALBUFF	SCH	Calibration buffer schematic – instantiates calibbuf.	N/A	C.13
	Calibration Buffer calibbuf	SM	Calibration buffer implementation.	194	C.14
	Camera Buffer CAMBUFF	SCH	Camera buffer schematic – instantiates cambuf.	N/A	C.15
	Camera Buffer cambuf	SM	Camera buffer implementation.	195	C.16
	Range Buffer LASERBUF3	SCH	Laser range buffer schematic – instantiates lasbuf3.	N/A	C.17

Continued

Appendix C – FPGA design

System section	Name	Type	Description	Lines of code	Figure
	Range Buffer lasbuf3	SM	Laser range buffer implementation.	255	C.18
	Range Buffer clob_del	SM	Implementation of clobber avoidance delay. The need for this component is explained in Appendix B.	78	C.19
	PC-Output Buffer PCBUFF	SCH	PC output buffer schematic – instantiates PCBuf.	N/A	C.20
	PC-Output Buffer PCBuf	SM	PC output buffer implementation.	263	C.21
	Processing Input Buffer IPBUFF	SCH	Processing input buffer schematic – instantiates ipbuf and implements support functions.	N/A	C.22, C.23 and C.24
	Processing Input Buffer ipbuf	SM	Processing input buffer implementation.	287	C.25
	Procesing Output Buffer OPBUFFER	SCH	Processing output buffer schematic – instantiates outbuff.	N/A	C.26
	Procesing Output Buffer outbuff	SM	Processing output buffer implementation.	147	C.27
Camera data path	Camera_Interface CAMIO	SCH	Physical connection from FPGA to camera.	N/A	C.29
	Camera_Interface CameraCont	SM	Interface to FUGA camera.	181	C.30
	Dynamic_Scale_ Space ScaleSp	SM	Implementation of dynamic scale space.	86	C.31
	Filter FILT	SCH	Filter schematic – instantiates Filter.	N/A	C.32
	Filter Filter	SM	Implementation of image filtering.	252	C.33
	Filter CAMPIXCL	SCH	Auxiliary pixel clock used by filter.	N/A	C.34
Laser data path	Laser Data Path LASERIN	SCH	Physical connection from FPGA to laser range sensor.	N/A	C.36
	Laser Data Path Laserint	SM	Interface to laser range sensor.	70	C.37

Continued

304 *Motion vision: design of compact motion sensing solutions*

System section	Name	Type	Description	Lines of code	Figure
PC data path	PC Data Path PCOUT	SCH	Physical connection from FPGA to PC.	N/A	C.39
Processing data path	Motion Estimation Data Path processing	SCH	Schematic for the motion processing data path.	N/A	C.41
	Derivative_Filter COLFILTM	SCH	Filtering of derivatives – schematic instantiating ColFilt.	N/A	C.42
	Derivative_Filter ColFilt	SM	Implementation of derivative filtering.	245	C.43
	Optical_Flow APPVELM	SCH	Calculation of optical flow – schematic instantiating AppVel.	N/A	C.44
	Optical_Flow AppVel	SM	Implementation of optical flow calculation.	142	C.45
	Velocity VCALCM	SCH	Conversion from optical flow to velocity – schematic instantiating vCalc.	N/A	C.46
	Velocity vCalc	SM	Implementation of conversion from optical flow to velocity.	168	C.47
	Robust_Column-wise_Average ROBAVGM2	SCH	Schematic instantiating RobAvg2.	N/A	C.48
	Robust_Column-wise_Average RobAvg2	SM	Implementation of robust column-wise averaging using LTSV.	202	C.49
	Robust_Column-wise_Average Robavgp	SM	Interface block for RobAvg2.	89	C.50
	Robust_Smoothing ROBAVGSEGM2	SCH	Schematic instantiating RobAvgSeg2.	N/A	C.51
	Robust_Smoothing RobAvgSeg2	SM	Implementation of robust smoothing (between range edges) using LTSV.	216	C.52
	Zero Crossings RangeEdges	SM	Implementation of range data segmentation using zero crossings.	156	C.53
	Temporal_Integration TEMPINTM	SCH	Schematic instantiating TempInt.	N/A	C.54
	Temporal_Integration TempInt	SM	Implementation of temporal integration.	178	C.55

Continued

System section	Name	Type	Description	Lines of code	Figure
Misc. components	GenClock	SCH	Biphase clock generation.	N/A	C.57
	SysReset	SM	System reset generation – waits 5 s after boot before removing the reset state.	66	C.58
	BufsDone	SM	Signals completion of I/O buffering.	77	C.59
	ScaleChange	SM	Monitor the changeScale bit to indicate if processing should begin.	76	C.60
	Udcnt	SM	Monitor the state of FIFOs/buffers.	136	C.61
	BUFE32	SCH	32 bit tristate buffer with active low tristate signal.	N/A	C.62
	BUF16EJ	SCH	16 bit tristate buffer with active low tristate signal.	N/A	C.63

C.2 Top level schematic

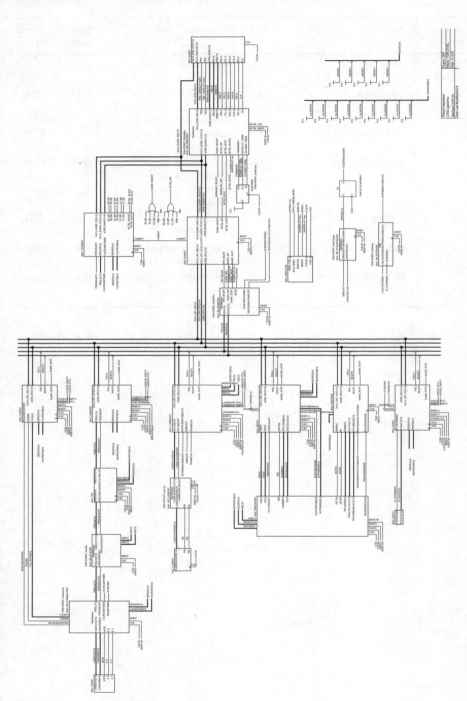

Figure C.1 Top level schematic. See Figure 7.1 (Section 7.2) for a simplified block diagram (see following schematics for a clearer view of each section)

Appendix C – FPGA design 307

C.3 RAMIC

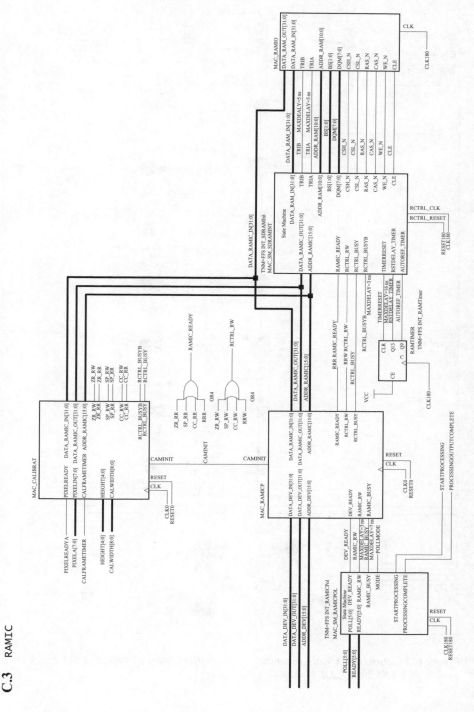

Figure C.2 Top level schematic – focus on the RAMIC. See Figure 7.7 (Section 7.2.4) for a simplified block diagram

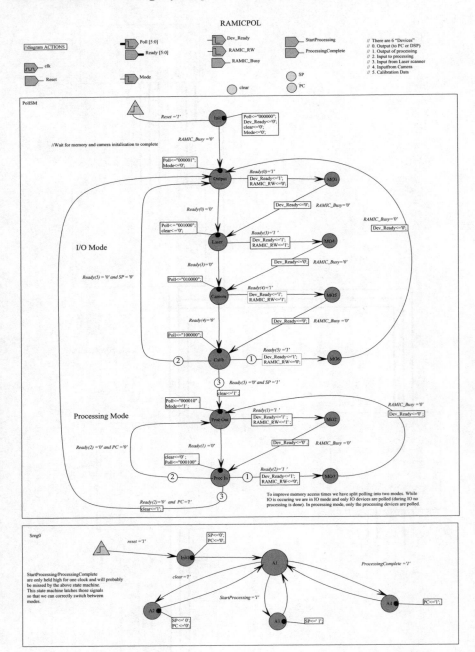

Figure C.3 RAMICPol *state machine – polls buffers (part of* RAMIC_Poll *block in Section 7.2.4.4)*

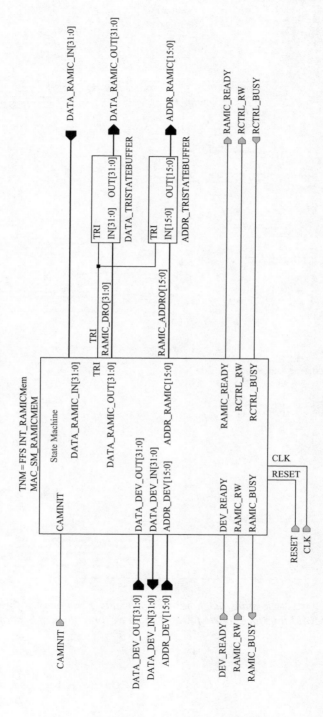

Figure C.4 RAMICP schematic – high level RAM interface (part of RAMIC_operations block in Section 7.2.4.2)

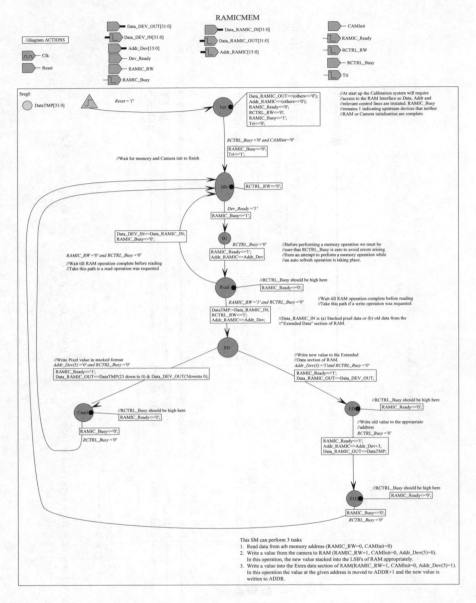

Figure C.5 RAMICMem *state machine – implements functionality for high level RAM interface (part of* RAMIC_Operations *block in Section 7.2.4.2)*

Appendix C – FPGA design 311

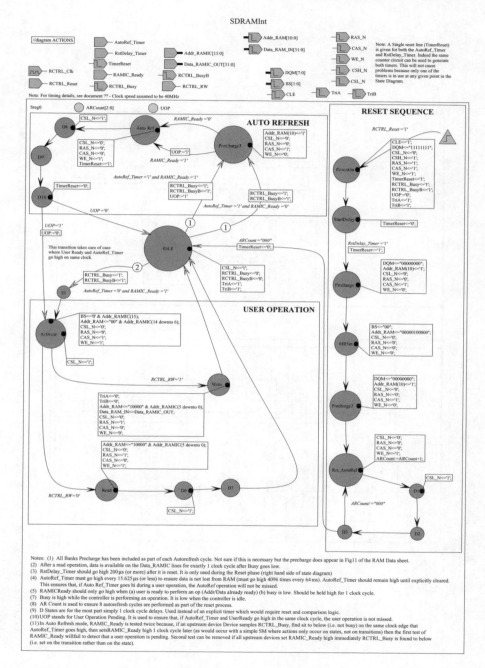

Figure C.6 SDRAMInt *state machine – low level RAM interface (part of* SDRAM_Interface *block in Section 7.2.4.1)*

312 *Motion vision: design of compact motion sensing solutions*

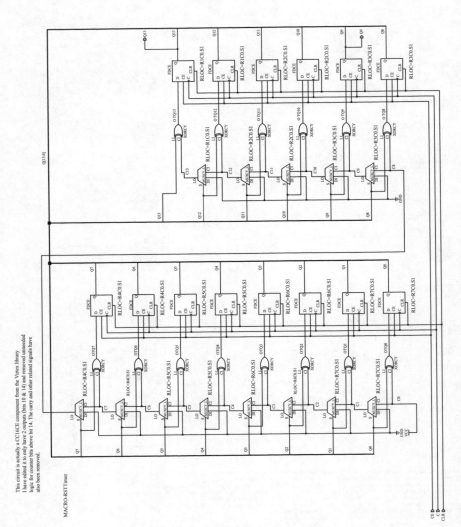

Figure C.7 `RstTimer` *schematic – timer used for* SDRAM *refresh cycles and at system boot (part of* `SDRAM_Interface` *block in Section 7.2.4.1)*

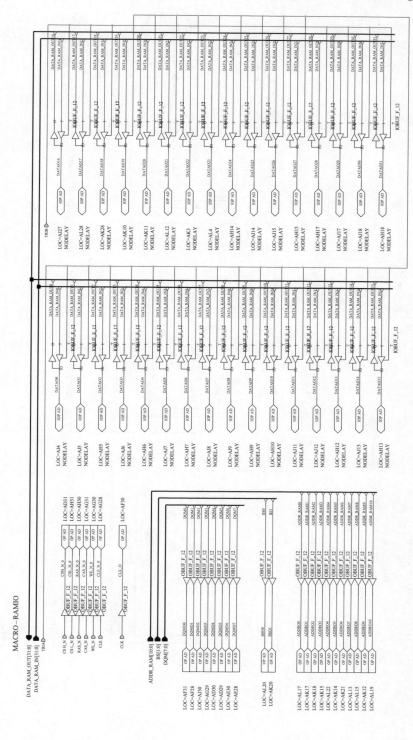

Figure C.8 RAMIO schematic – physical connection from SDRAMInt to SDRAM (part of SDRAM_Interface block in Section 7.2.4.1)

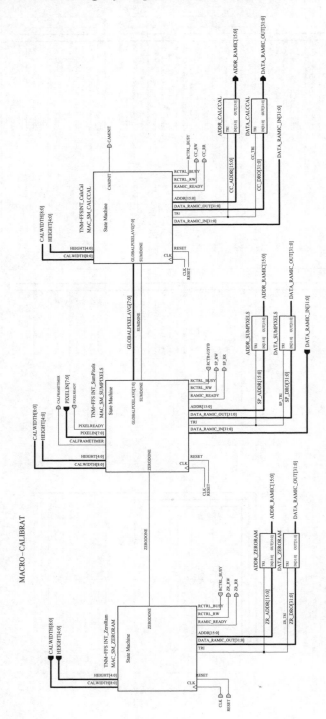

Figure C.9 CALIBRAT schematic – interconnection of components used to generate camera calibration image (part of calibration block in Section 7.2.4.3)

Appendix C – FPGA design

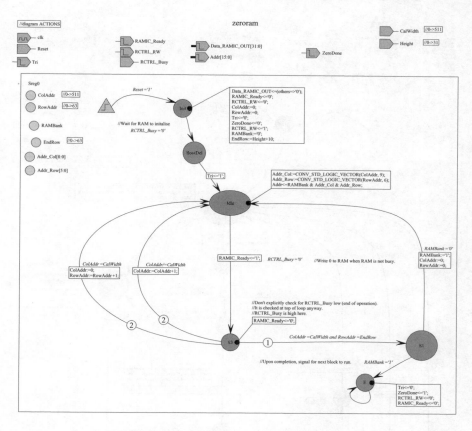

Figure C.10 **ZeroRam** *state machine – first part of calibration image generation – zero RAM (part of* **Calibration** *block in Section 7.2.4.3)*

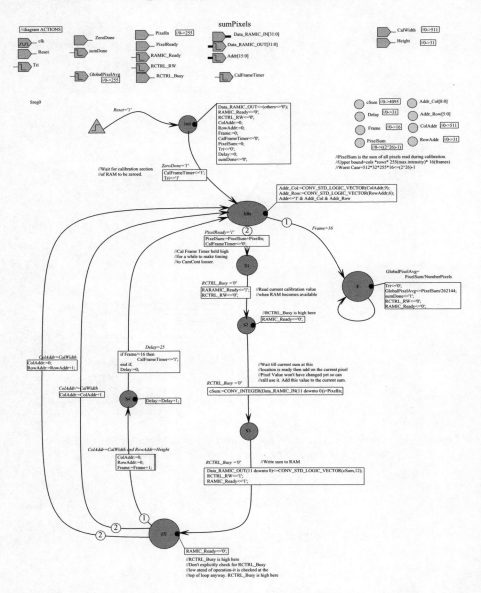

Figure C.11 SumPixels *state machine – second part of calibration image generation – sum 16 frames (part of* Calibration *block in Section 7.2.4.3)*

Appendix C – FPGA design 317

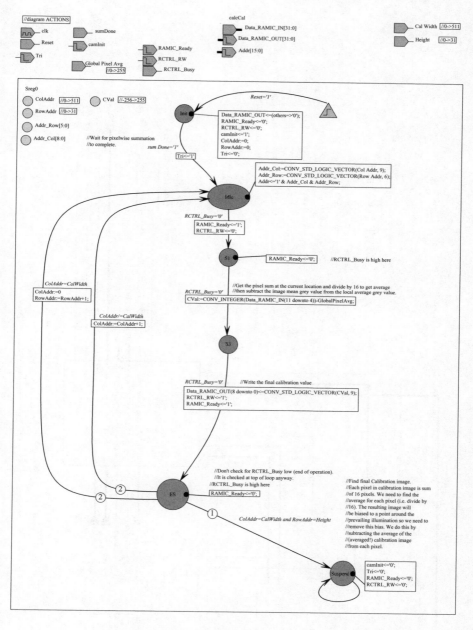

Figure C.12 calcCal *state machine – final part of calibration image generation – average pixel values and remove image bias (part of* Calibration *block in Section 7.2.4.3)*

C.4 Buffers

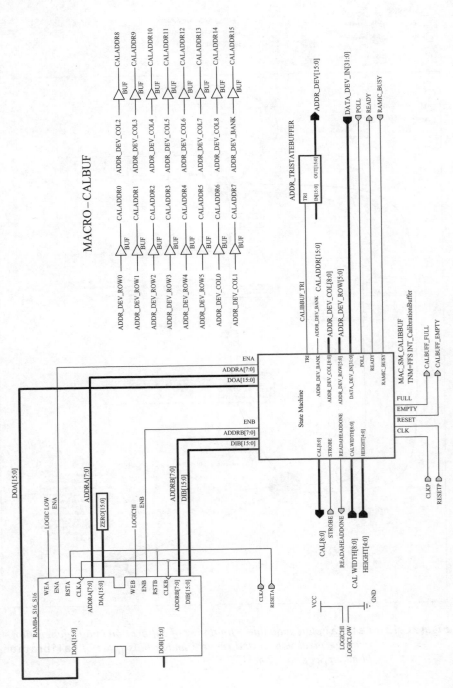

Figure C.13 Calibration buffer (CALBUFF) schematic (part of Calibration_Buffer in Section 7.2.5.1)

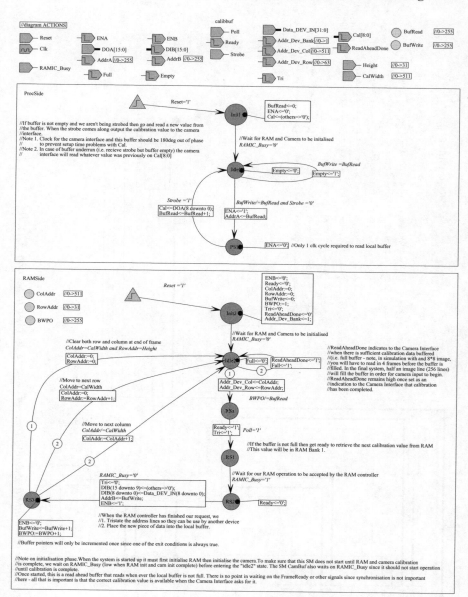

Figure C.14 Calibration buffer (CALIBBUF) state machine (part of Calibration_Buffer in Section 7.2.5.1)

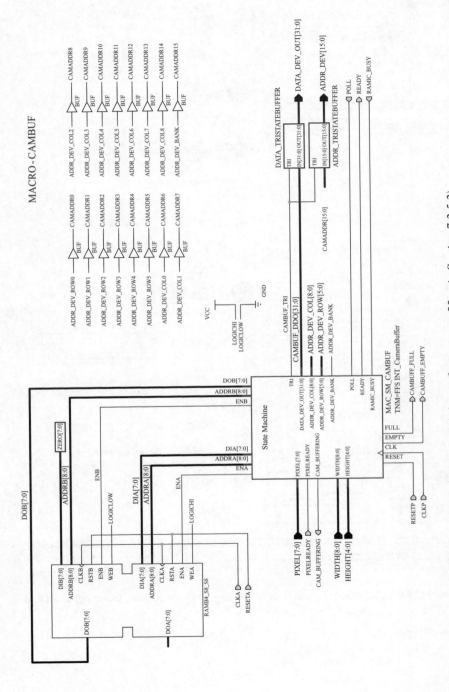

Figure C.15 Camera buffer schematic (CAMBUFF) (part of Camera_Buffer in Section 7.2.5.2)

Appendix C – FPGA design 321

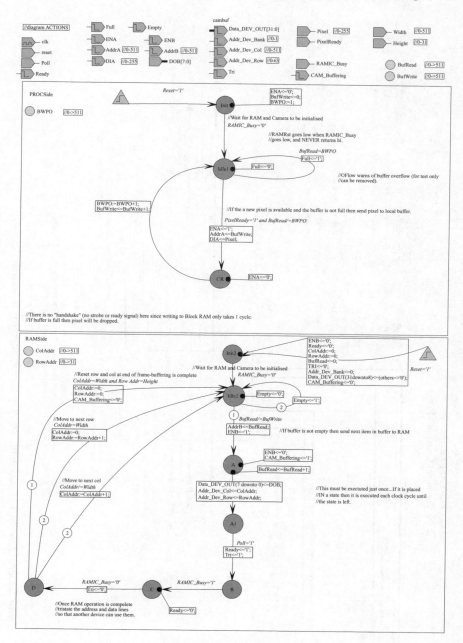

Figure C.16 Camera buffer (CAMBUF) state machine (part of Camera_Buffer in Section 7.2.5.2)

322 *Motion vision: design of compact motion sensing solutions*

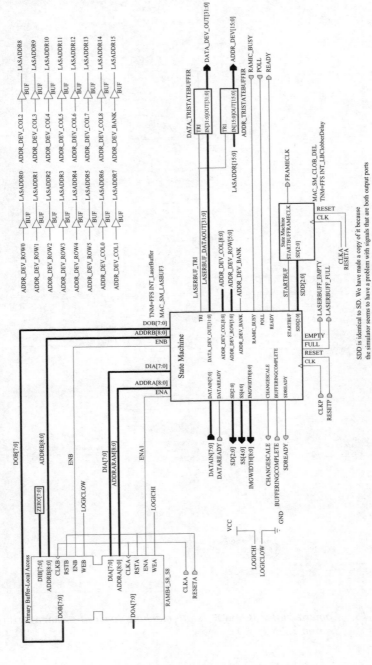

Figure C.17 *Laser buffer* (LASERBUF3) *schematic (part of* Range_Buffer *in Section 7.2.5.3)*

Appendix C – FPGA design 323

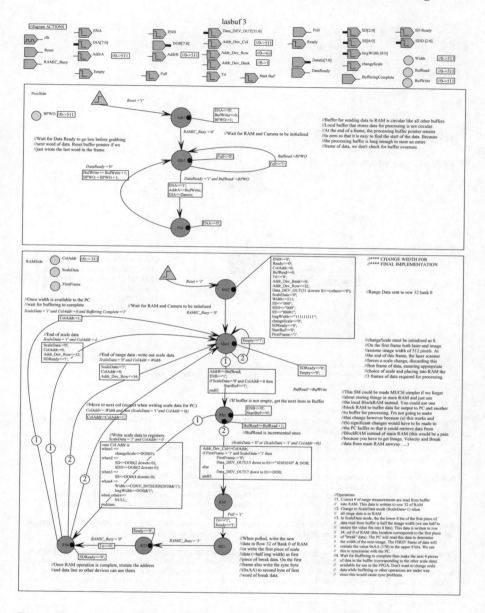

Figure C.18 Laser buffer (LASBUF3) state machine (part of Range_Buffer in Section 7.2.5.3)

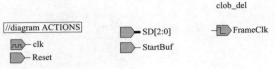

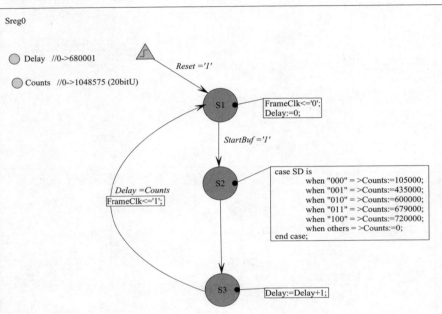

Output of range data to PC MUST occur after all new range data is read in. That is, output of range data must start a min of 19 ms after reading of new range data begins.

For SD=0, it takes 512*32*1μs = 16.384 ms for output of img data to PC, so output to PC should start 19-16.384=2.61 ms after range data reading begins. Rounding up slightly, this gives 105000 counts.

Similarly for   SD=1, Output time = 256*32*1us = 8.192ms ->>> 19-8.912 = 10.808ms ->>> 435000 counts
SD=2, Output time = 128*32*1us = 4.096ms ->>> 19-4.096 = 14.904ms ->>> 600000 counts
SD=3, Output time = 64*32*1us = 2.048ms ->>> 19-2.048 = 16.952ms ->>> 679000 counts
SD=4, Output time = 32*32*1us = 1.024ms ->>> 19-1.024 = 17.976ms ->>> 720000 counts

Note that extending/reducing these times does not change the amount of time available for processing. It simply changes where processing starts relative to the start of range data input (o in timing diagram).

Note that we assume that the time for the PC to start reading is from when it is signalled is 0 ns (see f in timing diagram). This time can be as high as 1.7 ms, but we assume here that it is zero for safety (don't want to assume some time and then have it violated). Besides, as I mentioned above, even if we take this time into account it will not change the amount of time available for processing. It will only change the relative delay between the start of range data input and the start of processing.

Figure C.19   Clob_Delay (clob_del) *state machine – implementation of clobber avoidance delay (part of* Range_Buffer *in Section 7.2.5.3). See Appendix B for more detail on the need for this component*

Appendix C – FPGA design 325

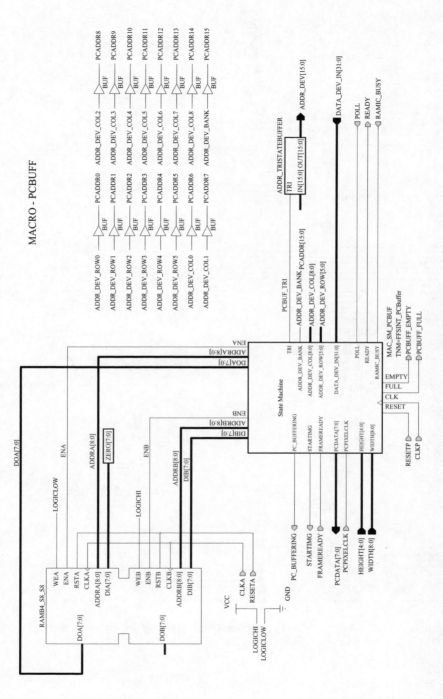

Figure C.20 PC buffer (PCBUFF) schematic (part of PC-Output_Buffer in Section 7.2.5.6)

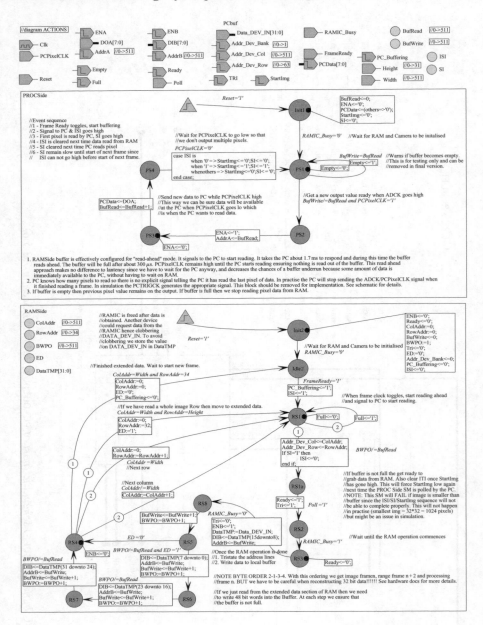

Figure C.21 PC buffer (PCBuf) state machine (part of PC-Output_Buffer in Section 7.2.5.6)

Appendix C – FPGA design 327

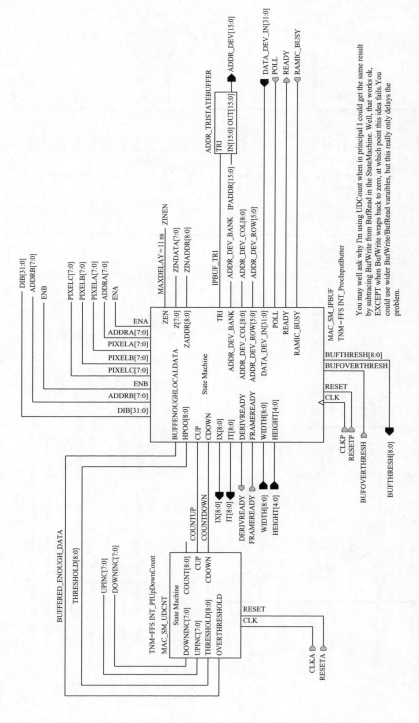

Figure C.22 Processing input buffer (IPBUFF) schematic – part 1 (part of Processing_Input_Buffer *in Section 7.2.5.4)*

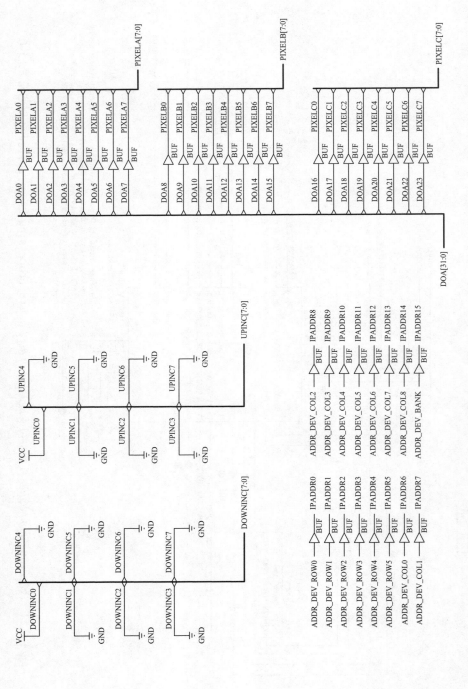

Figure C.23 Processing input buffer (IPBUFF) schematic – part 2 (part of Processing_Input_Buffer in Section 7.2.5.4)

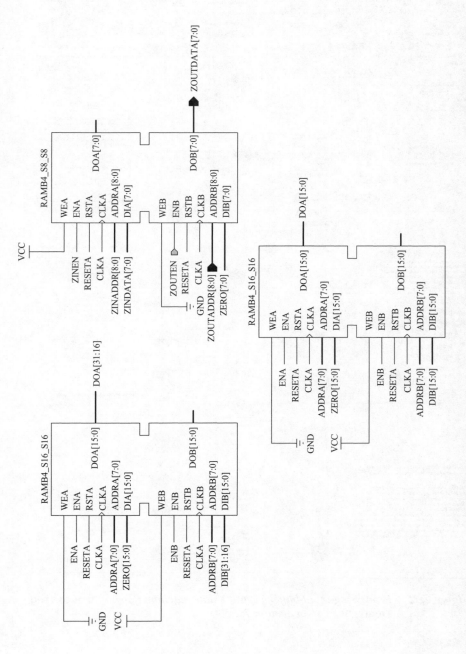

*Figure C.24   Processing input buffer (*IPBUFF*) schematic – part 3 (part of* Processing_Input_Buffer *in Section 7.2.5.4)*

330 *Motion vision: design of compact motion sensing solutions*

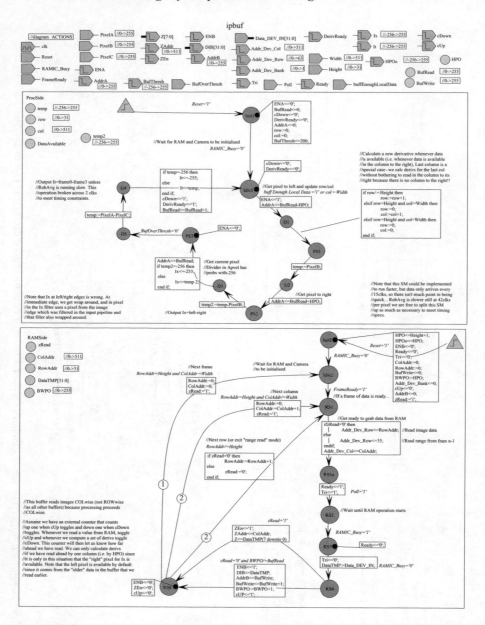

Figure C.25   *Processing input buffer (*IPBUF*) state machine (part of* Processing_Input_Buffer *in Section 7.2.5.4)*

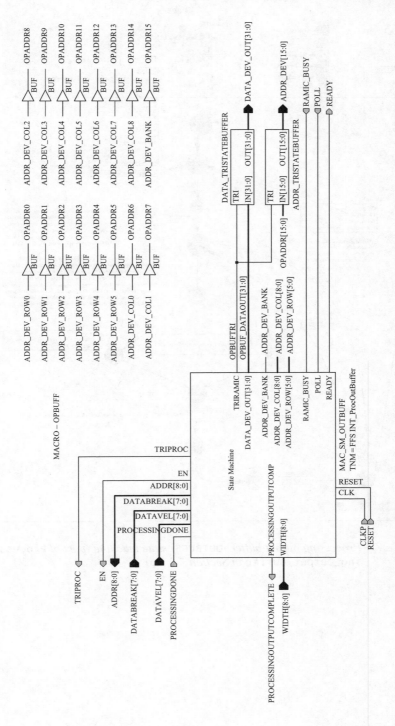

Figure C.26 Processing output buffer (`OPBUFFER`) schematic (part of `Processing_Output_Buffer` in Section 7.2.5.5)

332 *Motion vision: design of compact motion sensing solutions*

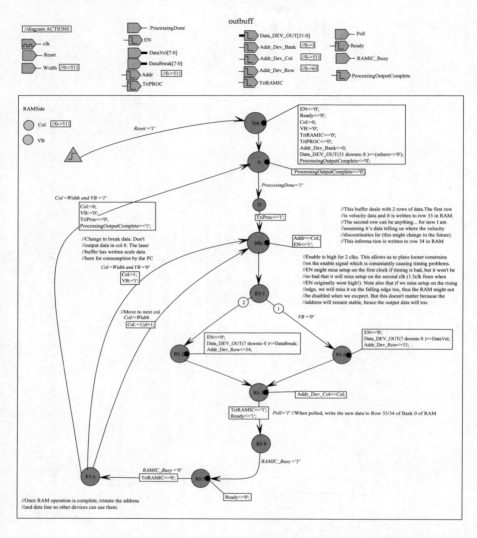

Figure C.27 *Processing output buffer (*OUTBUFF*) state machine (part of* Processing_Output_Buffer *in Section 7.2.5.5)*

C.4.1 Camera data path

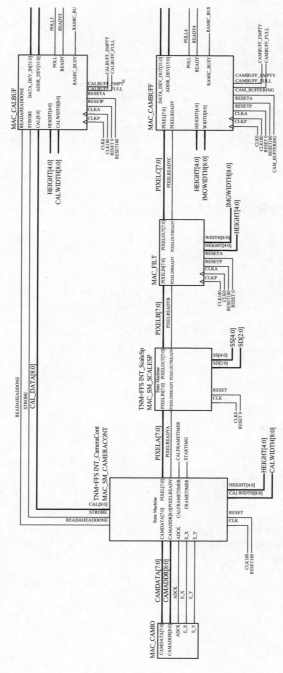

Figure C.28 Top level schematic – focus on the camera data path. See Figure 7.13 (Section 7.2.6.1) for a simplified block diagram

MACRO-CAMIO

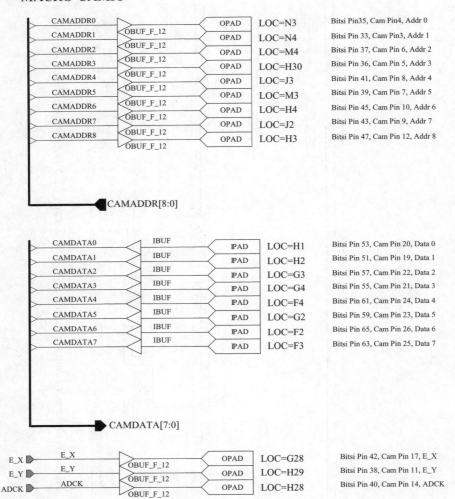

Figure C.29   Camera I/O (`CAMIO`*) schematic – physical connection from FPGA (*`CameraCont`*) to the FUGA camera (Part of* `Camera_Interface` *in Section 7.2.6.1)*

Appendix C – FPGA design 335

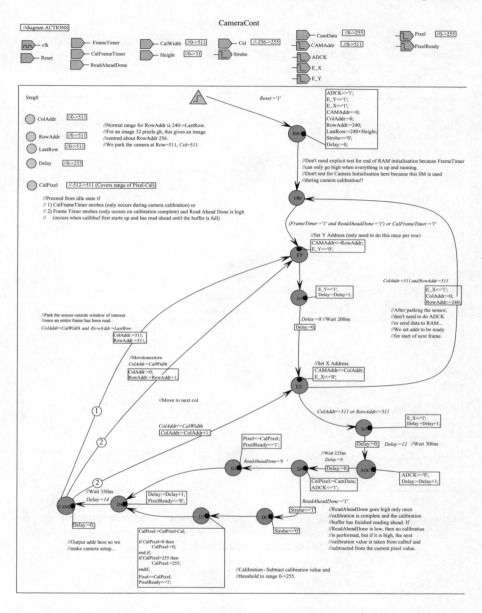

Figure C.30 Camera controller (`CameraCont`) state machine – implementation of an interface to the FUGA camera (part of `Camera_Interface` in Section 7.2.6.1).

336 *Motion vision: design of compact motion sensing solutions*

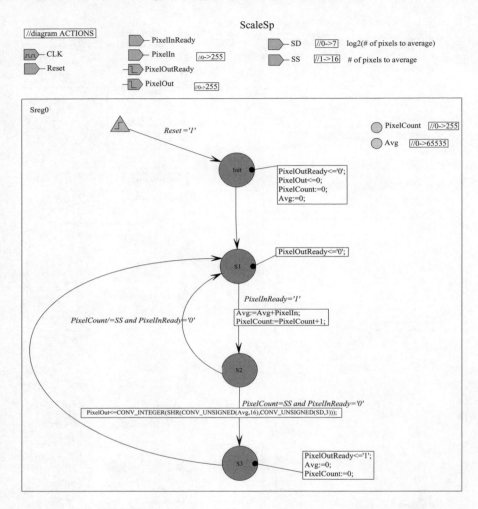

Figure C.31 *Scale space (*ScaleSp*) state machine – implementation of the image subsampling required for dynamic scale space (called* Dynamic_Scale_Space *in Section 7.2.6.1)*

Appendix C – FPGA design 337

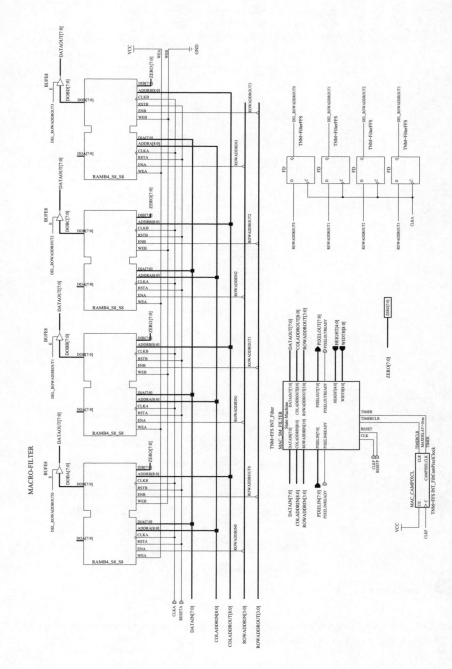

Figure C.32 Filter (FILT) schematic – schematic illustrating the implementation of a simple image filter (part of Filter block in Section 7.2.6.1)

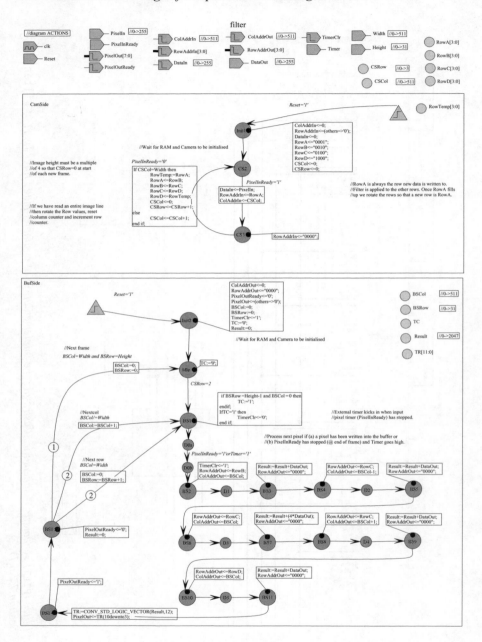

Figure C.33   Filter (filter) state machine – core of the image filter implementation (part of Filter block in Section 7.2.6.1)

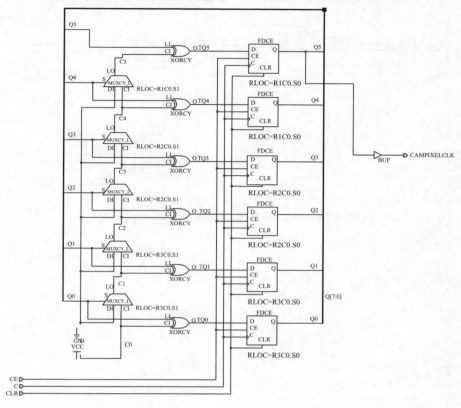

Figure C.34 Pixel clock (`CAMPIXCL`) schematic – auxiliary pixel clock used by the image filter (part of `Filter` block in Section 7.2.6.1)

C.5 Laser and PC data paths

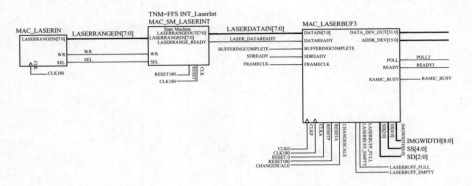

Figure C.35   Top level schematic – focusing on laser data path

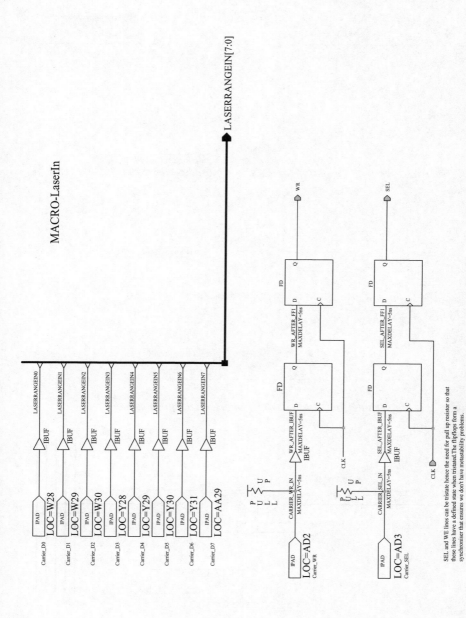

Figure C.36 *Laser input (LASERIN) schematic – physical connection from FPGA (laserint) to range sensor*

342 *Motion vision: design of compact motion sensing solutions*

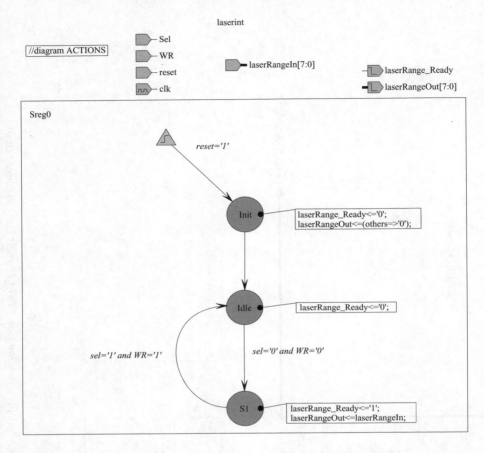

*Figure C.37   Laser interface (*LASERINT*) – interface to range sensor*

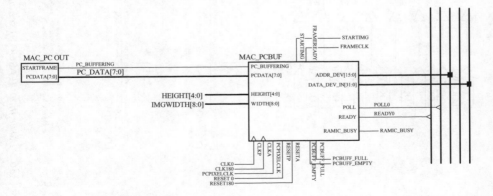

Figure C.38   Top level schematic – focus on PC output data path

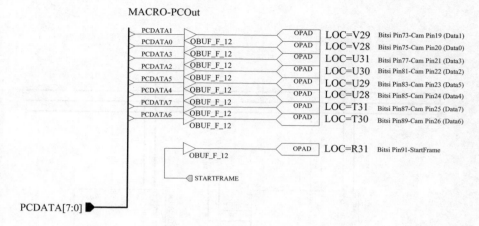

Figure C.39 PC output (PCOUT) schematic – physical connection from FPGA (PCBUFF) to PC

C.6 Processing

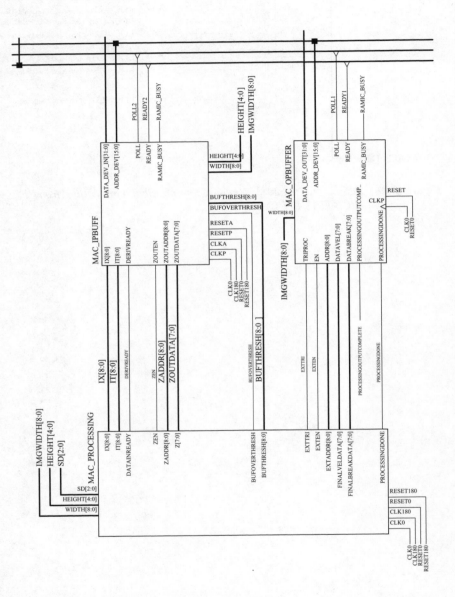

Figure C.40 *Top level schematic – focus on processing block (called* Motion Estimation Data Path *in Section 7.2.6.2)*

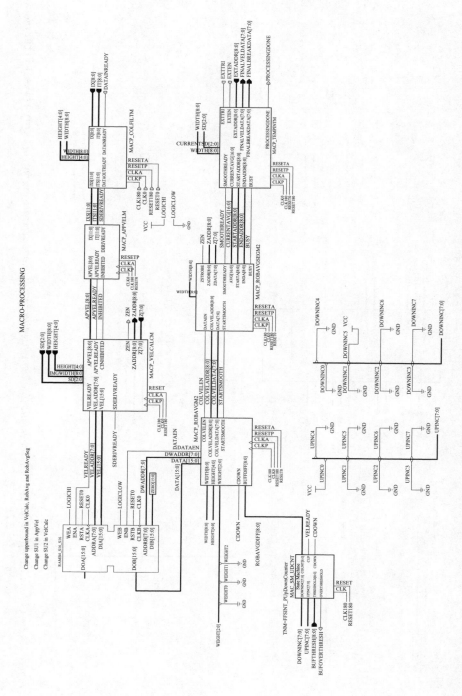

Figure C.41 *Processing block top level schematic (called* Motion Estimation Data *Path in Section 7.2.6.2)*

346 Motion vision: design of compact motion sensing solutions

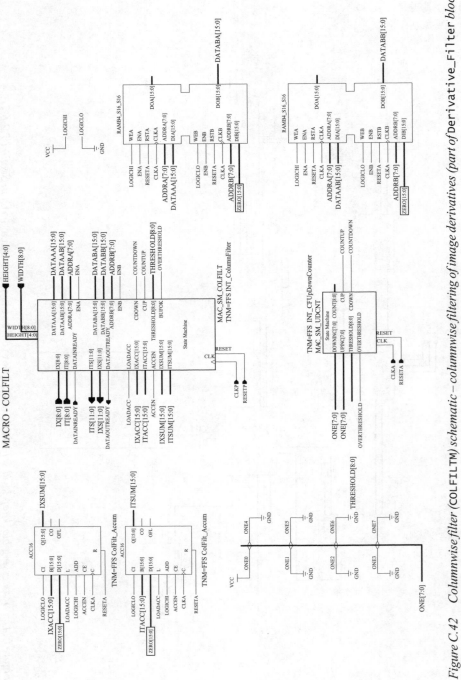

Figure C.42 Columnwise filter (COLFLTM) schematic – columnwise filtering of image derivatives (part of Derivative_Filter block in Section 7.2.6.2.1)

Appendix C – FPGA design 347

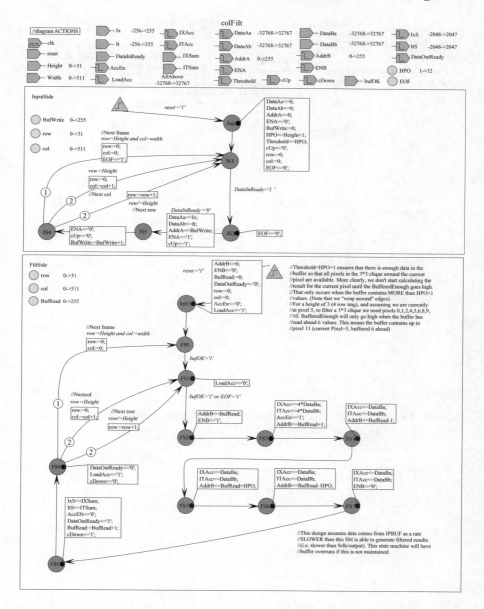

Figure C.43 Columnwise filter (colFilt) state machine – implementation of image derivative filtering (part of Derivative_Filter block in Section 7.2.6.2.1)

348 *Motion vision: design of compact motion sensing solutions*

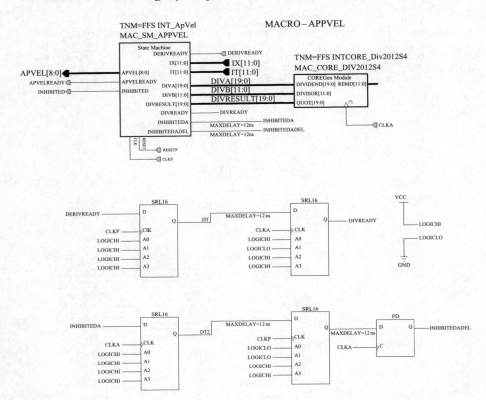

Figure C.44 *Apparent velocity calculation (`APPVELM`) schematic (part of `Optical_Flow` block in Section 7.2.6.2.2)*

Appendix C – FPGA design 349

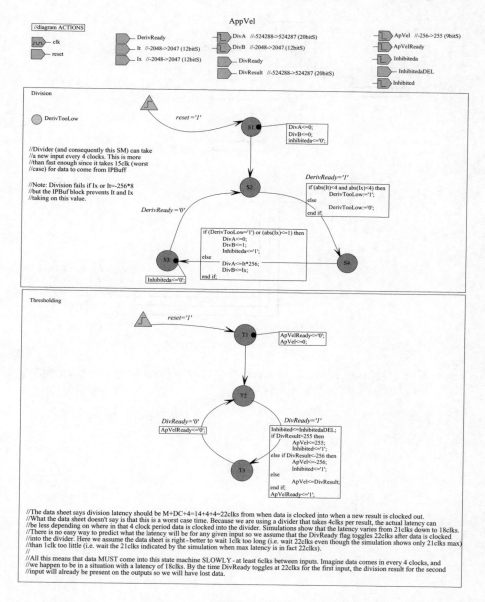

Figure C.45 *Apparent velocity calculation (*`APPVEL`*) state machine (part of* `Optical_Flow` *block in Section 7.2.6.2.2)*

350 *Motion vision: design of compact motion sensing solutions*

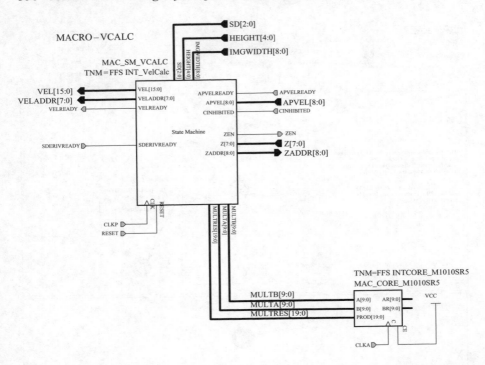

Figure C.46   *Velocity calculation block (*VCALCM*) schematic (part of* velocity *block in Section 7.2.6.2.2)*

Appendix C – FPGA design

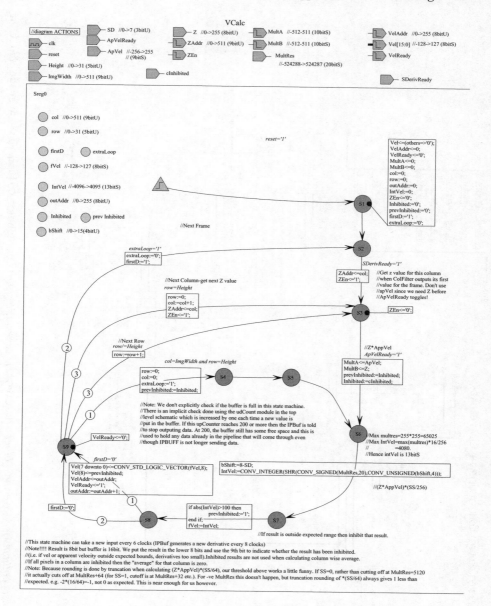

Figure C.47 *Velocity calculation block (*`VCalc`*) state machine (part of* `Velocity` *block in Section 7.2.6.2.2)*

352 Motion vision: design of compact motion sensing solutions

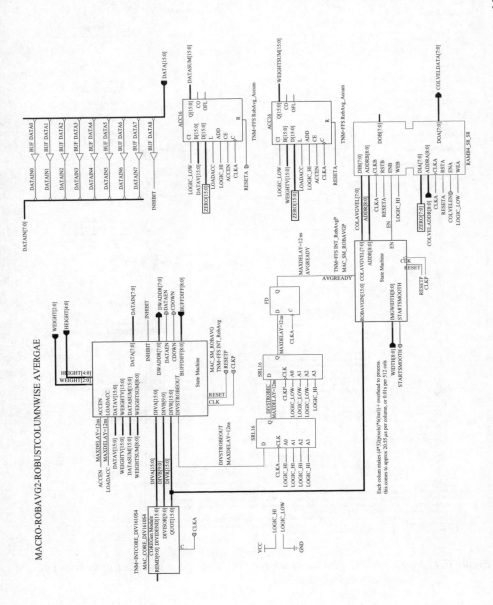

Figure C.48 Robust columnwise average block (ROVAVGM2) schematic (part of Robust_Column-wise_Average block in Section 7.2.6.2.3)

Appendix C – FPGA design 353

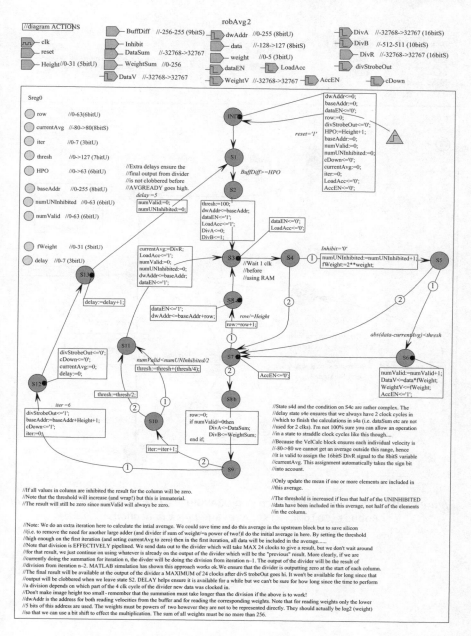

Figure C.49 Robust columnwise average block (`ROBAVG2`) state machine (part of `Robust_Column-wise_Average` block in Section 7.2.6.2.3)

354 *Motion vision: design of compact motion sensing solutions*

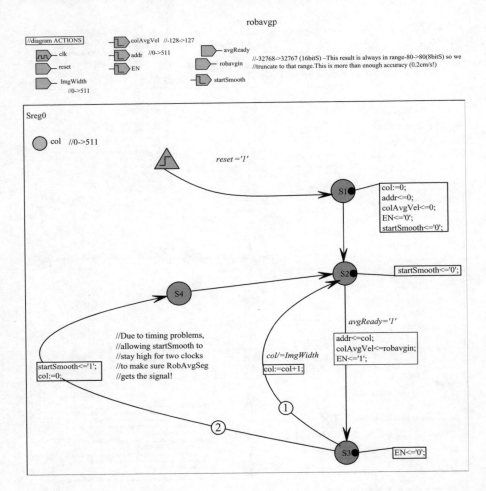

*Figure C.50 Interblock interface (*ROBAVGP*) state machine (part of* `Robust_Column-wise_Average` *block in Section 7.2.6.2.3)*

Appendix C – FPGA design 355

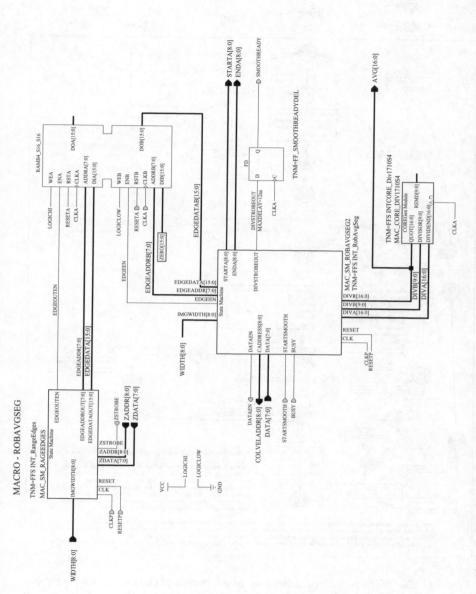

Figure C.51 Blobwise robust averaging block (ROBAVGSEGM2) schematic (part of Robust_smoothing block in Section 7.2.6.2.3)

356 *Motion vision: design of compact motion sensing solutions*

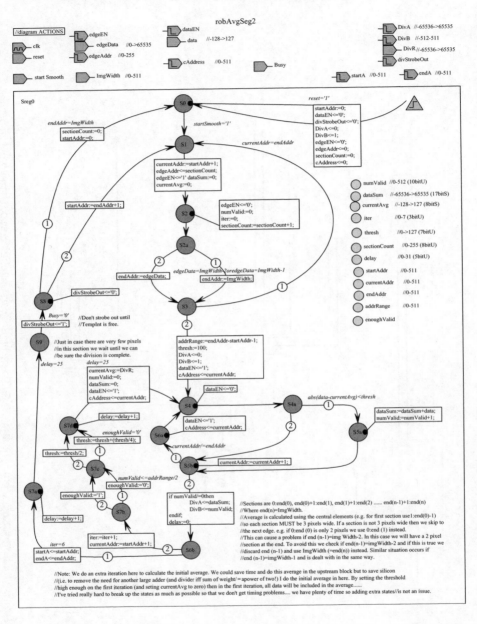

Figure C.52 Blobwise robust averaging block (ROBAVGSEG2) state machine (part of Robust_Smoothing block in Section 7.2.6.2.3)

Appendix C – FPGA design 357

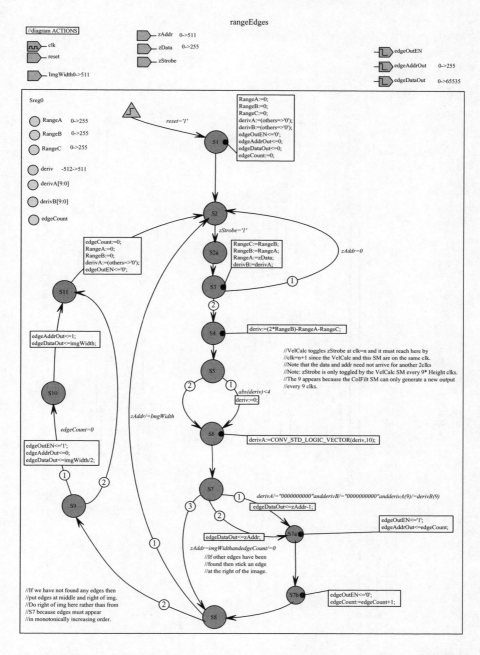

Figure C.53 Calculation of range edges (RangeEdges) state machine (called Zero_Crossings in Section 7.2.6.2.3)

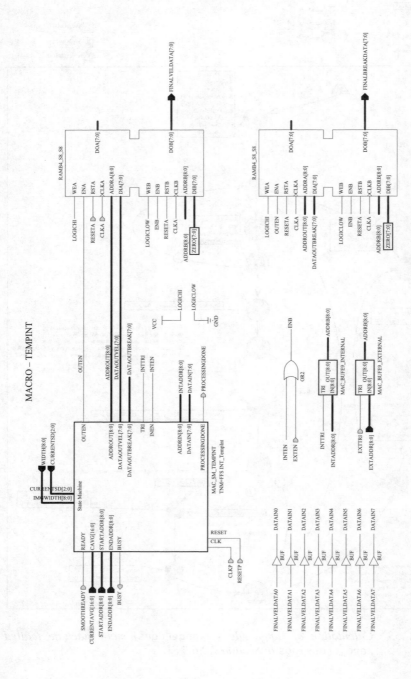

Figure C.54 Temporal integration block (TEMPINTM) schematic (part of Temporal_Integration block in Section 7.2.6.2.4)

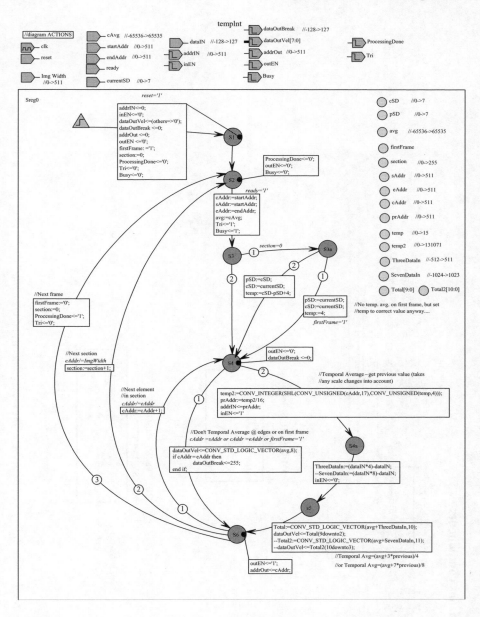

Figure C.55   Temporal integration block (`TempInt`) state machine (part of `Temporal_Integration` block in Section 7.2.6.2.4)

C.7 Miscellaneous components

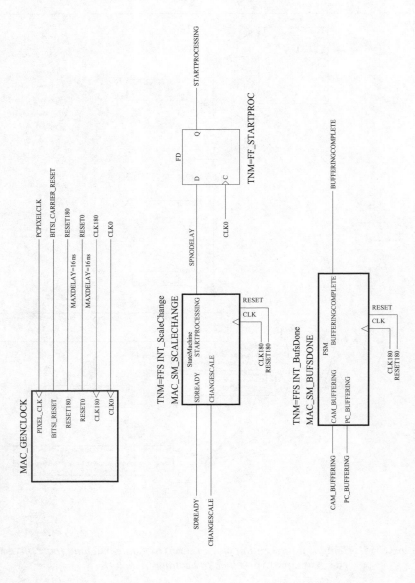

Figure C.56 Top level schematic – focus on some of the miscellaneous components

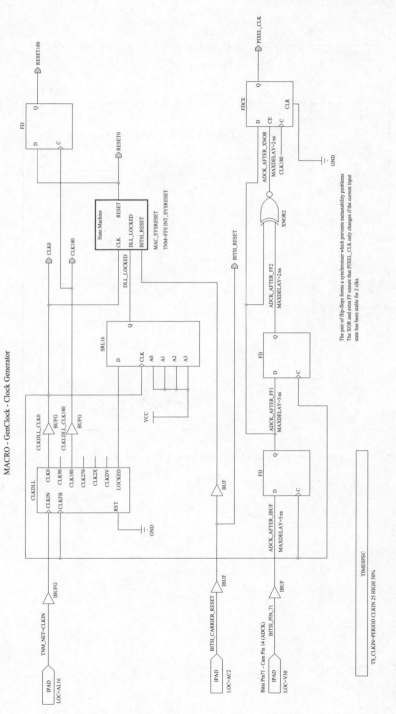

Figure C.57 Clock generation (GENCLK) schematic

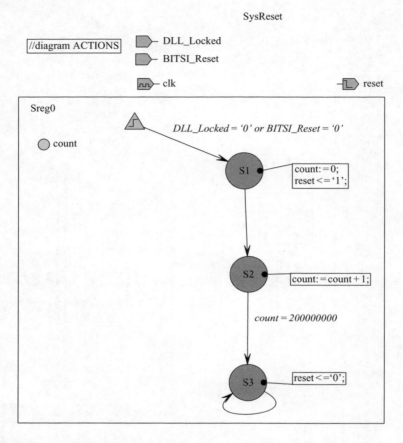

Figure C.58 Systems reset generation (SysReset) state machine

Appendix C – FPGA design   363

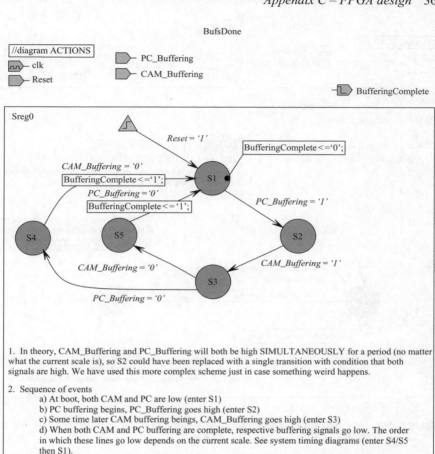

Figure C.59   BufsDone state machine – signal completion of I/O buffering

364 *Motion vision: design of compact motion sensing solutions*

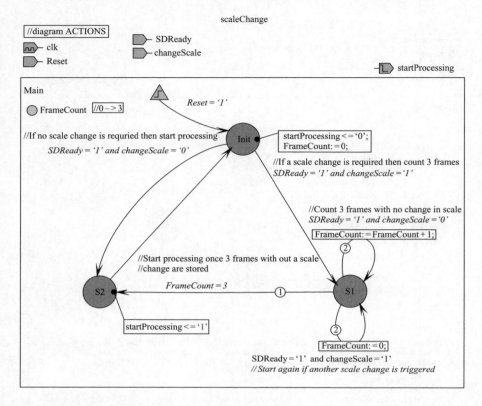

Figure C.60 scaleChange *state machine – monitor the* ChangeScale *bit to indicate if processing should begin*

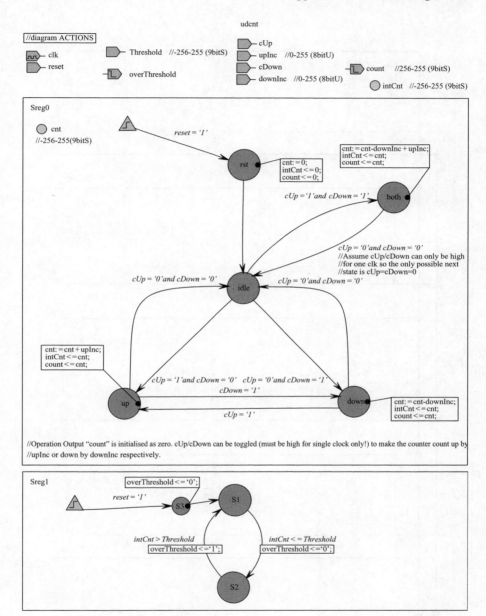

Figure C.61   UDCNT *state machine – block used to monitor the state of local buffers*

366 *Motion vision: design of compact motion sensing solutions*

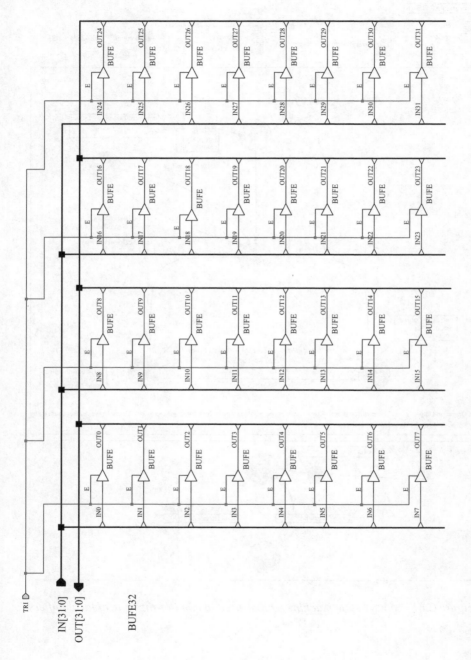

Figure C.62   *BUFE32 schematic – 32 bit tristate buffer, active low tristate signal*

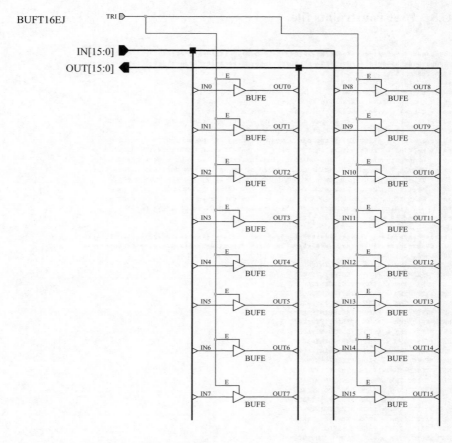

Figure C.63 BUF16EJ *schematic – 16 bit tristate buffer, active low tristate signal*

C.8 User constraints file

```
################################################################################
################################################################################
## Notes on the design
##
## 1) SDRAM - Inputs have 1.5ns (min) setup time, 1ns (min) hold time requirement
##             Output data hold time 2.5ns (min)
## 2) CAMERA - Asyncronous, timing as per camera data sheet
## 3) PC CAPTURE - Timing as per camera data sheet. All timing relative to ADCK.
## 4) DSP - Timing as per DSP RAM interface.
## 5) There are effectively two parts to the timing of the design. For a state
##    machine we must ensure that it does not take longer than 1clk for data to
##    propagate from flip-flop to flip-flop. Externally we must ensure the delay
##    from a SM output flip-flop to BlockRAM or to another SM does not exceed 0.5clk.
## 6) Input and ouput pads must be dealt with explicitly. For SDRAM we must
##    meet setup/hold times etc. relative to our 40MHz system clk. For PC we
##    base IO timing on the ADCK signal. For the camera we explicitly generate
##    all timing in accordance with the data sheet. DSP timing is as per the
##    DSP data sheet RAM IO specification.


################################################################################
################################################################################
## GLOBAL CONSTRAINTS
NET "TP0"  TNM = PADS "TESTPOINTPADS";
NET "TP1"  TNM = PADS "TESTPOINTPADS";
NET "TP2"  TNM = PADS "TESTPOINTPADS";
NET "TP3"  TNM = PADS "TESTPOINTPADS";
NET "TP4"  TNM = PADS "TESTPOINTPADS";
NET "TP5"  TNM = PADS "TESTPOINTPADS";
NET "TP6"  TNM = PADS "TESTPOINTPADS";
NET "TP7"  TNM = PADS "TESTPOINTPADS";
NET "TP8"  TNM = PADS "TESTPOINTPADS";
NET "TP9"  TNM = PADS "TESTPOINTPADS";
NET "TP10" TNM = PADS "TESTPOINTPADS";
NET "TP11" TNM = PADS "TESTPOINTPADS";
NET "TP12" TNM = PADS "TESTPOINTPADS";
NET "TP13" TNM = PADS "TESTPOINTPADS";
NET "TP14" TNM = PADS "TESTPOINTPADS";
NET "TP15" TNM = PADS "TESTPOINTPADS";
NET "TP16" TNM = PADS "TESTPOINTPADS";
NET "TP17" TNM = PADS "TESTPOINTPADS";
NET "TP18" TNM = PADS "TESTPOINTPADS";
NET "TP19" TNM = PADS "TESTPOINTPADS";
NET "TP20" TNM = PADS "TESTPOINTPADS";
NET "TP21" TNM = PADS "TESTPOINTPADS";
NET "TP22" TNM = PADS "TESTPOINTPADS";
NET "TP23" TNM = PADS "TESTPOINTPADS";
NET "TP24" TNM = PADS "TESTPOINTPADS";
NET "TP25" TNM = PADS "TESTPOINTPADS";
NET "TP26" TNM = PADS "TESTPOINTPADS";
NET "TP27" TNM = PADS "TESTPOINTPADS";
NET "TP28" TNM = PADS "TESTPOINTPADS";
NET "TP29" TNM = PADS "TESTPOINTPADS";
NET "TP30" TNM = PADS "TESTPOINTPADS";
NET "TP31" TNM = PADS "TESTPOINTPADS";
NET "TP32" TNM = PADS "TESTPOINTPADS";
NET "TP33" TNM = PADS "TESTPOINTPADS";
NET "TP34" TNM = PADS "TESTPOINTPADS";
NET "TP35" TNM = PADS "TESTPOINTPADS";
NET "TP36" TNM = PADS "TESTPOINTPADS";
NET "TP37" TNM = PADS "TESTPOINTPADS";
NET "TP38" TNM = PADS "TESTPOINTPADS";
NET "TP39" TNM = PADS "TESTPOINTPADS";
NET "TP40" TNM = PADS "TESTPOINTPADS";
NET "MAC_PROCESSING/TPPROC0" TNM = PADS "TESTPOINTPADS";
NET "MAC_PROCESSING/TPPROC1" TNM = PADS "TESTPOINTPADS";
NET "MAC_PROCESSING/TPPROC2" TNM = PADS "TESTPOINTPADS";
NET "MAC_PROCESSING/TPPROC3" TNM = PADS "TESTPOINTPADS";
```

Appendix C – FPGA design

```
NET "MAC_PROCESSING/TPPROC4"  TNM = PADS "TESTPOINTPADS";
NET "MAC_PROCESSING/TPPROC5"  TNM = PADS "TESTPOINTPADS";
NET "MAC_PROCESSING/TPPROC6"  TNM = PADS "TESTPOINTPADS";
NET "MAC_PROCESSING/TPPROC7"  TNM = PADS "TESTPOINTPADS";
NET "MAC_PROCESSING/TPPROC8"  TNM = PADS "TESTPOINTPADS";
NET "MAC_PROCESSING/TPPROC9"  TNM = PADS "TESTPOINTPADS";
NET "MAC_PROCESSING/TPPROC10" TNM = PADS "TESTPOINTPADS";
NET "MAC_PROCESSING/TPPROC11" TNM = PADS "TESTPOINTPADS";
NET "MAC_PROCESSING/TPPROC12" TNM = PADS "TESTPOINTPADS";

# NOTE - AT END OF FILE ALL ABOVE NETS ARE SET TO BE IGNORED FOR TIMING
#        DOING THIS IS SIMPLER THAN EXPLICITLY EXCLUDING THE PADS IN INDIVIDUAL
#            CONSTRAINTS.

TIMEGRP "USEFULPADS" = PADS except "TESTPOINTPADS";

# Constrain the RESET nets. Note that there is a reset net for each clk to allow max propagation time
NET "RESET0" TNM = "RESET0_SINKS";
TIMESPEC "TS_RESET0" =           TO  "RESET0_SINKS"          24.9 ns;

NET "RESET180" TNM = "RESET180_SINKS";
TIMESPEC "TS_RESET180" =         TO  "RESET180_SINKS"        24.9 ns;

################################################################################
################################################################################
##CONSTRAINTS FOR INSIDE STATE MACHINES
## Note - I allow just a bit less than 1clk for FF to FF delay within a state
##        machine. I could have made it 25ns, but I use 24.9ns for added confidence.

## -- SDRAM Interface and Calibration
TIMESPEC "TSRAMICMem" =       FROM  "INT_RAMICMem"      TO  "INT_RAMICMem"         24.9 ns;
TIMESPEC "TSZeroRam" =        FROM  "INT_ZeroRam"       TO  "INT_ZeroRam"          24.9 ns;
TIMESPEC "TSSumPixels" =      FROM  "INT_SumPixels"     TO  "INT_SumPixels"        24.9 ns;
TIMESPEC "TSCalcCal" =        FROM  "INT_CalcCal"       TO  "INT_CalcCal"          24.9 ns;
TIMESPEC "TSRAMICPol" =       FROM  "INT_RAMICPol"      TO  "INT_RAMICPol"         24.9 ns;
TIMESPEC "TSRAMTimer" =       FROM  "INT_RAMTimer"      TO  "INT_RAMTimer"         24.9 ns;
TIMESPEC "TSSDRAMInt" =       FROM  "INT_SDRAMInt"      TO  "INT_SDRAMInt"         24.9 ns;

## -- Buffers
TIMESPEC "TSCalBuffer" =      FROM  "INT_CalibrationBuffer" TO  "INT_CalibrationBuffer" 24.9 ns;
TIMESPEC "TSCameraBuffer" =   FROM  "INT_CameraBuffer"  TO  "INT_CameraBuffer"     24.9 ns;
TIMESPEC "TSLaserBuffer" =    FROM  "INT_LaserBuffer"   TO  "INT_LaserBuffer"      24.9 ns;
TIMESPEC "TSLBClobberDelay" = FROM  "INT_LBClobberDelay" TO  "INT_LBClobberDelay"  24.9 ns;
TIMESPEC "TSProcInputBuffer" = FROM "INT_ProcInputBuffer" TO "INT_ProcInputBuffer" 24.9 ns;
TIMESPEC "TSPIupDownCounter" = FROM "INT_PIUpDownCount" TO  "INT_PIUpDownCount"    24.9 ns;
TIMESPEC "TSProcOutBuffer" =  FROM  "INT_ProcOutBuffer" TO  "INT_ProcOutBuffer"    24.9 ns;
TIMESPEC "TSPCBuffer" =       FROM  "INT_PCBuffer"      TO  "INT_PCBuffer"         24.9 ns;

## -- Camera data path
TIMESPEC "TSCamCont" =        FROM  "INT_CameraCont"    TO  "INT_CameraCont"       24.9 ns;
TIMESPEC "TSScaleSp" =        FROM  "INT_ScaleSp"       TO  "INT_ScaleSp"          24.9 ns;
TIMESPEC "TSFiltCamPixelClk" = FROM "INT_FiltCamPixelClock" TO "INT_FiltCamPixelClock" 24.9 ns;
TIMESPEC "TSFilter" =         FROM  "INT_Filter"        TO  "INT_Filter"           24.9 ns;

## -- Laser data path
TIMESPEC "TSLaserInt" =       FROM  "INT_LaserInt"      TO  "INT_LaserInt"         24.9 ns;

## -- Processing data path
TIMESPEC "TSPUDCnt" =         FROM  "INT_PUpDownCounter" TO "INT_PUpDownCounter"   24.9 ns;
TIMESPEC "TSColFilt" =        FROM  "INT_ColumnFilter"  TO  "INT_ColumnFilter"     24.9 ns;
TIMESPEC "TSColFiltUPCnt" =   FROM  "INT_CFUpDownCounter" TO "INT_CFUpDownCounter" 24.9 ns;
TIMESPEC "TSApVel" =          FROM  "INT_ApVel"         TO  "INT_ApVel"            24.9 ns;
TIMESPEC "TSVelCalc" =        FROM  "INT_VelCalc"       TO  "INT_VelCalc"          24.9 ns;
TIMESPEC "TSRobAvg" =         FROM  "INT_RobAvg"        TO  "INT_RobAvg"           24.9 ns;
TIMESPEC "TSRobAvgP" =        FROM  "INT_RobAvgP"       TO  "INT_RobAvgP"          24.9 ns;
TIMESPEC "TSRangeEdges" =     FROM  "INT_RangeEdges"    TO  "INT_RangeEdges"       24.9 ns;
TIMESPEC "TSRobAvgSeg" =      FROM  "INT_RobAvgSeg"     TO  "INT_RobAvgSeg"        24.9 ns;
TIMESPEC "TSTempInt" =        FROM  "INT_TempInt"       TO  "INT_TempInt"          24.9 ns;

## -- Misc
TIMESPEC "TSReset" =          FROM  "INT_SYSRESET"      TO  "INT_SYSRESET"         24.9 ns;
```

```
TIMESPEC "TSBufsDone" =            FROM  "INT_BufsDone"        TO  "INT_BufsDone"           24.9 ns;
TIMESPEC "TSScaleChange" =         FROM  "INT_ScaleChange"     TO  "INT_ScaleChange"        24.9 ns;

TIMESPEC "TS_ApVel_Div" =          FROM  "INTCORE_Div2012S4"   TO  "INTCORE_Div2012S4"      24.9 ns;
TIMESPEC "TS_VelCalc_Mult" =       FROM  "INTCORE_M1010SR5"    TO  "INTCORE_M1010SR5"       24.9 ns;
TIMESPEC "TS_RobAvg_Div" =         FROM  "INTCORE_DIV1610S4"   TO  "INTCORE_DIV1610S4"      24.9 ns;
TIMESPEC "TS_RobAvgSeg_Div" =      FROM  "INTCORE_Div1710S4"   TO  "INTCORE_Div1710S4"      24.9 ns;

##############################################################################
##############################################################################
## Constraints between state machines
## Explicitly define constraints in both directions
## Do not use the syntax with missing too because that would apply to delays within
## the SM, which have already been constrained above.

###############################
####### SDRAMInt Connections

NET "RAMIC_READY" TNM = "SDRAMInt_Control";
NET "RCTRL_RW" TNM = "SDRAMInt_Control";
TIMEGRP "SDRAMInt_except_control" = "INT_SDRAMInt" except "SDRAMInt_Control";
TIMEGRP "CalibrationBits" = "INT_ZeroRam" "INT_SumPixels" "INT_CalcCal";

# SDRAMInt and SDRAM are on same clock. We use a constraint significantly less than 25 ns to ensure
# net delays and outside the FPGA don't cause problems.

# Allow for propagation outside FPGA
TIMESPEC "TS_SDRAMInt_SDRAM" =         FROM  "INT_SDRAMInt"    TO  "USEFULPADS"             16 ns;
TIMESPEC "TS_SDRAM_RAMIC" =            FROM  "USEFULPADS"      TO  "INT_RAMICMem"           16 ns;
TIMESPEC "TS_SDRAM_CALIB" =            FROM  "USEFULPADS"      TO  "CalibrationBits"        16 ns;

# Timer and SDRAMInt are on same clk
TIMESPEC "TS_RAMTimer_SDRAMINT"=       FROM  "INT_RAMTimer"    TO  "INT_SDRAMInt"           24.9 ns;
# RCTRL_Busy
TIMESPEC "TS_SDRAMInt_RAMICMEM"=       FROM  "INT_SDRAMInt"    TO  "INT_RAMICMem"           12.4 ns;
# RAMIC_Ready/RCTRL_RW - 0.5clk
TIMESPEC "TS_RAMICMEM_SDRAMIntCont"=   FROM  "INT_RAMICMem"    TO  "SDRAMInt_Control"       12.4 ns;
# Data/addr lines - not used for a bit - 1clk.
TIMESPEC "TS_RAMICMEM_SDRAMIntAddr"=   FROM  "INT_RAMICMem"    TO  "SDRAMInt_except_control" 24.9 ns;
TIMESPEC "TS_SDRAMInt_Calib" =         FROM  "INT_SDRAMInt"    TO  "CalibrationBits"        12.4 ns;
# RAMIC_Ready/RCTRL_RW - 0.5clk
TIMESPEC "TS_Calib_Control_SDRAMInt" = FROM  "CalibrationBits" TO  "SDRAMInt_Control"       12.4 ns;
# Data/addr lines - not used for a bit - 1clk.
TIMESPEC "TS_Calib_AddrData_SDRAMInt" = FROM "CalibrationBits" TO  "SDRAMInt_except_control" 24.9 ns;

###############################
####### Calibration Internal

# Same clk
TIMESPEC "TS_Zero_Sum" =               FROM  "INT_ZeroRam"     TO  "INT_SumPixels"          24.9 ns;
# Same clk
TIMESPEC "TS_Sum_Calc" =               FROM  "INT_SumPixels"   TO  "INT_CalcCal"            24.9 ns;
# Same clk
TIMESPEC "TS_Calc_RAMICMEM" =          FROM  "INT_CalcCal"     TO  "INT_RAMICMem"           24.9 ns;

###############################
####### RAMICMem/RAMICPol

TIMESPEC "TS_RAMICMem_RAMICPol"=       FROM  "INT_RAMICMem"    TO  "INT_RAMICPol"           12.4 ns;
TIMESPEC "TS_RAMICPol_RAMICMem"=       FROM  "INT_RAMICPol"    TO  "INT_RAMICMem"           12.4 ns;

TIMEGRP "Buffers" = "INT_CalibrationBuffer" "INT_CameraBuffer" "INT_LaserBuffer" "INT_ProcInputBuffer"
"INT_ProcOutBuffer" "INT_PCBuffer";

# Since both RAMICMem and Buffers run on same clock, a 25 ns constraint is appropriate
# RAMIC_Busy and DATA_DEV_IN Bus -
TIMESPEC "TS_RAMICMem_Buffers"=        FROM  "INT_RAMICMem"    TO  "Buffers"                24.9 ns;
# DATA_DEV_OUT and ADDR_DEV buses.
TIMESPEC "TS_Buffers_RAMICMem"=        FROM  "Buffers"         TO  "INT_RAMICMem"           24.9 ns;
```

```
# POLL bus
TIMESPEC "TS_POLLBUS" =                      FROM  "INT_RAMICPol"          TO  "Buffers"                    12.4ns;
# READY bus
TIMESPEC "TS_READYBUS" =                     FROM  "Buffers"               TO  "INT_RAMICPol"               12.4ns;


###############################
####### Buffers

INST "MAC_CALBUF/$RAM_FIFO" TNM = "CALBUFFERRAM";
TIMESPEC "TS_CALBUF_RAM" =         FROM  "INT_CalibrationBuffer"  TO  "CALBUFFERRAM"              12.4ns;
TIMESPEC "TS_RAM_CALBUF" =         FROM  "CALBUFFERRAM"           TO  "INT_CalibrationBuffer"     12.4ns;

INST "MAC_CAMBUFF/$RAM_FIFO" TNM = "CAMBUFFERRAM";
TIMESPEC "TS_CAMBUF_RAM" =         FROM  "INT_CameraBuffer"       TO  "CAMBUFFERRAM"              12.4ns;
TIMESPEC "TS_RAM_CAMBUF" =         FROM  "CAMBUFFERRAM"           TO  "INT_CameraBuffer"          12.4ns;

INST "MAC_LASERBUF3/$RAM_FIFO" TNM = "LASERBUFFERRAM";
TIMESPEC "TS_LASERBUF_RAM" =       FROM  "INT_LaserBuffer"        TO  "LASERBUFFERRAM"            12.4ns;
TIMESPEC "TS_RAM_LASERBUF" =       FROM  "LASERBUFFERRAM"         TO  "INT_LaserBuffer"           12.4ns;
TIMESPEC "TS_LaserBuf_ClobDel"=    FROM  "INT_LaserBuffer"        TO  "INT_LBClobberDelay"        12.4ns;

INST "MAC_IPBUFF/$RAM_FIFO_LSBDATA"    TNM = "INPUTBUFFERRAM";
INST "MAC_IPBUFF/$RAM_FIFO_MSBDATA"    TNM = "INPUTBUFFERRAM";
INST "MAC_IPBUFF/$RAM_FIFO_RANGEDATA"  TNM = "INPUTBUFFERRAM";
TIMESPEC "TS_INPUTBUF_RAM" =       FROM  "INT_ProcInputBuffer"    TO  "INPUTBUFFERRAM"            12.4ns;
TIMESPEC "TS_RAM_INPUTBUF" =       FROM  "INPUTBUFFERRAM"         TO  "INT_ProcInputBuffer"       12.4ns;
TIMESPEC "TS_InputBuf_PUDCount"=   FROM  "INT_ProcInputBuffer"    TO  "INT_PIUpDownCount"         12.4ns;
TIMESPEC "TS_PUDCount_InputBuf"=   FROM  "INT_PIUpDownCount"      TO  "INT_ProcInputBuffer"       12.4ns;

# Processing input buffer to column filter (Ix/It connection)
TIMESPEC "TS_PIBuffer_ColFilt"=    FROM  "INT_ProcInputBuffer"    TO  "INT_ColumnFilter"          12.4ns;

# Range data buffer - IPBUFF
# NOTE THAT ALL SM USING THE RANGE BUFFER ARE ON THE SAME CLOCK AS THE BUFFER SO 1clk CONSTRAINT IS
APPROPRIATE
# ZData is used by both VelCalc and Processing/RobAvgSeg/RangeEdges
# ZAddr/Zen are used by both IPBUFF/RangeBuffer and Processing/RobAvgSeg/RangeEdges
INST "MAC_IPBUFF/$RAM_FIFO_RANGEDATA" TNM = "IPBUFF_RANGEDATA_RAM";
TIMESPEC "TS_RANGEBUF_VELCALC"=    FROM  "IPBUFF_RANGEDATA_RAM"   TO  "INT_VelCalc"               24.9ns;
TIMESPEC "TS_RANGEBUF_RANGEDGES"=  FROM  "IPBUFF_RANGEDATA_RAM"   TO  "INT_RangeEdges"            24.9ns;
TIMESPEC "TS_VELCALC_RANGEBUF" =   FROM  "INT_VelCalc"            TO  "IPBUFF_RANGEDATA_RAM"      24.9ns;
TIMESPEC "TS_VELCALC_RANGEEDGES" = FROM  "INT_VelCalc"            TO  "INT_RangeEdges"            24.9ns;

# Processing input buffer to processing/udcnt
# The UDCNT and InputBuffer are on different clks so there is only 0.5clk for the BUFOVERTHRESH signal
# to propagate. No need to constrain BUFTHRESH since it is set once and never changed.
TIMESPEC "TS_BUFOVERTHRESH"=       FROM  "INT_PUpDownCounter"     TO  "INT_ProcInputBuffer"       12.4ns;

# Path from ProcessingOut buffer to RAMS in TEMPInt
# NOTE: EXTEN is high for 2 clocks, so we can allow a 1clk delay from EXTEN/EXTADDR to the RAMS.
INST "MAC_PROCESSING/MACP_TEMPINTM/$RAM_FIFO_VELOCITY" TNM = "TEMPINT_RAMS";
INST "MAC_PROCESSING/MACP_TEMPINTM/$RAM_FIFO_BREAK"    TNM = "TEMPINT_RAMS";
#ADDR/EN
TIMESPEC "TS_PROCOUTBUF_TEMPINTRAMS" = FROM  "INT_ProcOutBuffer"  TO  "TEMPINT_RAMS"              22.9ns;
#DATA
TIMESPEC "TS_TEMPINTRAMS_PROCOUTBUF" = FROM  "TEMPINT_RAMS"       TO  "INT_ProcOutBuffer"         12.4ns;

INST "MAC_PCBUF/$RAM_FIFO" TNM = "PCOUTBUFFERRAM";
TIMESPEC "TS_PCBUF_RAM" =          FROM  "INT_PCBuffer"           TO  "PCOUTBUFFERRAM"            12.4ns;
TIMESPEC "TS_RAM_PCBUF" =          FROM  "PCOUTBUFFERRAM"         TO  "INT_PCBuffer"              12.4ns;


###############################
####### Camera Data Path

#### CAMERA - CAMERA INTERFACE
## The CameraCont SM outputs data well before toggling address stobes so we should not
## have addr setup problems. Hold shouldn't be a problem either because the address is
## not changed until a little while after the address strobe is complete.
## Since we can be sure there are not setup/hold problems, the actual delays on the
## nets are not very relevant unless they get wildly out of hand.
## This constraint effects ADCK, E_X, E_Y, CAMADDR
```

372 Motion vision: design of compact motion sensing solutions

```
TIMESPEC "TS_CameraOutput" =            FROM  "INT_CameraCont"      TO  "USEFULPADS"         24.9 ns;

## Time for the data to travel to the SM is not too relevant. The SM waits for 9clk
## or so for the data to stabilise anyway...
TIMESPEC "TS_CameraInput" =             FROM  "USEFULPADS"          TO  "INT_CameraCont"     24.9 ns;

NET "CALFRAMETIMER" TNM = "CalFrameTimer_Sinks";
TIMEGRP "CamCont_X_CalFrameTimer" = "INT_CameraCont" except "CalFrameTimer_Sinks";
# PixelReadyA/PixelA
TIMESPEC "TS_CamCont_Calib" =           FROM  "INT_CameraCont"      TO  "INT_SumPixels"      12.4 ns;
# CamFrameTimer (only signal from Calib to CameraCont) - Stays high for a while - don't care if
we miss rising edge
TIMESPEC "TS_CalFrameTimer" =           FROM  "INT_SumPixels"       TO  "CalFrameTimer_Sinks" 24.9 ns;

NET "READAHEADDONE" TNM = "ReadAheadDone_Sinks";
TIMEGRP "CamCont_X_ReadAheadDone" = "INT_CameraCont" except "ReadAheadDone_Sinks";

# Strobe
TIMESPEC "TS_CamCont_CalBuf" =          FROM  "INT_CameraCont"      TO  "INT_CalibrationBuffer"  12.4 ns;
# Calibration Data - not used until 1 clk after strobe goes low.
TIMESPEC "TS_CalBuf_CamCont" =          FROM  "INT_CalibrationBuffer" TO  "CamCont_X_ReadAheadDone"  24.9 ns;
# ReadAheadDone toggles once - don't care if we miss rising edge
TIMESPEC "TS_ReadAheadDone" =           FROM  "INT_CalibrationBuffer" TO  "ReadAheadDone_Sinks"  24.9 ns;

NET "STARTIMG" TNM = "StartImg_Sinks";
# StartImg stays high for a while - don't care if we miss rising edge
TIMESPEC "TS_StartImg" =                FROM  "INT_PCBuffer"        TO  "StartImg_Sinks"     24.9 ns;

TIMESPEC "TS_CamCont_ScaleSp" =         FROM  "INT_CameraCont"      TO  "INT_ScaleSp"        12.4 ns;
TIMESPEC "TS_ScaleSp_Filter" =          FROM  "INT_ScaleSp"         TO  "INT_Filter"         12.4 ns;
TIMESPEC "TS_Filter_CamBuf" =           FROM  "INT_Filter"          TO  "INT_CameraBuffer"   12.4 ns;

# Filter and filter pixel clock are on the same clock....
TIMESPEC "TS_PixelClk_Filter" =         FROM  "INT_FiltCamPixelClock" TO  "INT_Filter"       24.9 ns;

# Constrain ROWADDROUT/DELROWADDROUT lines
TIMESPEC "TS_Filter_FFS" =              FROM  "INT_Filter"          TO  "FilterFFS"          12.4 ns;

# Constrain ADDR, and Data Input lines
INST "MAC_FILT/$RAM_FILTER_0" TNM = "FILTER_RAMS";
INST "MAC_FILT/$RAM_FILTER_1" TNM = "FILTER_RAMS";
INST "MAC_FILT/$RAM_FILTER_2" TNM = "FILTER_RAMS";
INST "MAC_FILT/$RAM_FILTER_3" TNM = "FILTER_RAMS";
TIMESPEC "TS_Filter_RAMS" =             FROM  "INT_Filter"          TO  "FILTER_RAMS"        12.4 ns;
# Constrain Data output lines (that pass through tristate buffers under control of DEL_ROWADDROUT)
# Filter is designed such that output of RAM is not use until 2clks after the address is sent out
# so we can safely allow 24ns on the data path from RAM to Filter.
TIMESPEC "TS_RAMS_Filter" =             FROM  "FILTER_RAMS"         TO  "INT_Filter"         24.9 ns;
# From FFS to Filter SM via 3 state buffers
TIMESPEC "TS_FILTERFFS_RAMS" =          FROM  "FilterFFS"           TO  "INT_Filter"         24.9 ns;

###############################
####### Laser Data Path

#### LASER INPUT - LASER INTERFACE
## WR and SEL are generated using a 40 MHz clock and to prevent metastability
## are run through a pair of flip-flops. Since these signals comes from FFs
## we can give them a 12 ns delay. We can also associate a 12 ns delay with
## the data bus from the DSP since the data bus is only read when WR and SEL
## are high.

NET "WR" TNM = "LASERDATAINPUT";
NET "SEL" TNM = "LASERDATAINPUT";

NET "LASERRANGEIN<0>" TNM = "LASERDATAINPUT";
NET "LASERRANGEIN<1>" TNM = "LASERDATAINPUT";
NET "LASERRANGEIN<2>" TNM = "LASERDATAINPUT";
NET "LASERRANGEIN<3>" TNM = "LASERDATAINPUT";
NET "LASERRANGEIN<4>" TNM = "LASERDATAINPUT";
NET "LASERRANGEIN<5>" TNM = "LASERDATAINPUT";
NET "LASERRANGEIN<6>" TNM = "LASERDATAINPUT";
```

Appendix C – FPGA design

```
NET "LASERRANGEIN<7>" TNM = "LASERDATAINPUT";

TIMESPEC "TS_LASERDATA" =            FROM  "USEFULPADS"           TO  "LASERDATAINPUT"         12.4ns;
TIMESPEC "TS_LaserInt_LaserBuf"=     FROM  "INT_LaserInt"         TO  "INT_LaserBuffer"        12.4ns;

###############################
####### PC Data Path

NET "PCPIXELCLK"    MAXDELAY = 7ns;    # Runs from CLK/5, not sure which clk phase this corresponds to

# The PC capture card expects data to be available on the camera
# data lines about 25ns after the ADCK line goes low.
# To make sure there are no timing problems, we actually
# place data on the camera data bus when ADCK returns high
# so that data are available long before ADCK goes low
# again. Thus the constraint on this output pin is
# not too critical.
TIMESPEC "TS_PCOUTPUT" =             FROM  "INT_PCBuffer"         TO  "USEFULPADS"             24.9ns;

###############################
####### Misc Top Level Signals

# Same clk!
TIMESPEC "TS_FRAMECLK" =             FROM  "INT_LaserBuffer"      TO  "INT_PCBuffer"           24.9ns;
TIMESPEC "TS_CAM_BUFFERING" =        FROM  "INT_CameraBuffer"     TO  "INT_BufsDone"           12.4ns;
TIMESPEC "TS_PC_BUFFERING" =         FROM  "INT_PCBuffer"         TO  "INT_BufsDone"           12.4ns;
TIMESPEC "TS_BUFFERINGCOMPLETE"=     FROM  "INT_BufsDone"         TO  "INT_LaserBuffer"        12.4ns;
TIMESPEC "TS_SDREADY" =              FROM  "INT_LaserBuffer"      TO  "INT_ScaleChange"        12.4ns;
TIMESPEC "TS_PROCOUTCOMPLETE" =      FROM  "INT_ProcOutBuffer"    TO  "INT_RAMICPol"           12.4ns;
TIMESPEC "TS_SPNODELAYA" =           FROM  "INT_ScaleChange"      TO  "FF_STARTPROC"           12.4ns;
TIMESPEC "TS_SPNODELAYB" =           FROM  "INT_ScaleChange"      TO  "INT_ProcInputBuffer"    12.4ns;
TIMESPEC "TS_STARTPROCESSING" =      FROM  "FF_STARTPROC"         TO  "INT_RAMICPol"           12.4ns;

# AT END OF FILE SCALEDATA, POLLMODE ARE EXPLICITLY SET TO BE IGNORED FOR TIMING PURPOSES.

#######################################
####### PROCESSING - ColFilt (ApVel)

INST "MAC_PROCESSING/MACP_COLFILTM/$RAM_FIFO_IX" TNM = "COLFILT_RAMS";
INST "MAC_PROCESSING/MACP_COLFILTM/$RAM_FIFO_IT" TNM = "COLFILT_RAMS";
TIMESPEC "TS_COLFILT_RAM" =          FROM  "INT_ColumnFilter"     TO  "COLFILT_RAMS"           12.4ns;
TIMESPEC "TS_RAM_COLFILT" =          FROM  "COLFILT_RAMS"         TO  "INT_ColumnFilter"       12.4ns;

TIMESPEC "TS_COLFILT_UDCNT" =        FROM  "INT_ColumnFilter"     TO  "INT_CFUpDownCounter"    12.4ns;
TIMESPEC "TS_UDCNT_COLFILT" =        FROM  "INT_CFUpDownCounter"  TO  "INT_ColumnFilter"       12.4ns;

TIMESPEC "TS_COLFILT_ACCUM" =        FROM  "INT_ColumnFilter"     TO  "ColFilt_Accum"          12.4ns;
TIMESPEC "TS_ACCUM_COLFILT" =        FROM  "ColFilt_Accum"        TO  "INT_ColumnFilter"       12.4ns;

TIMESPEC "TS_COLFILTRAM_ACCUM"=      FROM  "COLFILT_RAMS"         TO  "ColFilt_Accum"          12.4ns;
TIMESPEC "TS_ACCUM_COLFILTRAM"=      FROM  "ColFilt_Accum"        TO  "COLFILT_RAMS"           12.4ns;

TIMESPEC "TS_COLFILT_APVEL" =        FROM  "INT_ColumnFilter"     TO  "INT_ApVel"              12.4ns;
TIMESPEC "TS_COLFILTACUM_APVEL"=     FROM  "ColFilt_Accum"        TO  "INT_ApVel"              12.4ns;

# Constrain NET SDERIVREADY from ColFilt to VelCalc - the state machines are on same clk so 24ns
delay is appropriate
TIMESPEC "TS_SDERIVREADY" =          FROM  "INT_ColumnFilter"     TO  "INT_VelCalc"            24.9ns;

#######################################
####### PROCESSING - APVel (VelCalc)

# Note - all other paths between SRLs and flip-flops are constrained on the schematic to 12ns

TIMESPEC "TS_APPVEL_DIV" =           FROM  "INT_ApVel"            TO  "INTCORE_Div2012S4"      12.4ns;
TIMESPEC "TS_DIV_APPVEL" =           FROM  "INTCORE_Div2012S4"    TO  "INT_ApVel"              12.4ns;
TIMESPEC "TS_APPVEL_VELCALC" =       FROM  "INT_ApVel"            TO  "INT_VelCalc"            12.4ns;
#Added just in case there is a direct path from the divider to Velcalc.....
TIMESPEC "TS_DIV_VELCALC" =          FROM  "INTCORE_Div2012S4"    TO  "INT_VelCalc"            12.4ns;
#Added just in case there is a direct path from the divider to Velcalc.....
TIMESPEC "TS_DIV_VELCALCMULT" =      FROM  "INTCORE_Div2012S4"    TO  "INTCORE_M1010SR5"       12.4ns;
```

374 Motion vision: design of compact motion sensing solutions

```
########################################
####### PROCESSING - VelCalc

TIMESPEC "TS_VELCALC_MULT" =         FROM   "INT_VelCalc"           TO   "INTCORE_M1010SR5"     12.4 ns;
TIMESPEC "TS_MULT_VELCALC" =         FROM   "INTCORE_M1010SR5"      TO   "INT_VelCalc"          12.4 ns;

INST "MAC_PROCESSING/$RAM_FIFO_VELCALC-ROBAVG" TNM = "VELCALC_ROBAVG_RAM";
TIMESPEC "TS_VELCALC_RAM" =          FROM   "INT_VelCalc"           TO   "VELCALC_ROBAVG_RAM"   12.4 ns;
TIMESPEC "TS_MULT_RAM" =             FROM   "INTCORE_M1010SR5"      TO   "VELCALC_ROBAVG_RAM"   12.4 ns;

# Connection to IPBuff/RangeBuffer is constrained in section "buffers" above.

########################################
####### PROCESSING - UDPCnt

# PROCESSING/UDCNT is on the same clk as RobAvg, so we have 1clk for the ROBAVGDIFF signal to
propogate
TIMESPEC "TS_ROBAVGDIFF_ROBAVG"=     FROM   "INT_PUpDownCounter"    TO   "INT_RobAvg"           24.9 ns;
# VELREADY/CUP
TIMESPEC "TS_VELCALC_UDCNT" =        FROM   "INT_VelCalc"           TO   "INT_PUpDownCounter"   12.4 ns;
# CDOWN
TIMESPEC "TS_ROBAVG_UDCNT" =         FROM   "INT_RobAvg"            TO   "INT_PUpDownCounter"   12.4 ns;

########################################
####### PROCESSING - RobAvg

# Note - paths between SRLs and flip-flops are constrained on the schematic to 12 ns

TIMESPEC "TS_ROBAVG_RAM" =           FROM   "INT_RobAvg"            TO   "VELCALC_ROBAVG_RAM"
                   12.4 ns;
# Data from RAM is stable for 1clk before it is read, so can use 25 ns contraint.
TIMESPEC "TS_RAM_ROBAVG" =           FROM   "VELCALC_ROBAVG_RAM"    TO   "INT_RobAvg"           24.9 ns;

TIMESPEC "TS_ROBAVG_DIV" =           FROM   "INT_RobAvg"            TO   "INTCORE_DIV1610S4"    12.4 ns;
TIMESPEC "TS_DIV_ROBAVG" =           FROM   "INTCORE_DIV1610S4"     TO   "INT_RobAvg"           12.4 ns;
TIMESPEC "TS_ROBAVG_ACCUM" =         FROM   "INT_RobAvg"            TO   "RobAvg_Accum"         12.4 ns;
TIMESPEC "TS_ACCUM_ROBAVG" =         FROM   "RobAvg_Accum"          TO   "INT_RobAvg"           12.4 ns;

TIMESPEC "TS_DIV_ROBAVGP" =          FROM   "INTCORE_DIV1610S4"     TO   "INT_RobAvgP"          12.4 ns;

INST "MAC_PROCESSING/MACP_ROBAVGM2/$RAM_FIFO" TNM = "ROBAVG_RAM";
TIMESPEC "TS_ROBAVGDIV_RAM" =        FROM   "INTCORE_DIV1610S4"     TO   "ROBAVG_RAM"           12.4 ns;
TIMESPEC "TS_ROBAVGP_RAM" =          FROM   "INT_RobAvgP"           TO   "ROBAVG_RAM"           12.4 ns;

# Startsmooth is held high for 2clks
TIMESPEC "TS_STARTSMOOTH" =          FROM   "INT_RobAvgP"           TO   "INT_RobAvgSeg"        24.9 ns;

########################################
####### PROCESSING - RobAvgSeg (TempInt)

INST "MAC_PROCESSING/MACP_ROBAVGSEGM2/$RAM_FIFO" TNM = "ROBAVGSEG_RAM";
TIMESPEC "TS_RangeEdges_RAM"=        FROM   "INT_RangeEdges"        TO   "ROBAVGSEG_RAM"        12.4 ns;

TIMESPEC "TS_ROBAVGSEG_DIV" =        FROM   "INT_RobAvgSeg"         TO   "INTCORE_Div1710S4"    12.4 ns;
TIMESPEC "TS_DIV_ROBAVGSEG" =        FROM   "INTCORE_Div1710S4"     TO   "INT_RobAvgSeg"        12.4 ns;

# Constraints to/from Range Edge RAM (in ROBAVGSEG macro)
TIMESPEC "TS_ROBAVGSEG_RANGERAM"=    FROM   "INT_RobAvgSeg"         TO   "ROBAVGSEG_RAM"        12.4 ns;
# Data from RAM is stable for 1clk before it is read, so can use 25 ns contraint.
TIMESPEC "TS_RANGERAM_ROBAVGSEG"=    FROM   "ROBAVGSEG_RAM"         TO   "INT_RobAvgSeg"        24.9 ns;

# Constraints to/from Velocity RAM (in ROBAVG macro)
TIMESPEC "TS_ROBAVGSEG_VELRAM" =     FROM   "INT_RobAvgSeg"         TO   "ROBAVG_RAM"           12.4 ns;
# Data from RAM is stable for 1clk before it is read, so can use 25 ns contraint.
TIMESPEC "TS_VELRAM_ROBAVGSEG" =     FROM   "ROBAVG_RAM"            TO   "INT_RobAvgSeg"        24.9 ns;

TIMESPEC "TS_ROBAVGSEG_TEMPINT"=     FROM   "INT_RobAvgSeg"         TO   "INT_TempInt"          24.9 ns;
TIMESPEC "TS_DIV_TEMPINT" =          FROM   "INTCORE_Div1710S4"     TO   "INT_TempInt"          24.9 ns;
TIMESPEC "TS_SMOOTHREADY" =          FROM   "FF_SMOOTHREADYDEL"     TO   "INT_TempInt"          24.9 ns;
# Connections to the range buffer in IPBuff are constrained above in section "Processing top level"
```

```
#####################################
####### PROCESSING - TempInt

# TEMPINT_RAMS is defined in buffers section above
# Constrains tristate lines too

TIMESPEC "TS_TEMPINT_TEMPINTRAMS" = FROM "INT_TempInt"    TO "TEMPINT_RAMS"      12.4 ns;
TIMESPEC "TS_TEMPINTRAMS_TEMPINT" = FROM "TEMPINT_RAMS"   TO "INT_TempInt"       12.4 ns;
TIMESPEC "TS_ProcessingComplete"  = FROM "INT_TempInt"    TO "INT_ProcOutBuffer" 12.4 ns;
TIMESPEC "TS_BUSY" =                FROM "INT_TempInt"    TO "INT_RobAvgSeg"     24.9 ns;

#####################################
####### IGNORED PATHS

NET "TP0" TIG;
NET "TP1" TIG;
NET "TP2" TIG;
NET "TP3" TIG;
NET "TP4" TIG;
NET "TP5" TIG;
NET "TP6" TIG;
NET "TP7" TIG;
NET "TP8" TIG;
NET "TP9" TIG;
NET "TP10" TIG;
NET "TP11" TIG;
NET "TP12" TIG;
NET "TP13" TIG;
NET "TP14" TIG;
NET "TP15" TIG;
NET "TP16" TIG;
NET "TP17" TIG;
NET "TP18" TIG;
NET "TP19" TIG;
NET "TP20" TIG;
NET "TP21" TIG;
NET "TP22" TIG;
NET "TP23" TIG;
NET "TP24" TIG;
NET "TP25" TIG;
NET "TP26" TIG;
NET "TP27" TIG;
NET "TP28" TIG;
NET "TP29" TIG;
NET "TP30" TIG;
NET "TP31" TIG;
NET "TP32" TIG;
NET "TP33" TIG;
NET "TP34" TIG;
NET "TP35" TIG;
NET "TP36" TIG;
NET "TP37" TIG;
NET "TP38" TIG;
NET "TP39" TIG;
NET "TP40" TIG;
NET "MAC_PROCESSING/TPPROC0" TIG;
NET "MAC_PROCESSING/TPPROC1" TIG;
NET "MAC_PROCESSING/TPPROC2" TIG;
NET "MAC_PROCESSING/TPPROC3" TIG;
NET "MAC_PROCESSING/TPPROC4" TIG;
NET "MAC_PROCESSING/TPPROC5" TIG;
NET "MAC_PROCESSING/TPPROC6" TIG;
NET "MAC_PROCESSING/TPPROC7" TIG;
NET "MAC_PROCESSING/TPPROC8" TIG;
NET "MAC_PROCESSING/TPPROC9" TIG;
NET "MAC_PROCESSING/TPPROC10" TIG;
NET "MAC_PROCESSING/TPPROC11" TIG;
NET "MAC_PROCESSING/TPPROC12" TIG;

# NO CONSTRAINT ON SCALE DATA - IT IS NOT USED FOR QUITE SOME TIME AFTER IT CHANGES
NET "IMGWIDTH<0>" TIG;
```

```
NET "IMGWIDTH<1>" TIG;
NET "IMGWIDTH<2>" TIG;
NET "IMGWIDTH<3>" TIG;
NET "IMGWIDTH<4>" TIG;
NET "IMGWIDTH<5>" TIG;
NET "IMGWIDTH<6>" TIG;
NET "IMGWIDTH<7>" TIG;
NET "IMGWIDTH<8>" TIG;

NET "SS<0>" TIG;
NET "SS<1>" TIG;
NET "SS<2>" TIG;
NET "SS<3>" TIG;
NET "SS<4>" TIG;

NET "SD<0>" TIG;
NET "SD<1>" TIG;
NET "SD<2>" TIG;

NET "POLLMODE" TIG; # POLL MODE IS ONLY USED FOR TEST PURPOSES. NO NEED FOR CONSTRAINT

#### NOTE - FOR IGNORED TRISTATE CONTROL LINES I ASSUME THAT THE TIMING ANALYSER WILL ONLY IGNORE
####        PATHS ORIGINATING FROM THE TRISTATE CONTROL LINE. I ASSUME PATHS THAT RUN THROUGH THE
####        TRISTATE BUFFER FROM INPUT TO OUTPUT (AND ONTO A REGISTER) ARE STILL ANALYSED.

# MAC_RAMICP/Tri only changes when calibration finishes.
NET "MAC_RAMICP/TRI" TIG;

# MAC_CALIBRAT/ZR_Tri low at reset - high when clearing - low at completion. Timing not critical
NET "MAC_CALIBRAT/ZR_TRI" TIG;

# MAC_CALIBRAT/SP_Tri low at reset - high during operation (after zeroing complete) - low at completion.
Timing not critical
NET "MAC_CALIBRAT/SP_TRI" TIG;

# MAC_CALIBRAT/CC_Tri low at reset - high during operation (after sumpixels complete) - low at
completion. Timing not critical
NET "MAC_CALIBRAT/CC_TRI" TIG;

#EXTTRI/INTTRI - These signals control access to the processing result buffer. During processing
EXTTRI is low and
#INTRI is high so that processing has exclusive access to buffer via the INTEN lines (See
MAC_PROCESSING/MAC_TEMPINT).
#When processing is complete the processing logic releases the buffer (INTTRI goes low) and the
processing output
#buffer is signalled. The output buffer now takes control of the buffer (EXTTRI goes high) for
as long as it takes
#to ouput the data to RAM. This is completed well before the next frame of data starts to be
read so there is no
#chance conflict of access to this buffer.
NET "EXTTRI" TIG;
NET "MAC_PROCESSING/MACP_TEMPINTM/INTTRI" TIG;
```

Appendix D
Simulation of range data

In this appendix we present the code we used to simulate range data. This is essentially a ray-tracing program where the user can specify a simple environment. This is C code targeted at the ADSP Sharc 21062 DSP on the SignalMaster prototype platform used in our work and the code uses libraries targeted at this design environment. Despite this, the core algorithm should be portable to any C compiler. Please remember this is 'prototype code' – it worked for our purposes but it is not necessarily fully documented and it may not represent an 'ideal' implementation.

Use of this code (or any part thereof) is entirely at your risk. We cannot provide support and will not accept any liability for your use of this code (in part or whole). This code (or any part thereof) must not be used to maintain the safety or security of people or equipment.

```
#include <21060.h>            // Functions for the internal timer
#include <def21062.h>
#include <sport.h>
#include <signal.h>           // Function and macros for interrupts
#include <macros.h>
#include <stdlib.h>
#include <stdio.h>
#include <math.h>
#include ".\external\dspbios.h"
#include ".\external\bitsi.h"
#include ".\external\gatesm.h"
#include ".\external\lcd.h"

#define JWAIT_STATE 1E6
#define PI                      3.1415926

//-------------------------------------------------------------------
// *** MACROS ***
//-------------------------------------------------------------------
#define TRANSLATE_POINT(X, Z, a, b)    {   (X) = (X)+(a);           \
                                           (Z) = (Z)+(b);    }
#define ROTATE_POINT(X, Z, ca, sa)     {   float xt,zt;             \
                                           xt = (X)*(ca) - (Z)*(sa);  \
                                           zt = (X)*(sa) + (Z)*(ca);  \
                                           X = xt;   Z = zt;         }
```

```
#define RADIANS2DEGREES(r, d)         { d = r * (180.0/PI);         }
#define DEGREES2RADIANS(d, r)         { r = d * (PI/180.0);         }

//------------------------------------------------------------------------------
// *** ENVIRONMENT CONFIGURATION ***
//------------------------------------------------------------------------------

// ---IMAGE INTERPOLATION DATA---
// We assume the camera optical axis and the range finder 0 degree scan
// direction to be coincident. Note that the camera optical axis passes through
// the middle of the image which is BETWEEN pixels ((n/2)-1) and (n/2)
#define CAM_FOCAL_LENGTH    0.0048          // Camera focal length in metres
#define CAM_PIXEL_PITCH     0.0000125       // Camera pixel pitch in metres
#define CAM_XPIXELS         512             // Number of pixels in the
                                            // X direction.


// ---TIME---
#define CLKRATE             40E6            // 40 MHz clock rate
#define TIME_PERSAMPLE      29E-6           // Output time per sample in
                                            // seconds
#define COUNT_PERSAMPLE     TIME_PERSAMPLE*CLKRATE
#define TIME_PERSCAN        0.0625          // Time to complete a 360 deg
                                            // revolution of scan head in
                                            // seconds
#define COUNT_PERSCAN       TIME_PERSCAN*CLKRATE

// ---RANGE SCANNER----
// Although the vehicle carrying the range scanner can only move in
// the direction it is facing, we have provided a more flexible scheme
// for defining the scanner's position. Its position at time T is given
// by (SCANNER_X(T),SCANNER_Z(T)) and its orientation (direction in which
// the scanners 0 deg scan is made) is given by SCANNER_ORIENTATION(T).
// We assume the scan is made in a plane parallel to the floor

#define SCANNER_DIST_RES        0.05            // Range scanner distance resolution
                                                // in metres
#define SCANNER_DIST_MAX        10.0            // Maximum scan distance

#define SCANNER_X(T)            ((T)*0.00 + 0.00)   // Start X position in metres
#define SCANNER_Z(T)            ((T)*0.1 + 0.00)    // Start Z position in metres
#define SCANNER_ORIENTATION(T)  ((T)*0.00 + 0.00)   // Angle which the scanner
                                                    // facing (measured from
                                                    // Z axis) degrees

// ---OBJECTS---
// An object can either be a circle or a polygon (or can be ignored). The
// positions of the object vertices (in the case of a polygon) or the
// position of the centre (in the case of a circle) are all defined in the
// VERTEX_LOCATIONS array. The first element in each row of the array
// defines the object type and subsequent values in the row provide information
// defining the object. If the object is to be ignored (allowing
// simple addition/removal of objects) then first value in the row will be 0.
// For circles this value is 1 and for polygons this first value will be the
// number of vertices in the object. The number of vertices in a polygon
// cannot be more than MAX_VERTICES.
// For a circle, the following 3 values define the X and Z locations of the
// centre and the circles radius respectively. All other values are ignored.
// For a polygon, subsequent values are taken pairwise to form the location
// of each vertex.
```

```
//    {4, 10,    10,   10,   -10,  -10,  -10,  -10,  10 }, \ //Large square vertices
//    {4, 0.5,   0.5,  0.5,  -0.5, -0.5, -0.5, -0.5, 0.5}, \ //smaller square
//                                                           //vertices
//    {1, 0,     0,    1,    0,    0,    0,    0,    0  }, \ //Circle, centre (0,0),
//                                                           //radius 1
//    {0, 0,     0,    0,    0,    0,    0,    0,    0  }, \ //Ignored
//    {0, 0,     0,    0,    0,    0,    0,    0,    0  };   //Ignored
//
// All transformations of object positions are done WRT the origin, thus
// if you don't centre your objects on the origin you will get some odd results.
// centre your objects on the origin then use the appropriate position function
// to set its initial position.
// The location of each object's origin is given by (OBJ?_XPOS(T), OBJ?_ZPOS(T)).
// The orientation of the object is given by OBJ?_ORIENTATION(T). The 0 degrees
// orientation is along the +ve Z axis. Note that changing the orientation of a
// circle makes no difference - a circle always looks the same no matter its
// orientation. The number of objects to use is defined by the NUMBER_OBJECTS
// parameter. This number SHOULD NOT include objects whose first entry in
// VERTEX_LOCATIONS=0 because these objects are skipped.
// There are 5 objects defined below. If you want more or less, then you
// should update the definition list appropriately. You'll have to update the
// code too... (see functions initObjectPos and updateObjectPos)
// No collision checking between objects is performed.
// Don't over do it with objects - each circle and each edge of a polygon
// increases the time it takes to compute the range. Remember you only
// have TIME_PERSCAN-(CAMXPIXELS*TIME_PERSAMPLE) to calculate the scan.
// If you run overtime you'll get a message on the LCD. See the interrupt
// handler to see the message you'll get.

#define NUMBER_OBJECTS          2                   // Number of non-ignored
                                                    // objects
#define MAX_VERTICES            4                   // Max number of vertices
                                                    // for an object
#define VERTEX_ARRAY_LEN        ((MAX_VERTICES*2)+1)

#define OBJ1_XPOS(T)            ((T)*0.00 + 0.00)   // X pos of object centre
#define OBJ1_ZPOS(T)            ((T)*0.00 + 0.00)   // Z pos of object centre
#define OBJ1_ORIENTATION(T)     ((T)*0.00 + 0.00)   // "Forward" direction in
                                                    // degrees from Z axis

#define OBJ2_XPOS(T)            ((T)*0.00 + 0.00)   // X pos of object centre
#define OBJ2_ZPOS(T)            ((T)*0.00 + 0.00)   // Z pos of object centre
#define OBJ2_ORIENTATION(T)     ((T)*0.00 + 0.00)   // "Forward" direction in
                                                    // degrees from Z axis

#define OBJ3_XPOS(T)            ((T)*0.10 + 0.00)   // X pos of object centre
#define OBJ3_ZPOS(T)            ((T)*0.00 + 2.13)   // Z pos of object centre
#define OBJ3_ORIENTATION(T)     ((T)*0.00 + 0.00)   // "Forward" direction in
                                                    // degrees from Z axis

#define OBJ4_XPOS(T)            ((T)*0.00 + 0.00)   // X pos of object centre
#define OBJ4_ZPOS(T)            ((T)*0.00 + 0.00)   // Z pos of object centre
#define OBJ4_ORIENTATION(T)     ((T)*0.00 + 0.00)   // "Forward" direction in
                                                    // degrees from Z axis

#define OBJ5_XPOS(T)            ((T)*0.00 + 0.00)   // X pos of object centre
#define OBJ5_ZPOS(T)            ((T)*0.00 + 0.00)   // Z pos of object centre
#define OBJ5_ORIENTATION(T)     ((T)*0.00 + 0.00)   // "Forward" direction in
                                                    // degrees from Z axis
```

```c
float    VERTEX_LOCATIONS[5][VERTEX_ARRAY_LEN] = {                    \
                {4,   3,   3,   3,   -3,  -3,  -3,  -3, 3  }, \
                {0,   0,   0,   0,    0,   0,   0,   0, 0  }, \
                {1,   0,   0,   0.13, 0,   0,   0,   0, 0  }, \
                {0,   0,   0,   0,    0,   0,   0,   0, 0  }, \
                {0,   0,   0,   0,    0,   0,   0,   0, 0  }};

#define ARRAYLOC(O, P)          ( ((O)*VERTEX_ARRAY_LEN) + (P) )

//------------------------------------------------------------------------------
// GLOBAL VARIABLES
//------------------------------------------------------------------------------
// Scale choice table.
float    lowerCutoff[7]    =    {0.0,  0.25,  0.5,   1.00,  2.00,  4.00,  10000.0};
float    upperCutoff[7]    =    {0.0,  0.25,  0.5,   1.00,  2.00,  4.00,  10000.0};
float    SSValue[7]        =    {16,   16  ,  8    , 4    , 2    , 1    , 1    };
float    SDValue[7]        =    {4,    4   ,  3    , 2    , 1    , 0    , 0    };
float    ImgValue[7]       =    {32,   32  ,  64   , 128  , 256  , 512  , 512  };

float    *SINE_lookuptable;
float    *COS_lookuptable;
int      TABLE_length;
volatile int timerIRQOccured;
int      currentlyBusy;
int      frame;
int      pixel;

//------------------------------------------------------------------------------
// DATA TYPES
//------------------------------------------------------------------------------
typedef struct _pointDummy {
        float X;                    // metres
        float Z;                    // metres
} _point;

typedef struct _scaleDataDummy {
        int imgWidth;               // Image width (in pixels)
        int SD;                     // 2^SD = SS
        int SS;                     // CAM_XPIXELS/SS = imgWidth
        int changeScale;            // 1 if scale change pending, else 0
} _scaleData;

//------------------------------------------------------------------------------
// FUNCTION PROTOTYPES
//------------------------------------------------------------------------------
void     updateObjectPos(float time, float *objects);
void     transformObject(int index, float rotObj, float cosScan, float sinScan,
         float xTrans, float zTrans, float *object);
float    performScan(float *scan, float *objects, _scaleData scale);
float    calcMinRange(int index, float *objects);
void     updateScaleData(float minRange, float prevMinRange, _scaleData *scale);

// Output of data
void     outputRangeToFPGA(float *scan, _scaleData scale);
void     outputScaleToFPGA(_scaleData scale);

// Generate various arrays
void     camBaseline(float *Angle, int SS);        // Generate camera angles
void     SINCOS_LUT(float *angles, int number);    // Generate sin/cos lookup tables
```

Appendix D – Simulation of range data 381

```c
// Intersections and range calculation
inline float     linelineIntersection(int alpha_index, float X3, float Z3,
                                      float X4, float Z4);
inline float     linecircleIntersection(int alpha_index, float Xc, float Zc,
                                        float radius);

void timer_irq_handler(int signal);


//-------------------------------------------------------------------------
// Function                : timer_irq_handler
// Description             : Deal with a timer interrupt
//-------------------------------------------------------------------------
void timer_irq_handler(int signal) {

        char msg[15];
        int i;

        timerIRQOccured=1;            // Flag that a timer interrupt occurred
        clear_interrupt(SIG_TMZ0);    // Clear the interrupt flag

        if (currentlyBusy!=0) {
                LcdGotoXY(0,0);
                LcdWriteString("Timing Error!");
                LcdGotoXY(0,1);
                sprintf(msg,"F %i P %i M %i",frame,pixel,currentlyBusy);
                LcdWriteString(msg);
                for (i=0; i<200000000; i++) {}
        }
}



//-------------------------------------------------------------------------
// Function    : updateObjectPos
//
// Parameters  : A time for which to calculate object positions. A pointer to
//               an array of object positions.
//
// Description : Calculate the positions of object centres (for circular
//               objects) or object corners (for rectangle objects) in the
//               scanner centred coordinate system at a given time
//-------------------------------------------------------------------------
void updateObjectPos(float time, float *objects) {

        float scannerX, scannerZ, scannerO;
        float objX, objZ, objO;
        float sinScan, cosScan;
        int objIndex;
        int i,t;

        // Scanner position
        scannerX = SCANNER_X(time);
        scannerZ = SCANNER_Z(time);
        DEGREES2RADIANS(SCANNER_ORIENTATION(time),scannerO);
        sinScan = sin(-scannerO);
        cosScan = cos(-scannerO);

        objIndex=0;
        t=0;
```

```
// Object 1 (note, copy this code for each object)
if (VERTEX_LOCATIONS[0][0]!=0) {
        objX = OBJ1_XPOS(time);
        objZ = OBJ1_ZPOS(time);
        DEGREES2RADIANS(OBJ1_ORIENTATION(time),objO);
        for(i=0; i<VERTEX_ARRAY_LEN; i++) {
                objects[t]=VERTEX_LOCATIONS[0][i];
                t++;
        }
        transformObject(objIndex,objO, cosScan, sinScan, objX-scannerX,
                        objZ-scannerZ, objects);
        objIndex++;
}

// Object 2
if (VERTEX_LOCATIONS[1][0]!=0) {
        objX = OBJ2_XPOS(time);
        objZ = OBJ2_ZPOS(time);
        DEGREES2RADIANS(OBJ1_ORIENTATION(time),objO);
        for(i=0; i<VERTEX_ARRAY_LEN; i++) {
                objects[t]=VERTEX_LOCATIONS[1][i];
                t++;
        }
        transformObject(objIndex,objO, cosScan, sinScan, objX-scannerX,
                        objZ-scannerZ, objects);
        objIndex++;
}

// Object 3
if (VERTEX_LOCATIONS[2][0]!=0) {
        objX = OBJ3_XPOS(time);
        objZ = OBJ3_ZPOS(time);
        DEGREES2RADIANS(OBJ1_ORIENTATION(time),objO);
        for(i=0; i<VERTEX_ARRAY_LEN; i++) {
                objects[t]=VERTEX_LOCATIONS[2][i];
                t++;
        }
        transformObject(objIndex,objO, cosScan, sinScan, objX-scannerX,
                        objZ-scannerZ, objects);
        objIndex++;
}

// Object 4
if (VERTEX_LOCATIONS[3][0]!=0) {
        objX = OBJ4_XPOS(time);
        objZ = OBJ4_ZPOS(time);
        DEGREES2RADIANS(OBJ1_ORIENTATION(time),objO);
        for(i=0; i<VERTEX_ARRAY_LEN; i++) {
                objects[t]=VERTEX_LOCATIONS[3][i];
                t++;
        }
        transformObject(objIndex,objO, cosScan, sinScan, objX-scannerX,
                        objZ-scannerZ, objects);
        objIndex++;
}

// Object 5
if (VERTEX_LOCATIONS[4][0]!=0) {
        objX = OBJ5_XPOS(time);
```

```
                    objZ = OBJ5_ZPOS(time);
                    DEGREES2RADIANS(OBJ1_ORIENTATION(time),objO);
                    for(i=0; i<VERTEX_ARRAY_LEN; i++) {
                            objects[t]=VERTEX_LOCATIONS[4][i];
                            t++;
                    }
                    transformObject(objIndex,objO, cosScan, sinScan, objX-scannerX,
                                    objZ-scannerZ, objects);
                    objIndex++;
            }
    }

    //--------------------------------------------------------------------------------
    // Function      : transformObject
    //
    // Parameters    : index - which object to transform in the object array
    //                 rotObj - Object orientation
    //                 cosScan, sinScan - Sin/Cos for scanner orientation
    //                 xTrans, zTrans - translation
    //                 object - an array of objects
    //
    // Description   : For each object vertex, transform that vertex to its correct
    //                 location in the scanner centred coordinate system.
    //--------------------------------------------------------------------------------
    void transformObject(int index, float rotObj, float cosScan, float sinScan,
                        float xTrans, float zTrans, float *object){
        float sinObj, cosObj;
        int numVert;
        int i,base;
        float x,z;

        sinObj = sin(rotObj);   // Compute sin/cos for object orientation once only
        cosObj = cos(rotObj);

        base=index*VERTEX_ARRAY_LEN;
        numVert = object[base];

        for (i=0; i<numVert; i++) {
            x = object[base+(i*2)+1];
            z = object[base+(i*2)+2];

            ROTATE_POINT(x, z, cosObj, sinObj);      // Get object orientation
                                                     // right
            TRANSLATE_POINT(x, z, xTrans, zTrans);   // Position object at right
                                                     // position
            ROTATE_POINT(x, z, cosScan, sinScan);    // Get scanner orientation
                                                     // right

            object[base+(i*2)+1] = x;
            object[base+(i*2)+2] = z;
        }
    }

    //--------------------------------------------------------------------------------
    // Function      : performScan
    // Parameters    :
    // Description   : Performs a range scan. The angles at which measurements are
    //                 made are the same as the angles used to generate the
    //                 SINE_lookuptable and COS_lookuptable. So, if you want to
```

384 *Motion vision: design of compact motion sensing solutions*

```
//                  change the angles at which range measurements occur, you must
//                  update SINE_lookuptable, COS_lookuptable and TABLE_length.
//                  Also takes scale into account so that correct subsampling
//                  is performed.
//-----------------------------------------------------------------------------
float performScan(float *scan, float *objects, _scaleData scale) {

        float minRange;
        float blockMin;
        int index;
        float currentAvgRange, cDist;
        int count;
        int t;

        minRange=SCANNER_DIST_MAX;
        index=0;
        pixel=0;


        for (t=0; t<(int)(TABLE_length/scale.SS); t++) {

                currentAvgRange=0;
                blockMin = SCANNER_DIST_MAX;

                for (count=0; count<scale.SS; count++) {
                    cDist = calcMinRange(index,objects);
                    if (cDist<blockMin) blockMin=cDist;      // Store lowest range
                                                             // in this BLOCK
                    index++;
                }

                if (blockMin<minRange) minRange=blockMin;    // Store lowest range
                                                             // in this SCAN
                scan[t] = blockMin;
                pixel++;
        }

        return(minRange);
}

//-----------------------------------------------------------------------------
// Function    : calcMinRange
// Parameters  : An index pointing to the angle in SIN_lookuptable/
//               COS_lookuptable which corresponds to the angle we are measuring
//               distance in. A list of objects.
// Description : Calculates the minimum distance to an object in the direction
//               defined by index. Distance is limited to SCANNER_DIST_MAX and
//               and is rounded to the nearest SCANNER_DIST_RES.
//-----------------------------------------------------------------------------
float calcMinRange(int index, float *objects) {

        int numVert;
        int i,j,t,tmp;
        int cloc;
        float minDist, cDist;
        float x1,x2,z1,z2,r;

        minDist = SCANNER_DIST_MAX;
        t=0;
```

```
        for (i=0; i<NUMBER_OBJECTS; i++) {      // For each object, find any
                                                // intersection

            numVert = objects[t];

            if (numVert==1) {                   // Circle
                    x1=objects[t+1];
                    z1=objects[t+2];
                    r = objects[t+3];
                    // Don't look for intersection if circle is completely
                    // behind scanner
                    if ( (z1+r) > 0) {
                            cDist=linecircleIntersection(index, x1,z1,r);
                            if (cDist!=-1 && cDist<minDist) minDist=cDist;
                    }
            } else {                            // Rectangle
                    for (j=0; j<numVert-1; j++) {   // For each vertex of the
                                                    // current object
                            cloc=t+((j*2)+1);
                            x1 = objects[cloc];
                            z1 = objects[cloc+1];
                            x2 = objects[cloc+2];
                            z2 = objects[cloc+3];
                            // Don't try to calculate range for edges
                            // completely behind the scanner
                            if (z1>0 && z2>0) {
                                    cDist=linelineIntersection(index, x1,z1,x2,z2);
                                    if (cDist!=-1.0 && cDist<minDist) minDist=cDist;
                            }
                    }
                    tmp=t+(((numVert-1)*2)+1);  // speed up calculation slightly.
                    x1 = objects[tmp];
                    z1 = objects[tmp+1];
                    x2 = objects[t+1];
                    z2 = objects[t+2];

                    // Don't try to calculate range for edges completely behind
                    // the scanner
                    if (z1>0 && z2>0) {
                            cDist=linelineIntersection(index, x1,z1,x2,z2);
                            if (cDist!=-1.0 && cDist<minDist) minDist=cDist;
                    }
            }
            t+=VERTEX_ARRAY_LEN;
        }
        // Limit range to the user defined max
        if (minDist>SCANNER_DIST_MAX) minDist=SCANNER_DIST_MAX;

        // Round range to the user defined resolution
        minDist=(float) ((int)(minDist/(float)SCANNER_DIST_RES)) *
                                (float)SCANNER_DIST_RES;
        return(minDist);
}

//-------------------------------------------------------------------------
// Function    :
// Parameters  :
// Description : Range is output in cm/4
//-------------------------------------------------------------------------
```

```
void outputRangeToFPGA(float *scan, _scaleData scale)

        int i;

        // Overall output time per frame must be CAM_XPIXELS*TIME_PERSAMPLE
        if (timer_set (COUNT_PERSAMPLE*scale.SS, COUNT_PERSAMPLE*scale.SS) != 1)
                timer_on();

        // For each range measurement
        for (i=0; i<TABLE_length/scale.SS; i++) {

                //Wait for timer interrupt then send data
                while(!timerIRQOccured) {}
                timerIRQOccured=0;
                //Output data in cm/4
                *(pMS2) = ((int)(scan[i]*100/4));
        }
        timer_off();
}

//-----------------------------------------------------------------------------
// Function    :
// Parameters  :
// Description :
//-----------------------------------------------------------------------------
void outputScaleToFPGA(_scaleData scale) {

            if (timer_set (1200, 1200) != 1) timer_on();

            // Wait for 30µs - output 1/2 img width
            while(!timerIRQOccured) {}
            timerIRQOccured=0;
            *(pMS2) = (int)(((scale.imgWidth)/2.0)-1);

            // Wait for 30µs - output changeScale Bit
            while(!timerIRQOccured) {}
            timerIRQOccured=0;
            if ((int)(scale.changeScale)==0) {
                       *(pMS2) = (int)0;
            } else {
                       *(pMS2) = (int)1;
            }

            // Wait for 30µs - output SD
            while(!timerIRQOccured) {}
            timerIRQOccured=0;
            *(pMS2) = (int)(scale.SD);

            // Wait for 30µs - output SS
            while(!timerIRQOccured) {}
            timerIRQOccured=0;
            *(pMS2) = (int)(scale.SS);

            // Wait for 30µs - output 1/2 img width
            while(!timerIRQOccured) {}
            timerIRQOccured=0;
            *(pMS2) = (int)(((scale.imgWidth)/2.0)-1);
            timer_off();
```

```
//-----------------------------------------------------------------------
// Function    : updateScaleData
//
// Parameters  : Takes the current and the previous minimum ranges as well as
//               the previous scale data.
//               Also uses the scale choice table (global variables) to
//               determine the boundaries at which scale changes should occur.
//
// Description : Using information about the nearest object, determine which
//               scale should currently be in use.
//
//               - If the nearest object gets closer between frames, we choose
//                 a scale so that lowerCutoff[i+1]>minDist>=lowerCutoff[i].
//               - If the nearest object gets further between frames we choose
//                 a scale so that upperCutoff[i]>=minDist>upperCutoff[i+1];
//               - If i=0 then the nearest object is too close, so we suspend
//                 processing by setting the changeScale paramter to 1.
//               - If this is the first frame (i.e. prevMinDist==0) then we flag
//                 a scale change so that we can get 3 frames of data at the
//                 correct scale
//               - If the new scale is different to the previous scale, then
//                 flag a scale change.
//-----------------------------------------------------------------------
void updateScaleData(float minRange, float prevMinRange, _scaleData *scale) {
    int i;
    int j;
    int oldWidth;

    scale->changeScale=0;          // Assume no scale change to start with
    oldWidth=scale->imgWidth;      // Remember what the current scale is
    i=0;                           // If no scale is found, then object is too
                                   // close!
    // Find appropriate scale
    for (j=0; j<6; j++) {
        // Note this may cause a bit of toggling as we cross a scale boundary
        // The range might fall below the cut-off, then move back over the
        // cut-off on the next frame. This usually settles quickly but will
        // cause a bit of "flicker" as scales change.
        if ((minRange>=upperCutoff[j]) && (minRange<upperCutoff[j+1])) i=j;
    }
    if (prevMinRange!=minRange) {
            scale->imgWidth = ImgValue[i];
            scale->SD = SDValue[i];
            scale->SS = SSValue[i];
    }

    // Flag a scale change if...
    if (scale->imgWidth!=oldWidth)  scale->changeScale = 1; // ...changed scale
    else if (prevMinRange==0)       scale->changeScale = 1; // ...this is
                                                            // the first frame
    else if (i==0)                  scale->changeScale = 1; // ...object is
                                                            // too close
}

//-----------------------------------------------------------------------
// Function    : camBaseline
//
// Parameters  : Angle contains angle of pixel centre in radians
//               SS is the current degree of subsampling (in pixels)
```

```
//                      Also uses CAM_XPIXELS, CAM_PIXEL_PITCH,
//                      CAM_FOCAL_LENGTH
// Description : Work out the angle from the centre of each pixel through the
//                      pin hole taking subsampling into account. We assume an EVEN
//                      number of pixels and that the optical axis passes between the
//                      middle two pixels.
//
//                                  X Direction (for camera)
//                      Pixel 4  |\   |
//                               \ |
//                      Pixel 3  |  \|Pinhole
//                      ---------\---------- Optical Axis (Z direction for camera)
//                      Pixel 2  |   |\  a
//                               |    \
//                      Pixel 1  |    |  \
//
//                         ->---<- Focal Length
//
//                      The above diagram illustrates our pin hole camera model.
//                      The angle we are computing is marked with 'a'. An intermediate
//                      part of the calculation determines the physical distance of
//                      a pixel's centre from the optical axis.
//-----------------------------------------------------------------------------
void camBaseline(float *Angle, int SS) {

        int counter;
        float pos;
        float location;
        float effective_pitch = CAM_PIXEL_PITCH*SS;
        float CAM_PIXELRANGE=((CAM_XPIXELS/SS)-1)/2.0;

        counter = 0;
        for (pos=-CAM_PIXELRANGE; pos<=CAM_PIXELRANGE; pos++) {
                location=effective_pitch*pos;
                Angle[counter]=-atan2f(location,(float)CAM_FOCAL_LENGTH);
                counter++;
        }
}


//-----------------------------------------------------------------------------
// Function     : SINCOS_LUT
// Parameters   : A pointer to an array of floats containing the angles at
//                      which to compute sin/cos + a integer telling how many elements
//                      are in the array.
//                      Writes to global variables SINE_lookuptable, COS_lookuptable
//                      and TABLE_Length
// Description  : Generates a lookup table for the SIN and COS functions to
//                      speed up operation.
//-----------------------------------------------------------------------------
void SINCOS_LUT(float *angles, int number) {
        int index;

        for (index=0; index<number; index++) {
                SINE_lookuptable[index]=sin(angles[index]);
                COS_lookuptable[index] =cos(angles[index]);
        }

        TABLE_length = number;
}
```

```
//----------------------------------------------------------------------
// Function     : linelineIntersection
//
// Parameters   : alpha_index - index to angle between line and the Z axis in
//                radians index=0          corresponds to angle SCANNER_ANG_MIN
//                index=SCANNER_POINTS-1 corresponds to angle SCANNER_ANG_MAX
//                We use this form so that we can use LUT rather than
//                the expensive SIN/COS functions.
//                P3, P4 are two points defining a LINE SEGMENT in the
//                environment
//
// Description  : Determines the intersection of two lines using the algorithm
//                http://astronomy.swin.edu.au/~pbourke/geometry/sphereline
//
//                Assumes the first line (defined by points P1 and P2) starts
//                from the origin (i.e. P1=(0,0)), that the line is of unit
//                length and that the line forms angle alpha (with corresponding
//                alpha_index) with the Z axis. This way of solving for the
//                intersection automatically gives the distance from the origin
//                to the intersection.
//
//                Will return -1 if the ray from the range scanner and the
//                LINE SEGMENT defined by P3 and P4 do not intersect otherwise
//                returns the distance from the origin to the point of
//                intersection.
//----------------------------------------------------------------------
inline float linelineIntersection(int alpha_index, float X3, float Z3,
                                  float X4, float Z4) {
    float term1, term2;
    float den;
    float ua,ub;

    term1 = X4-X3;
    term2 = Z4-Z3;

    den = term2*SINE_lookuptable[alpha_index] -
          term1*COS_lookuptable[alpha_index];

    if (den==0) return(-1.0);                  // Lines are parallel
    else {
        ua = (-term1*Z3 + term2*X3)/den;
        ub = (-Z3*SINE_lookuptable[alpha_index] +
              X3*COS_lookuptable[alpha_index])/den;

        if (ub<0 || ub>1) return (-1.0);       // No intersection in line
                                               // segment
        else if (ua<0) return (-1.0);          // Forward only
        else return(ua);
    }
}


//----------------------------------------------------------------------
// Function     : linecircleIntersection
//
// Parameters   : alpha_index - index to angle between line and the Z axis in
//                radians index=0          corresponds to angle SCANNER_ANG_MIN
//                index=SCANNER_POINTS-1 corresponds to angle SCANNER_ANG_MAX
//                We use this form so that we can use LUT rather than
//                the expensive SIN/COS functions.
```

```
//                centre - centre of the circle in metres
//                radius - radius of the circle
//
// Description : Determines the point of intersection between a circle and
//               a line using the algorithm from
//               http://astronomy.swin.edu.au/~pbourke/geometry/sphereline
//
//               Assumes the line (defined by points P1 and P2) starts from
//               the origin (i.e. P1 = (0,0)), that the line is of unit
//               length and that the line forms angle alpha with the Z axis.
//               This way solving for u effectively gives the distance
//               of the circle from the origin.
//
//               Will return -1 if the line does not intersect the circle,
//               otherwise returns u for the "near side" intersection with
//               the circle.
//-----------------------------------------------------------------------------
inline float linecircleIntersection(int alpha_index, float Xc, float Zc,
                                    float radius) {
        float u;
        float b,c;
        float det;

        b = -2.0 * ((Xc*SINE_lookuptable[alpha_index]) +
                    (Zc*COS_lookuptable[alpha_index]));
        c = (Xc*Xc) + (Zc*Zc) - (radius*radius);

        det = b*b - 4.0*c;

        if (det<0) return(-1.0);                    // No intersection
        else {
                det=sqrt(det);
                u=MIN((-b+det), (-b-det))/2.0;
        }

        if (u<0) return (-1.0);                     // Forward only
        else return(u);
}


//-----------------------------------------------------------------------------
// Function    : hwInit
//
// Description : Do all the hardware initialisation stuff
//-----------------------------------------------------------------------------
void hwInit(void) {
        int i;
        char msg[15];

        // Setup Signalmaster
        for(i=0;i<JWAIT_STATE;i++);                 // Wait before using display
        DspInit();                                  // Initialise DSP
        for(i=0;i<JWAIT_STATE;i++);                 // Wait before using display
        IpcInit();                                  // Initialise comms
        for(i=0;i<JWAIT_STATE;i++);                 // Wait before using display
        LcdInit();                                  // Initialise LCD
        for(i=0;i<JWAIT_STATE;i++);                 // Wait before using display
        LcdClear();                                 // Clear LCD
```

```
            // Setup Gatesmaster
            for(i=0;i<2000000;i++) {}
            LcdGotoXY(0,0); LcdWriteString("Initialising    ");
            GatesMasterInit();                  // Initialise GM

            for(i=0;i<2000000;i++) {}
            LcdGotoXY(0,0); LcdWriteString("Bypassing       ");
            GatesMasterSetBypass(1);            // Turn off bypass

            // Setup timer interrupt
            timerIRQOccured=0;                  // Clear interrupt flag
            interrupt(SIG_TMZ0, timer_irq_handler); // Register timer
                                                // interrupt handler
            // Setup interrupt for 1s delay
            if (timer_set (40000000, 40000000) != 1) timer_on();

            // FPGA commences Calibration 5s after being booted. This gives
            // us enough time to turn off bypass after booting. We wait a little
            // more than 5s just to be sure that the FPGA is really running
            // before starting processing/output from the DSP.
            for (i=0;i<5;i++){                  // 10 second countdown
                    while(!timerIRQOccured) {}
                    timerIRQOccured=0;
                    LcdGotoXY(0,1);
                    sprintf(msg,"DSP Starts in %i ",5-i-1);
                    LcdWriteString(msg);
            }
            timer_off();                        // Turn timer off
            LcdClear();                         // Clear LCD

            LcdGotoXY(0,1);
            LcdWriteString("DSP Running     ");
}

//*******************************************************************************
// ****************************** MAIN *****************************************
//*******************************************************************************
void main(void) {

    int iJK;
    float calctime;
    char msg[15];

    float *pixelAngles;         // Pixel angles
    float *finalRanges;         // Camera range measurements taking
                                // subsampling to account.
    float currentTime;
    float minRange, prevMinRange;
    _scaleData scale;           // Scale information
    float *objectPos;           // Position of each object

    //----------------
    // Initialisation
    //----------------

    // Initialise scale parameters - assume no subsampling to start with
    scale.imgWidth=CAM_XPIXELS;
    scale.SD=0;
    scale.SS=1;
```

```
scale.changeScale=0;
minRange = 0;                       // This value will force a scale change after
                                    // the first frame

// Allocate memory
pixelAngles      = (float *)malloc(sizeof(float)*CAM_XPIXELS);
if (pixelAngles==NULL) exit(0);
finalRanges      = (float *)malloc(sizeof(float)*CAM_XPIXELS);
if (finalRanges==NULL) exit(0);
SINE_lookuptable = (float *)malloc(sizeof(float)*CAM_XPIXELS);
if (SINE_lookuptable==NULL) exit(0);
COS_lookuptable  = (float *)malloc(sizeof(float)*CAM_XPIXELS);
if (COS_lookuptable==NULL) exit(0);
objectPos        = (float *)malloc(sizeof(float)*NUMBER_OBJECTS *
                                   VERTEX_ARRAY_LEN);
if (objectPos==NULL) exit(0);

camBaseline(pixelAngles, 1);         // Generate pixel angles
SINCOS_LUT(pixelAngles,CAM_XPIXELS); // Generate lookup tables

// Setup timing parameters
calctime = ((float)TIME_PERSCAN-((float)CAM_XPIXELS *
                  (float)TIME_PERSAMPLE));
calctime = calctime*(float)CLKRATE;

hwInit();            // Setup the GatesMaster and Signal Master Hardware

//-----------
// Main Loop
//-----------
frame=0;
iJK=0;
currentTime=0;
currentlyBusy=0;    // Not busy yet!
while (1) {
        // Set timer to interrupt after TIME_PERSCAN -
        // TIME_PERSAMPLE*CAM_XPIXELS
        // I.e. time it takes for a scan - time it takes to output data
        // This way we can be sure to get one complete scan every
        // TIME_PERSCAN second

    if (timer_set ((int)calctime, (int)calctime) != 1) timer_on();
    currentlyBusy=1; // Busy now - starting to process data. If we get
                    // interrupted (i.e. timer interrupt) while processing
                    // then obviously we can't process as quickly as user
                    // wants. The interrupt handler function will display
                    // a message on the LCD to inform user.

    // Update environment - find object positions in the camera centred
    // coordinate system
    updateObjectPos(currentTime, objectPos);

    // Generate range scan using the camera angles
    prevMinRange=minRange;
    minRange=performScan(finalRanges, objectPos, scale);

    currentlyBusy=0;              // Processing complete - no longer busy
    while(!timerIRQOccured) {}    // Wait for timer interrupt
    timerIRQOccured=0;            // Clear timer interrupt flag
```

```c
        timer_off();                        // Turn timer off so it can be used for
                                            // data output

    // Output Range data (using old scale data)
    outputRangeToFPGA(finalRanges, scale);

    // Update scale data for next frame
    updateScaleData(minRange, prevMinRange, & scale);

    // Output scale data for next frame
    outputScaleToFPGA(scale);

    // Output current status to DSP occasionally.
    // Output Time, current scale and current minimum range.
    if (iJK==10) {
                sprintf(msg,"T%i S%i R%i CS%i",(int)(currentTime), //
                        scale.SS,(int)(minRange*100),scale.changeScale);
                LcdGotoXY(0,0);
                LcdWriteString(msg);
                iJK=0;
    }
    frame++;
    iJK++;
    currentTime+=(float)TIME_PERSCAN;                        // Update time.
}   // end main loop

//----------
// Clean up
//----------

// Won't usually get here since the main loop is an infinite loop.
// Included for completeness
//free(pixelAngles);
//free(finalRanges);
//free(SINE_lookuptable);
//free(COS_lookuptable);
//free(objectPos);
}
```

Bibliography

1 'Bus arbitration, I/O interface to CPU, memory' http://www.cs.umt.edu/CS/COURSES/CS232-votavap/Slides/lecture22x2.pdf, cited 20 April 2004, last update 2001
2 'Discriminant analysis, the mahalanobis distance', http://www.galactic.com/Algorithms/discrim_mahaldist.htm, cited 15 April 2004, last updated 2002
3 'Gate count capacity metrics for FPGAs', Xilinx Application Note XAPP 059, V1.1, 1 February 1997. Available from http://www.xilinx.com/bvdocs/appnotes/xapp059.pdf, cited 10 August 2004
4 'Mahalanobis distance', www.wu-wien.ac.at/usr/h99c/h9951826/distance.pdf, cited 15 April 2004
5 'Virtex 2.5V field programmable gate arrays product specification', 2 April 2001. Available from http://direct.xilinx.com/bvdocs/publications/ds003.pdf, cited 10 April 2004
6 ADELSON, E. H., ANDERSON, C. H., BERGEN, J. R., BURT, P. J., and OGDEN, J. M.: 'Pyramid methods in image processing', *RCA Engineer*, 1984, **29**(6), pp. 33–41
7 ADORNI, G., BROGGI, A., CONTE, G., and D'ANDREA, V.: 'A self-tuning system for real-time optical flow detection', *Proc. IEEE System, Man, and Cybernetics Conference*, 1993, **3**, pp. 7–12
8 AGULLO, J.: 'New algorithms for computing the least trimmed squares regression estimator', *Computational Statistics & Data Analysis*, 2001, **36**, pp. 425–439
9 ALFKE, P., and PHILOFSKY, B.: 'Metastable recovery', Xilinx Application Note 94, XAPP 094, November 24, 1997. Available from www.xilinx.com
10 ALFKE, P., and PHILOFSKY, B.: 'Metastable recovery', Xilinx Application Note 94, XAPP 094, November 24, 1997. Available from www.xilinx.com
11 ALOUPIS, G.: 'On computing geometric estimators of location', Masters Thesis, School of Computer Science, McGill University, Canada, March 2001. Also available from http://cgm.cs.mcgill.ca/~athens/
12 ALOUPIS, G.: 'Robust estimators of location', http://cgm.cs.mcgill.ca/~athens/Geometric-Estimators/location.html, cited 14 April 2004

13 AL-REGIB, G., ALTUNBASAK, Y., and MERSEREAU, R. M.: 'Hierarchical motion estimation with content-based meshes', *IEEE Trans. Circuits and Systems for Video Technology*, 2003, **13**(10), pp. 1000–1005
14 ANADAN, P.: 'A computational framework and an algorithm for the measurement of visual motion', *International Journal of Computer Vision*, 1989, **2**, pp. 283–310
15 ANADAN, P., BERGEN, J. R., HANNA, K. J., and HINGORANI, R.: 'Hierarchical model-based motion estimation', Chapter 1 in 'Motion analysis and image sequence processing', SEZAN, M. I., and LAGENDIJK, R. L. (Eds) (Kluwer Academic Publishers, Boston, USA, 1993)
16 ANANDAN, P.: 'Motion and stereopsis', Chapter 4 in 'Computer vision: theory and industrial applications', TORRAS, C. (Ed.) (Springer Verlag, New York, USA, 1992)
17 ANCONA, N., and POGGIO, T.: 'Optical flow from 1D correlation: application to a simple time-to-crash detector', AI Memo 1375, MIT Artificial Intelligence Laboratory, October 1993
18 ANDREWS, D. F. *et al.*: 'Robust estimates of location: survey and advances' (Princeton University Press, Princeton, New Jersey, 1972)
19 ANSCOMBE, F. J.: 'Rejection of outliers', *Technometrics*, 1960, **2**(2), pp. 123–165
20 ARGYROS, A. A., LOURAKIS, M. I. A., TRAHANIAS, P. E., and ORPHANOUDAKIS, S. C.: 'Independent 3D motion detection through robust regression in depth layers', British Machine Vision Conference (BMVC '96). Available for download from http://www.ics.forth.gr/~argyros/research/imd.htm, 1996
21 ARGYROS, A. A., and ORPHANOUDAKIS, S. C.: 'Independent 3D motion detection based on depth elimination in normal flow fields', Conference on *Computer Vision and Pattern Recognition* (CVPR97), San Juan, Puerto Rico, June 1997, pp. 672–677
22 ASKEY, P.: 'Depth of field', http://www.dpreview.com/learn/Glossary/Optical/Depth_of_field_01.htm, cited 13 January 2004
23 BAB-HADIASHAR, A., and SUTER, D.: 'Robust optic flow computation', *International Journal of Computer Vision*, 1998, **29**(1), pp. 59–77
24 BARATOFF, G., TOEPFER, C., WENDE, M., and NEUMMAN, H.: 'Real-time navigation and obstacle avoidance from optical flow on a space-variant map', Proceedings of the 1998 IEEE International Symposium on *Intelligent Control (ISIC)* held jointly with IEEE International Symposium on *Computational Intelligence in Robotics and Automation (CIRA) Intelligent Systems and Semiotics (ISAS)*, Gaithersbury, Maryland, USA, September 1998, pp. 289–294
25 BARRON, J. L., FLEET, D. J., and BEAUCHEMIN, S. S.: 'Systems and experiments: performance of optical flow techniques', *International Journal of Computer Vision*, 1994, **12**(1), pp. 43–77
26 BARRON, J. L., and SPIES, H.: 'The fusion of image and range flow', Multi-Image Analysis: 10th Int. Workshop on *Theoretical Foundations*

of Computer Vision, Lecture Notes in Computer Science, 2001, **2032**, pp. 171–189
27 BATTITI, R., AMALDI, E., and KOCH, C.: 'Computing optical flow across multiple scales: an adaptive coarse to fine strategy', International Journal of Computer Vision, 1991, **6**(2), pp. 133–145
28 BEAUCHEMIN, S. S., and BARRON, J. L.: 'The computation of optical flow', ACM Computing Surveys, 1995, **27**(3), pp. 433–467
29 BENOIT, S. M.: 'Monocular optical flow for real-time vision systems', Masters Thesis, Dept. of Electrical Engineering, McGill University, April 1999
30 BERALDIN, J. A., EL-HAKIM, S. F., and BLAIS, F.: 'Performance evaluation of three active vision systems built at the nation research council of Canada', Conf. on Optical 3D Measurement Techniques III, pp. 352–361, October 1995
31 BERTERO, M., POGGIO, T., and TORRE, V.: 'Ill-posed problems in early vision', A.I. Memo 924, MIT A.I. Lab, May 1987
32 BERTHOZ, A.: 'The brain's sense of movement' (Harvard University Press, Cambridge, MA, 2000)
33 BERTOZZI, M., BROGGI, A., and FASCIOLI, A.: 'Vision-based intelligent vehicles: state of the art and perspectives', Robotics and Autonomous Systems, 2000, **32**, pp. 1–16
34 BETKE, M., HARITAOGLU, E., and DAVIS, L. S.: 'Real-time multiple vehicle detection and tracking from a moving vehicle', Machine Vision and Applications, 2000, **12**, pp. 69–83
35 BLACK, M. J.: 'Robust incremental optical flow', Ph.D. Thesis, Yale, 1992
36 BLAKE, A., and ZISSERMAN, A.: 'Visual reconstruction' (MIT Press, Cambridge, Massachusetts, USA, 1987)
37 BLOOMFIELD, P., and STEIGER, W. L.: 'Least absolute deviations: theory, applications and algorithms' (Birkhauser, Boston, USA, 1983)
38 BOBER, M., and KITTLER, J.: 'Combining the Hough transform and multiresolution MRFs for the robust motion estimation', in 'Recent developments in computer vision', 2nd Asian Conference on Computer Vision: Invited Session Papers, LI, S. K., MITAL, D. P., TEOH, E. K., and WAN, H. (Eds) pp. 91–100, 1995
39 BOBER, M., PETROU, M., and KITTLER, J.: 'Nonlinear motion estimation using the supercoupling approach', IEEE Trans. Pattern Analysis and Machine Intelligence, 1998, **20**(5), pp. 550–555
40 BOLDUC, M., and LEVINE, M. D.: 'A review of biologically motivated space-variant data reduction models for robotic vision', Computer Vision and Image Understanding, 1998, **69**(2), pp. 170–184
41 BOLDUC, M., and LEVINE, M. D.: 'A review of biologically motivated space-variant data reduction models for robotic vision', Computer Vision and Image Understanding, 1998, **69**(2), pp. 170–184
42 BOLUDA, J. A., DOMINGO, J., PARDO, F., and PELECHANO, J.: 'Motion detection independent of the camera movement with a log-polar sensor', Computer Analysis of Images and Patterns, Lecture Notes in Computer Science, 1997, **1296**, pp. 702–709

43 BOURKE, P.: 'Intersection of a sphere and a line (or a circle)', http://astronomy.swin.edu.au/~pbourke/geometry/sphereline, cited 4 January 2004, last updated November 1992

44 BROGGI, A.: 'Robust real-time lane and road detection in critical shadow conditions', *Proc. IEEE International Symposium on Computer Vision*, pp. 353–358, 1995

45 BROGGI, A., and BERTOZZI, M.: 'ARGO prototype vehicle', Chapter 14 in 'Intelligent vehicle technologies: theory and applications', VLACIC, L., PARENT, M., and HARASHIMA, F. (Eds) (Butterworth Heinemann, Reading, Massachusetts, USA, 2001)

46 BROGGI, A., BERTOZZI, M., and FASCIOLI, A.: 'Critical analysis of a stereo vision-based guidance system', Proceedings ISATA – International Symposium on *Automotive Technology and Automation*, Wien, Austria, 14–17 June, 1999, pp. 161–168

47 BROGGI, A., CONTE, G., and BURZIO, G. *et al.*: 'PAPRICA-3: a real-time morphological image processor', Proc. ICIP – First IEEE International Conference on *Image Processing*, Austin, Texas, USA, 13–16 November, 1994, pp. 654–658

48 BROWN, C. (Ed.): 'Tutorial on filtering, restoration, and state estimation', Technical Report 534, The University of Rochester, Computer Science Department, June 1995

49 BROWN, M. Z., BURSCHKA, D., and HAGER, G. D.: 'Advances in computational stereo', *IEEE Transactions on Pattern Analysis and Machine Intelligence*, 2003, **25**(8), pp. 993–1008

50 BRUSS, A. R., and HORN, B. K. P.: 'Passive navigation', *Computer Vision, Graphics, and Image Processing*, 1983, **21**, pp. 3–20

51 BURT, P. J., YEN, C., and XU, X.: 'Local correlation measures for motion analysis: a comparative study', IEEE Computer Society Conference on *Pattern Recognition and Image Processing* (PRIP82), Las Vegas, USA, June, 1982, pp. 269–274

52 CANTONI, V., and LEVIALDI, S.: 'Note: matching the task to an image processing architecture', *Computer Vision, Graphics and Image Processing*, 1983, **22**, pp. 301–309

53 CASTELLANO, G., BOYCE, J., and SANDLER, M.: 'Regularized CDWT optical flow applied to moving target detection in IR imagery', *Machine Vision and Applications*, 2000, **11**, pp. 277–288

54 CHALIDABHONGSE, J., and KUO, C.-C. J.: 'Fast motion vector estimation using multiresolution-spatio-temporal correlations', *IEEE Trans. Circuits and Systems for Video Technology*, 1997, **7**(3), pp. 477–488

55 CHATILA, R.: 'Deliberation and reactivity in autonomous mobile robots', *Robotics and Autonomous Systems*, 1995, **16**(2), pp. 197–211

56 CHAUDHURY, K., MEHROTRA, R., and SRINIVASAN, C.: 'Detecting 3D flow', Proc. IEEE Int. Conf. *Robotics and Automation*, San Diego, USA, 1994, **2**, pp. 1073–1078

57 CHEN, P.-Y.: 'A fuzzy search block-matching chip for motion estimation', *Integration, The VLSI Journal*, 2002, **32**, pp. 133–147

58   CHEN, V. C.: 'Gabor-wavelet pyramid for the extraction of image flow', *SPIE Mathematical Imaging*, 1993, **2034**, pp. 128–136
59   CHIN, T. M., KARL, W. C., and WILLSKY, A. S.: 'Probabilistic and sequential computation of optical flow using temporal coherence', *IEEE Trans. on Image Processing*, 1994, **3**(6), pp. 773–788
60   CIPOLLA, R., and BLAKE, A.: 'Surface orientation and time to contact from image divergence and deformation', European Conference on *Computer Vision*, ECCV92, *Lecture Notes in Computer Science*, 1992, **588**, pp. 187–202
61   COHEN, I., and HERLIN, I.: 'Non uniform multiresolution method for optical flow and phase portrait models: environmental applications', *International Journal of Computer Vision*, 1999, **33**(1), pp. 29–49
62   COMANICIU, D., and MEER, P.: 'Mean shift: a robust approach towards feature space analysis', *IEEE Trans. Pattern Analysis and Machine Intelligence*, 2002, **24**(5), pp. 603–619. Also available from http://www.caip.rutgers.edu/riul/research/e_mnshft.pdf
63   COOMBS, D., HERMAN, M., HONG, T., and MASHMAN, M.: 'Real-time obstacle avoidance using central flow divergence and peripheral flow', NIST Internal Report (NISTIR) 5605, 1995
64   COOPER, G., and McGILLEM, C.: 'Probabilistic methods of signal and system analysis' (Oxford University Press, Oxford, 1999, 3rd edn)
65   CORNELIUS, N., and KANADE, T.: 'Adapting optical flow to measure object motion in reflectance and X-ray image sequences', ACM SIGGRAPH/SIGART Interdisciplinary Workshop on *Motion: Representation and Perception*, Toronto, Canada, April, 1983
66   CRICK, F.: 'The astonishing hypothesis' (Touchstone Books, Simon & Schuster, London, 1995)
67   CUCCHIARA, R., PICCARDI, M., and PRATI, A.: 'Detecting moving objects, ghosts and shadows in video streams', *IEEE Trans. Pattern Anal. and Machine Vision*, 2003, **25**(10), pp. 1337–1342
68   DARRELL, T., and SIMONCELLI, E.: 'Separation of transparent motion into layers using velocity-tuned mechanisms', MIT Media Laboratory Vision and Modelling Group Technical Report #244, October 1993
69   DARRELL, T., and SIMONCELLI, E.: 'On the use of "Nulling" filters to separate transparent motions', MIT Media Laboratory Vision and Modelling Group Technical Report, No. 198, 1992. This paper can also be downloaded from http://cgi.media.mit.edu/vismod/tr_pagemaker.cgi?range=99-199
70   de GROEN, P. P. N.: 'An introduction to total least squares', http://arxiv.org/abs/math.RA/9805076, last updated Mon, 18 May 1998, cited 24 April 2004
71   DENNEY, T. S. JR., and PRINCE, J. L.: 'Optimal brightness patterns for 2-D optical flow', ICASSP-93, *IEEE Int. Conf. Acoustics, Speech and Signal Proc.*, 1993, **5**, pp. 225–228
72   DERICHE, R., KORNPROBST, P., and AUBERT, G.: 'Optical-flow estimation while preserving its discontinuities: a variational approach', Recent Developments in Computer Vision, 2nd Asian Conference on *Computer*

Vision: Invited Session Papers, LI, S. K., MITAL, D. P., TEOH, E. K., and WAN, H. (Eds), 1995, pp. 71–80
73 DEV, A.: 'Visual navigation on optical flow', Ph.D. Thesis, University of Amsterdam, 1998
74 Development System Reference Guide, Xilinx Foundation 2.1i Documentation, 1999. Available from http://toolbox.xilinx.com/docsan/2_1i/
75 DRON, L. G.: 'Computing 3D motion in custom analog and digital VLSI', Ph.D. Thesis, Dept. Electrical Engineering and Computer Science, MIT, 1994
76 DUBOIS, E., and KONRAD, J.: 'Estimation of 2D motion fields from image sequences with application to motion compensated processing', Chapter 3 in 'Motion analysis and image sequence processing', SEZAN, M. I., and LAGENDIJK, R. L. (Eds) (Kluwer Academic Publishers, Boston, USA, 1993)
77 DUNCAN, J. H., and CHOU, T.-C.: 'On the detection of motion and the computation of optical flow', *IEEE Trans. Pattern Analysis and Machine Intelligence*, 1992, **14**(3), pp. 346–352
78 DURIC, Z., ROSENFELD, A., and DUNCAN, J.: 'The applicability of Green's theorem of computation of rate of approach', *International Journal of Computer Vision*, 1999, **31**(1), pp. 83–98
79 ELOUARDI, A., BOUAZIZ, S., DUPRET, A., KLIEN, J. O., and REYNAUD, R.: 'On chip vision system architecture using a CMOS retina', 2004 IEEE Intelligent Vehicles Symposium (IV2004), University of Parma, Parma, Italy, 14–17 June, 2004, pp. 206–211
80 ENKELMANN, W.: 'Investigation of multigrid algorithms for the estimation of optical flow', *CVGIP*, 1988, **43**, pp. 150–177
81 ETIENNE-CUMMINGS, R., Van der SPIEGEL, J., and MUELLER, P.: 'A focal plane visual motion measurement sensor', *IEEE Trans. Circuits and Systems I: Fundamental Theory and Applications*, 1997, **44**(1), pp. 55–66
82 ETIENNE-CUMMINGS, R., Van der SPIEGEL, J., MUELLER, P., and ZHANG, M.-Z.: 'A foveated silicon retina for two-dimensional tracking', *IEEE Trans. Circuits and Systems II: Analog and Digital Signal Processing*, 2000, **47**(6), pp. 504–517
83 EWALD, A., and WILLHOEFT, V.: 'Laser scanners for obstacle detection in automotive applications', Intelligent Vehicles Symposium 2000 (IV2000), Dearborn, Michigan, USA, 3–5 October, 2000, pp. 682–687
84 FANUCCI, L., SAPONARA, S., and BERTINI, L.: 'A parametric VLSI architecture for video motion estimation', *Integration, The VLSI Journal*, 2001, **31**, pp. 79–100
85 FARID, H., and SIMONCELLI, E. P.: 'Optimally rotation – equivariant directional derivative kernels', 7th Int. Conf. *Computer Analysis of Images and Patterns*, Kiel, German, 10–12 September, 1997
86 FARID, H., and SIMONCELLI, E. P.: 'Range estimation by optical differentiation', *Journal of the Optical Society of America*, 1998, **15**(5), pp. 1777–1786

87 FAWCETT, B.: 'PLD capacity and gate counting', *Xilinx XCELL* Magazine, 1996, **23**, Q4. Available from http://www.xilinx.com/xcell/xl23/xl23_2.pdf, cited 10 August 2004

88 FERMULLER, C., and ALOIMONOS, Y.: 'What is computed by structure from motion algorithms', Computer Vision Laboratory, Center for Automation Research, University of Maryland, Technical Report CAR-TR-863 & CS-TR-3809, June 1997

89 FERRARA, A.: 'Automatic pre-crash collision avoidance in cars', 2004 IEEE Intelligent Vehicles Symposium (IV2004), University of Parma, Parma, Italy, 14–17 June, 2004, pp. 133–138

90 FIORINI, P.: 'Robot motion planning among moving obstacles', Ph.D. Thesis, University of California, 1995

91 FISHER, R. B.: 'The RANSAC (Random Sample Consensus) algorithm', http://homepages.inf.ed.ac.uk/rbf/CVonline/LOCAL_COPIES/FISHER/RANSAC/, last updated 6 May 2002, cited 3 June 2004

92 FISHER, R.: 'Geometric feature extraction methods', http://www.dai.ed.ac.uk/CVonline/feature.htm, site last updated 26 August 2003, cited 11 December 2003

93 FLEET, D. J., and JEPSON, A. D.: 'Computation of component image velocity from local phase information', *International Journal of Computer Vision*, 1990, **5**(1), pp. 77–104

94 Foundation Series 2.1i User Guide, Xilinx Foundation 2.1i Documentation, 1999. Available from http://toolbox.xilinx.com/docsan/2_1i/

95 FRANKE, U., and HEINRICH, S.: 'Fast obstacle detection for urban traffic situations', *IEEE Trans. Intelligent Transportation Systems*, 2002, **3**(3), pp. 173–181

96 FREEMAN, W. T., PASZTOR, E. C., and CARMICHAEL, O. T.: 'Learning low-level vision', *International Journal of Computer Vision*, 2000, **40**(1), pp. 25–47

97 ftp://ftp.fu-berlin.de/pub/unix/graphics/polyray/, cited 5 January 2004, last updated 1997

98 GALVIN, B., McCAN, B., NOVINS, K., MASON, D., and MILLS, S.: 'Recovering motion fields: an evaluation of eight optical flow algorithms', British Machine Vision Conference, University of Southampton, UK, 1998, pp. 195–204

99 GAO, Q., YANG, J., and SUEMATSU, Y.: 'The improved algorithm to compute the optical flow in the wide angle high distortion lens system', Proc. ICAR2003, 11th Int. Conf. on *Advanced Robotics*, University of Coimbra, Portugal, 30 June–3 July 2003, pp. 1166–1171

100 GEIGER, D., and GIROSI, F.: 'Parallel and deterministic algorithms from MRFs: surface reconstruction', *IEEE Trans. Pattern Analysis and Machine Intelligence*, 1991, **13**(5), pp. 401–412

101 GERALD, C. F., and PATRICK, O. W.: 'Applied numerical analysis' (Addison-Wesley Publishing Company, Reading, Massachusetts, USA, 1994, 5th edn)

102 GERVINI, D.: 'A robust and efficient adaptive reweighted estimator of multivariate location and scatter', *Journal of Multivariate Analysis*, 2003, **84**, pp. 116–144
103 GERVINI, D., and YOHAI, V. J.: 'A class of robust and fully efficient regression estimators', *The Annals of Statistics*, 2002, **30**(2), pp. 583–616
104 GIACHETTI, A., and TORRE, V.: 'Refinement of optical flow estimation and detection of motion edges', European Conference on *Computer Vision* (ECCV96), Cambridge, UK, 1996, pp. 151–160
105 GIBSON, J.: 'The perception of the visual world' (Houghton Mifflin, Boston, 1950)
106 GIROD, B.: 'Motion compensation: visual aspects, accuracy, and fundamental limits', Chapter 5 in 'Motion analysis and image sequence processing', SEZAN, M. I., and LAGENDIJK, R. L. (Eds) (Kluwer Academic Publishers, Boston, USA, 1993)
107 GIROD, B.: 'Motion-compensating prediction with fractional-pel accuracy', *IEEE Transactions on Communications*, 1993, **41**(4) pp. 604–612
108 GOLLAND, P., and BRUCKSTEIN, A. M.: 'Motion from colour', *Computer Vision and Image Understanding*, 1997, **68**(3) pp. 346–362
109 GONZALEZ, R. C., and RICHARDS, E. W.: 'Digital image processing' (Addison Wesley, Reading, Massachusetts, USA, 1992)
110 GOVINDARAJULI, Z., and LESLIE, R. T.: 'Annotated bibliography on robustness studies of statistical procedures', U.S. Department of Health, Education and Welfare, DHEW Publication Number (HSM) 72-1051, Vital and Health Statistics Series 2, No. 51, April 1972. Available from http://www.cdc.gov/nchs/products/pubs/pubd/series/sr02/100-1/100-1.htm
111 GREGORETTI, F., REYNERI, L. M., SANSOÈ, C., BROGGI, A., and CONTE, G.: 'The PAPRICA SIMD array: critical reviews and perspectives', Proc. ASAP'93 – IEEE Computer Society International Conference on *Application Specific Array Processors*, Venice, Italy, 25–27 October, 1993, pp. 309–320
112 GRUYER, D., ROYERE, C., LABAYRADE, R., and AUBERT, D.: 'Credibilistic multi-sensor fusion for real time application. Application to obstacle detection and tracking', Proc. ICAR 2003, International Conference on *Advanced Robotics*, University of Coimbra, Portugal, 30 June–3 July, 2003, pp. 1462–1467
113 GUPTA, N. C.: 'Recovering shape and motion from a sequence of images', Ph.D. Thesis, University of Maryland, 1993
114 GUPTA, N., and KANAL, L.: 'Gradient based image motion estimation without computing gradients', *International Journal of Computer Vision*, 1997, **22**(1), pp. 81–101
115 GUPTA, S. N., and PRINCE, J. L.: 'Stochastic formulations of optical flow algorithms under variable brightness conditions', Proc. Int. Conf. on *Image Proc.*, Washington D.C., 1995, **III**, pp. 484–487
116 HAAG, M., FRANK, T., KOLLNIG, H., and NAGEL, H. H.: 'Influence of an explicitly modelled 3D scene on the tracking of partially

occluded vehicles', *Computer Vision and Image Understanding*, 1997, **65**(2), pp. 206–255

117 HAAG, M., and NAGEL, H.: 'Combination of edge element tracking and optical flow estimates for 3D-model-based vehicle tracking in traffic image sequences', *International Journal of Computer Vision*, 1999, **35**(3), pp. 295–319

118 HACKETT, J. K., and SHAH, M.: 'Multi-sensor fusion: a perspective', *Trends in Optical Engineering*, 1993, **1**, pp. 99–118

119 HADI, A. S., and LUCENO, A.: 'Maximum trimmed likelihood estimators: a unified approach, examples and algorithms', *Computational Statistics & Data Analysis*, 1997, **25**, pp. 251–272

120 HALPERIN, D., KAVRAKI, L., and LATOMBE, J.-C.: 'Robot algorithms', Chapter 21 in 'Algorithms and theory of computation handbook', ATALLAH, M. J. (Ed.) (CRC Press, Boca Raton, Florida, USA, 1999)

121 HAMMING, R. W.: 'Numerical methods for scientists and engineers' (McGraw-Hill Book Company, New York, USA, 1962)

122 HAMPEL, F. R. *et al.*: 'Robust statistics: the approach based on influence functions' (John Wiley & Sons, New York, USA, 1986)

123 HANCOCK, J. A.: 'Laser intensity-based obstacle detection and tracking', Ph.D. Thesis, The Robotics Institute, Carnegie Mellon University, 1999

124 HARRISON, R. R., and KOCH, C.: 'A neuromorphic visual motion sensor for real-world robots', Workshop on *Defining the Future of Biomorphic Robotics*, IROS 1998, Victoria, B.C., Canada, 1998

125 HARTER, H. L.: 'The method of least squares and some alternatives', *Int. Stat. Rev.*, in 6 Parts – Part I: **42**(2), pp. 147–174, 1974. Part II: **42**(3), pp. 235–264, 1974. Part III: **43**(1), pp. 1–44, 1975. Part IV: **43**(2), pp. 125–190, 1975. Part V: **43**(3), pp. 269–278, 1975. Part VI: **44**(1), pp. 113–159, 1976

126 HARVILLE, M. *et al.*: '3D pose tracking with linear depth and brightness constraints', Proc. 7th Int. Conf. *Computer Vision*, Kerkyra, Greece, 20–27 September, 1999, pp. 206–213

127 HAYES, M. H.: 'Statistical digital signal processing and modelling' (John Wiley & Sons, New York, USA, 1996)

128 HEEGER, D. J., SIMONCELLI, E. P., and MOVSHON, J. A.: 'Computational models of cortical visual processing', *Proc. Natl Acad. Sci. USA.*, 1996, **93**, pp. 623–627

129 HEEGER, D., and SIMONCELLI, E. P.: 'Model of visual motion sensing', in 'Spatial vision in humans and robots', HARRIS, L. and JENKIN, M. (Eds) (Cambridge University Press, Cambridge, 1994), pp. 367–392

130 HEEL, J.: 'Direct estimation of structure and motion from multiple frames', AI Memo #1190, MIT Artificial Intelligence Laboratory, March 1990

131 HEITZ, F., and BOUTHEMY, P.: 'Multimodal estimation of discontinuous optical flow using Markov random fields', *IEEE Trans. Pattern Analysis and Machine Intelligence*, 1993, **15**(12), pp. 1217–1232

132 von HELMHOLTZ, H.: 'Treatise on physiological optics', 1924 (First English Translation, Gryphon Edition, New Jersey, USA, 1985). Available from http://www.psych.upenn.edu/backuslab/helmholtz/

133 HILDRETH, E. C., and KOCH, C.: 'The analysis of visual motion: from computation theory to neuronal mechanisms', MIT Artificial Intelligence Lab, AI Memo 919, 1986
134 HORN, B., and SCHUNCK, B.G.: 'Determining optical flow', MIT Artificial Intelligence Lab, AI Memo 572, April 1980
135 HORN, B.: 'Robot vision' (MIT Press, Cambridge, Massachusetts, USA, 1986)
136 HOUGHTON, A. D., MAWER, J. R., and IVEY, P. A.: 'ASIC for high resolution motion sensing', *Electronics Letters*, 1995, **31**(8), pp. 635–636
137 HOUGHTON, A., REES, G., and IVEY, P.: 'A method for processing laser speckle images to extract high-resolution motion', *Measurement Science and Technology*, 1997, **8**(6), pp. 611–617
138 http://www.cast-inc.com/, cited 2 April 2004
139 http://www.free-ip.com/cores.htm, cited 2 April 2004
140 http://www.griffith.edu.au/centre/icsl, cited 4 August 2004
141 http://www.ibeo-as.de/, cited 5 January 2004
142 http://www.intelliga.co.uk/, cited 2 April 2004
143 http://www.lyrtech.com, cited 5 January 2004
144 http://www.opencores.org/, cited 2 April 2004
145 http://www.seti.org, cited 5 January 2004
146 http://www.vector-international.be/C-Cam/Cindex.html, cited 5 January 2004
147 http://www.xilinx.com, cited 10 August 2004
148 HUBER, P. J.: 'Robust statistics' (John Wiley & Sons, New York, USA, 1981)
149 ISARD, M., and BLAKE, A.: 'CONDENSATION – conditional density propagation for visual tracking', *Int. Journal of Computer Vision*, 1998, **29**(1), pp. 5–28
150 IU, S.-L.: 'Robust estimation of motion vector fields with discontinuity and occlusion using local outliers rejection', *Journal of Visual Communication and Image Representation*, 1995, **6**(2), pp. 132–141
151 IU, S.-L., and WU, E. C.: 'Noise reduction using multi-frame motion estimation, with outlier rejection and trajectory correction', IEEE International Conference on *Acoustics, Speech and Signal Processing* (ICASSP93), Minneapolis, MN, USA, 27–30 April, 1993, **5**, pp. 205–207
152 JEPSON, A. D., and BLACK, M.: 'Mixture models for optical flow computation', *Proc. Computer Vision and Pattern Rec.*, 1993, pp. 760–761
153 JEPSON, A., and BLACK, M.: 'Mixture models for optical flow computation', Univ. Toronto, Dept. Computer Science, Technical Report: RBCV-TR-93-44, April 1993
154 JIN, J. S., ZHU, Z., and XU, G.: 'A stable vision system for moving vehicles', *IEEE Transactions on Intelligent Transportation Systems*, 2000, **1**(1), pp. 32–39
155 JOHNSON, D.: 'Cramer-Rao bound', available from http://cnx.rice.edu/content/m11266/latest, cited 28 April 2004, last updated 22 Aug 2003
156 KACZMARCZYK, G.: 'Downhill simplex method for many (~20) dimensions', http://paula.univ.gda.pl/~dokgrk/simplex.html, cited 14 April 2004, last modified December 1999

157 KARR, C. R., CRAFT, M. A., and CISNEROS, J. E.: 'Dynamic obstacle avoidance', *SPIE Proc. Distributed Interactive Simulation Systems in the Aerospace Environment*, 1995, **CR58**, pp. 195–219
158 KATO, T., NINOMIYA, Y., and MASAKI, I.: 'An obstacle detection method by fusion of radar and motion stereo', *IEEE Trans. Intelligent Transportation Systems*, 2002, **3**(3), pp. 182–188
159 KATZ, R. H.: 'Contemporary logic design' (Benjamin/Cummings Publishing Company Inc., Redwood City, California, USA, 1994)
160 KEARNEY, J. K., THOMPSON, W. B., and BOLEY, D. L.: 'Optical flow estimation: an error analysis of gradient-based methods with local optimisation', *IEEE Trans. Pattern Analysis and Machine Intelligence*, 1987, **9**(2), pp. 229–244
161 KNUDSEN, C. B., and CHRISTENSEN, H. I.: 'On methods for efficient pyramid construction', Proc. 7th Scandinavian Conf. on *Image Anal.*, 13–16, August 1991, pp. 29–39
162 KOLLER, D., WEBER, J., and MALIK, J.: 'Robust multiple car tracking with occlusion reasoning', *ECCV '94, Lecture Notes in Computer Science*, 1994, **800**, pp. 189–196
163 KOLODKO, J., PETERS, L., and VLACIC, L.: 'Motion estimation hardware for autonomous vehicle guidance', IEEE International Conference on *Industrial Control, Electronics and Instrumentation* (IECON 2000), Dearborn, Michigan, USA, 3–5 October, 2000
164 KOLODKO, J., PETERS, L., and VLACIC, L.: 'On the use of motion as a primitive quantity for autonomous vehicle guidance', IEEE Intelligent Vehicles Symposium 2000 (IV2000), Nagoya, Japan, 22–28 October, 2000, pp. 64–69
165 KOLODKO, J., and VLACIC, L.: 'A motion estimation system for autonomous navigation', 11th International Conference on *Advanced Robotics* (ICAR2003), University of Coimbra, Portugal, 30 June–3 July, 2003, **3**, pp. 1474–1479
166 KOLODKO, J., and VLACIC, L.: 'A motion estimation system', IEEE International Symposium on *Industrial Electronics* (ISIE2003), Rio de Janeiro, Brazil, 9–11 June, 2003, Paper #BF-000314
167 KOLODKO, J., and VLACIC, L.: 'Cooperative autonomous driving at the intelligent control systems laboratory', *IEEE Intelligent Systems Magazine*, July–August, 2003, pp. 8–11
168 KOLODKO, J., and VLACIC, L.: 'Experimental system for real-time motion sensing, segmentation and tracking', IEEE/ASME International Conference on *Advanced Intelligent Mechatronics* (AIM2003), Kobe, Japan, 20–24 June, 2003, pp. 981–986
169 KOLODKO, J., and VLACIC, L.: 'From motion processing to autonomous navigation', FIRA Robot World Congress (FIRA2002), Seoul, Korea, 26–29 May, 2002, pp. 160–164
170 KOLODKO, J., and VLACIC, L.: 'From motion processing to autonomous navigation', Chapter 3.2 in 'Intelligent robots: vision, learning and interaction' (KIAST Press, Daejeor, Republic of Korea, 2003), pp. 107–122

171 KOLODKO, J., and VLACIC, L.: 'Fusing vision and range for motion estimation in hardware', The 29th Annual Conference of the *IEEE Industrial Electronics Society* (IECON2003), Roanoke, Virginia, USA, 2–6 November, 2003, pp. 1487–1492

172 KOLODKO, J., and VLACIC, L.: 'Real time motion processing for autonomous navigation', *International Journal of Control, Automation, and Systems*, 2003, **1**(1), pp. 156–161

173 KOLODKO, J., and VLACIC, L.: 'Real-time motion segmentation hardware', *Autonomous Minirobots for Research and Edutainment* (AMIRE2003), Brisbane, Australia, 18–29, February 2003, pp. 83–92

174 KONRAD, J.: 'Motion detection and estimation', Chapter 3.8 in 'Image and video processing handbook', BOVIC, A. (Ed.) (Academic Press, San Diego, California, USA, 1999)

175 KONRAD, J., and DUBOIS, E.: 'Bayesian estimation of motion vector fields', *IEEE Transactions on Pattern Analysis and Machine Intelligence*, 1992, **13**(9), pp. 910–927

176 KONRAD, J., and STILLER, C.: 'On Gibbs-Markov models for motion estimation', Chapter 4 in 'Video data compression for multimedia computing: statistically based and biologically inspired techniques', LI, H., SUN, S., and DERIN, H. (Eds) (Kluwer Academic Publishers, Boston, USA, 1997), pp. 121–154

177 KRAMER, J., SARPESHKAR, R., and KOCH, C.: 'Analog VLSI motion discontinuity detectors for image segmentation', IEEE Int. Symp. on *Circuits and Systems*, **2**, pp. 620–623, 1996

178 KROSE, B. *et al.*: 'Visual navigation on optic flow', Proceedings 1997 RWC Symposium, RWC Technical Report TR – 96001, pp. 89–95, 1997

179 KRUGER, S.: 'Motion analysis and estimation using multiresolution affine models', Ph.D. Thesis, Dept. Computer Science, University of Bristol, 1998

180 KUFFNER, J. JR.: 'Motion planning with dynamics', March 1998. Prepared as part of Ph.D. thesis. Available from citeseer.nj.nec.com/kuffner98motion.html

181 KUHN, P. M.: 'Fast MPEG-4 motion estimation: processor based and flexible VLSI implementations', *Journal of VLSI Signal Processing*, 1999, **23**, pp. 67–92

182 KUMAR, V. *et al.*: 'Introduction to parallel computing: design and analysis of algorithms' (Benjamin/Cummings Publishing Company Inc., Redwood City, California, USA, 1994)

183 KYRAIKOPOULOS, K. J., and SARIDIS, G. N.: 'An integrated collision prediction and avoidance scheme for mobile robots in non-stationary environments', *Automatica*, 1993, **29**(2), pp. 309–322

184 LANGER, M. S., and MANN, R.: 'Optical snow', *International Journal of Computer Vision*, 2003, **55**(1), pp. 55–81

185 LAPPE, M., and GRIGO, A.: 'How stereovision interacts with optic flow perception: neural mechanisms', *Neural Networks*, 1999, **12**, pp. 1325–1329

186 LARGE, F., VASQUEZ, D., FRAICHARD, T., and LAUGIER, C.: 'Avoid cars and pedestrians using velocity obstacles and motion prediction', IEEE

Intelligent Vehicles Symposium (IV2004), University of Parma, Parma, Italy, 14–17 June, 2004, pp. 375–379
187 LAUGIER, C., and FRAICHARD, T.: 'Decisional architectures for motion autonomy', Chapter 11 in 'Intelligent vehicle technologies: theory and applications', VLACIC, L., PARENT, M., and HARASHIMA, F. (Eds) (SAE International, Warrendale, USA, 2001)
188 LEE, J.-C., and FANG, W.-C.: 'VLSI neuroprocessors for video motion detection', *IEEE Trans. on Neural Networks*, 1993, **4**(2), pp. 178–191
189 LETANG, J. M., REBUFFEL, V., and BOUTHEMY, P.: 'Motion detection robust to perturbations: a statistical regularization and temporal integration framework', International Conference on *Computer Vision* (ICCV93), Berlin, Germany, 11–13 May, 1993, pp. 21–30
190 LETANG, J. M., and RUFFEL, V.: 'Motion detection based on a temporal multiscale approach', Int. Conf. *Pattern Recognition*, Den Haag, Netherlands, 30 August–3 September, 1992, pp. 65–68
191 LI, S. Z.: 'Markov random field modelling in computer vision' (Springer Verlag, London, UK, 1995)
192 LI, S. Z.: 'Robustizing robust M-estimation using deterministic annealing', *Pattern Recognition*, 1996, **29**(1), pp. 159–166
193 LIN, T., and BARRON, J. L.: 'Image reconstruction error for optical flow', in 'Vision interface', Banff National Park, Canada, pp. 73–80, May 1994
194 LINDEBERG, T.: 'Scale-space theory: a basic tool for analysing structures at different scales', *Journal of Applied Statistics*, 1994, **2**(2), pp. 225–270
195 LIU, H., CHELLAPPA, R., and ROSENFELD, A.: 'Accurate dense optical flow estimation using adaptive structure tensors and a parametric model', *IEEE Transactions on Image Processing*, 2003, **12**(10), pp. 1170–1180
196 LIU, H., HONG, T.-H., HERMAN, M., and CAMUS, T.: 'Accuracy vs efficiency trade-offs in optical flow algorithms', *Computer Vision and Image Understanding*, 1998, **72**(3), pp. 271–286
197 LUCAS, B. D., and KANADE, T.: 'An iterative image-registration technique with an application to stereo vision', Proc. IJCAI, Vancouver, B.C., August 1981, pp. 674–679
198 LUTHON, F., POPESCU, G. V., and CAPLIER, A.: 'An MRF based motion detection algorithm implemented on analog resistive network', *Lecture Notes in Computer Science*, **800**, European Conference on *Computer Vision* (ECCV94), pp. 167–174, 1994
199 MAE, Y., SHIRAI, Y., MIURA, J., and KUNO, Y.: 'Object tracking in cluttered background based on optical flow edges', Proc. ICPR'96, International Conference on *Pattern Recognition*, pp. 196–200, 1996
200 MAUREL, D., and DONIKIAN, S.: 'ACC systems – overview and examples', Chapter 13 in 'Intelligent vehicle technologies: theory and applications', VLACIC, L., PARENT, M., and HARASHIMA, F. (Eds) (SAE International, Warrendale, USA, 2001)
201 MEER, P.: 'Robust techniques for computer vision', in 'Emerging topics in computer vision', MEDIONI, G., and KANG, S. B. (Eds) (Prentice Hall,

New Jersey, USA, 2004). Also available from http://www.caip.rutgers.edu/riul/research/papers/pdf/rotechcv.pdf

202 MEER, P., MINTZ, D., and ROSENFELD, A.: 'Robust regression methods for computer vision: a review', *International Journal of Computer Vision*, 1991, **6**(1), pp. 59–70

203 MEYER, F., and BOUTHEMY, P.: 'Estimation of time-to-collision maps from first order motion models and normal flows', International Conference on *Pattern Recognition* (ICRP92), Den Haag, Netherlands, 30 August–3 September, 1992, pp. 78–82

204 MITICHE, A.: 'Computational analysis of visual motion' (Plenum Press, New York, 1994)

205 MITICHE, A., and BOUTHEMY, P.: 'Computational analysis of image motion: a synopsis of current problems and methods', *International Journal of Computer Vision*, 1996, **19**(1), pp. 29–55

206 MOINI, A.: 'Vision chips or seeing silicon', http://www.iee.et.tu-dresden.de/iee/eb/analog/papers/mirror/visionchips/vision_chips/vision_chips.html, cited 16 January 2004, last updated March 1997

207 MONTEIRO, D., and JOUVENCEI, B.: 'Visual servoing for fast mobile robot: adaptive estimation of kinematic parameters', IEEE International Conference on *Industrial Control, Electronics and Instrumentation* (IECON '93), **3**, pp. 1588–1593, 1993

208 MOORE, A., and KOCH, C.: 'A multiplication based analog motion detection chip', *Visual Information Processing: From Neurons to Chips, SPIE*, 1991, **1473**, pp. 66–75

209 MUKAWA, N.: 'Optical-model-based analysis of consecutive images', *Computer Vision and Image Understanding*, 1997, **66**(1) pp. 25–32

210 MURRAY, D. W., and BUXTON, B. F.: 'Experiments in the machine interpretation of visual motion' (MIT Press, Cambridge, Massachusetts, USA, 1990)

211 MURRAY, D. W., and BUXTON, B. F.: 'Scene segmentation from visual motion using global estimation', *IEEE Trans. on Pattern Analysis and Machine Intelligence*, 1987, **9**(2), pp. 220–228

212 MURRAY, R. M., LI, Z., and SASTRY, S. S.: 'A mathematical introduction to robotic manipulation' (CRC Press, Boca Raton, Florida, USA, 1994)

213 MYLES, Z., and LOBO, N. dV.: 'Recovering affine motion and defocus blur simultaneously', *IEEE Trans. Pattern Analysis and Machine Intelligence*, 1998, **20**(6), pp. 652–658

214 NAGEL, H. H.: 'Displacement vectors derived from second-order intensity variation in image sequences', *Computer Graphics and Image Processing*, 1983, **21**, pp. 85–117

215 NAKAYAMA, K.: 'Biological image motion processing: a review', *Vision Res.*, 1985, **25**(5), pp. 625–660

216 NASRAOUI, O.: 'A brief overview of robust statistics', Dept. Electrical and Computer Engineering, University of Memphis, last updated 08-10-2002. Available from http://archer.ee.memphis.edu/www.ee.memphis.edu/people/faculty/nasraoui/MY_TUTORIALS/RobustStatistics/RobustStatistics.html

217 NEGAHDARIPOUR, S.: 'Revised definition of optical flow: integration of radiometric and geometric cues for dynamic scene analysis', *IEEE Trans. on Pattern Analysis and Machine Intelligence*, 1998, **20**(9), pp. 961–979
218 NEGAHDARIPOUR, S., and HORN, B. K. P.: 'Direct passive navigation', *IEEE Trans. on Pattern Analysis and Machine Intelligence*, 1987, **9**(1), pp. 168–176
219 NESI, P., DEL BIMBO, A., and BEN-TZVI, D.: 'A robust algorithm for optical flow estimation', *Computer Vision and Image Understanding*, 1995, **62**(1), pp. 55–68
220 NEYKOV, N. M., and MULLER, C. H.: 'Breakdown point and computation of trimmed likelihood estimators in generalized linear models', in 'Developments in robust statistics', DUTTER, R., FILZMOSER, P., GATHER, U., and ROUSSEEUW, P. J. (Eds) (Physica-Verlag, Heidelberg, 2003) pp. 277–286. Also available from http://www.member.uni-oldenburg.de/ch.mueller/publikation.html
221 NIESSEN, W. J. *et al.*: 'A multiscale approach to image sequence analysis', *Computer Vision and Image Understanding*, 1997, **65**(2), pp. 259–268
222 NOSRATINIA, A., and ORCHARD, M. T.: 'Discrete formulation of pel-recursive motion compensation with recursive least squares updates', IEEE Int. Conf. on *Acoustics, Speech and Signal Processing* (ICASSP93), Minneapolis, MN, USA, 27–30 April, 1993, **5**, pp. 229–232
223 OGATA, M., and SATO, T.: 'Motion-detection model with two stages: spatiotemporal filtering and feature matching', *Journal of the Optical Society of America A*, 1992, **9**(3), pp. 377–387
224 ONG, E. P., and SPANN, M.: 'Robust optical flow computation based on least-median-of-squares regression', *International Journal of Computer Vision*, 1999, **31**(1), pp. 51–82
225 OSIANDER, R.: 'Terrahertz imaging and spectroscopy', http://www.jhuapl.edu/programs/rtdc/SensorTechnology/TerahertzImaging.html, cited 5 January 2004
226 PALIWAL K.: 'Advance signal processing lecture notes' (Griffith University, Brisbane, Australia, 1998)
227 PALL, G. A.: 'Introduction to scientific computing' (Meredith Corp, New York, USA, 1971)
228 PERRONE, J. A.: 'Simple technique for optical flow estimation', *Journal of the Optical Society of America A*, 1990, **7**(2), pp. 264–278
229 PRATI, A. *et al.*: 'Detecting moving shadows: algorithms and evaluation', *IEEE Trans. Pattern Anal. and Machine Vision*, 2003, **25**(7), pp. 918–923
230 PRESS, H. W., FLANNERY, B. P., REUOLSKY, S. A., and VETTERLING, W. T.: 'Numerical recipes: the art of scientific computing' (Cambridge University Press, Cambridge, 1986)
231 PRESS, W. H. *et al.*: 'Numerical recipes in C: the art of scientific computing' (Cambridge University Press, Cambridge, 2002, 2nd edn). Also available from http://www.nr.com

232 PRMIA, S., AYACHE, N., BARRICK, T., and ROBERTS, N.: 'Maximum likelihood estimation of the bias field in MR brain images: investigating different modelings of the imaging process', in 'MICCAI2001, Lecture Notes in Computer Science', NIESSEN, W., and VIERGEVER, M. (Eds), **2208**, pp. 811–819, 2001

233 PROAKIS, J. G., and MONOLAKIS, D. G.:'Digital signal processing: principals, algorithms and applications' (Prentice Hall International, New Jersey, USA, 1996, 3rd edn)

234 PROESMANS, M., VAN GOOL, L., PAUWELS, E., and OOSTERLINCK, A.: 'Determination of optical flow and its discontinuities using non-linear diffusion', *Lecture Notes in Computer Science*, **801**, European Conference on *Computer Vision* (ECCV94), pp. 296–304, 1994

235 QIAN, X., and MITCHIE, A.: 'Direct motion interpretation and segmentation based on the robust estimation of parametric models', *Vision Interface* '99, Trios-Rivieres, Canada, 19–21 May, pp. 552–558, 1999

236 RATH, G. B., and MAKUR, A.: 'Subblock matching-based conditional motion estimation with automatic threshold selection for video compression', *IEEE Trans. Circuits and Systems for Video Technology*, 2003, **13**(9), pp. 914–924

237 REKLEITIS, I. M.: 'Steerable filters and cepstral analysis for optical flow calculation from a single blurred image', *Vision Interface* '96, Toronto, May, pp. 159–166, 1996

238 RIPLEY, D.: 'Robust statistics', available from www.stats.ox.ac.uk/~ripley/StatMethods/Robust.pdf, cited 6 May 2004, last updated 2003

239 ROUSSEEUW, P. J., and LEROY, A. M.: 'Robust regression and outlier detection' (John Wiley & Sons, New York, USA, 1987)

240 ROUSSEEUW, P. J., and VAN DRIESSEN, K.: 'Computing LTS regression for large data sets', Technical Report, Antwerp Group on Robust & Applied Statistics, Department of Mathematics and Computer Sciences, University of Antwerp (UA), available from http://www.agoras.ua.ac.be/abstract/Comlts99.htm (cited 12 January 2004), 1999

241 ROWEKAMP, T.: 'A smart sensor system for real-time optical flow estimation', Ph.D. Thesis, Faculty for Mathematics, Science and Information Technology, University of Cottbus, Germany, 1997

242 SANTOS-VICTOR, J., and SANDINI, G.: 'Uncalibrated obstacle detection using normal flow', *Machine Vision and Applications*, 1996, **9**, pp. 130–137

243 SARPESHKAR, R., KRAMER, J., INDIVERI, G., and KOCH, C.: 'Analog VLSI architectures for motion processing: from fundamental limits to system applications', *Proceedings of the IEEE*, 1996, **84**(7), pp. 969–987

244 SATO, T.: 'DMax: relations to low & high-level motion processes', Chapter 4 in 'High level motion processing, computational, neurobiological and psychophical perspectives', WATANABE, T. (Ed.) (MIT Press, Cambridge, Massachusetts, USA, 1998)

245 SCHALKOFF, R. J.: 'Digital image processing and computer vision' (John Wiley & Sons, New York, USA, 1989)

246 SCHRATER, P. R., KNILL, D. C., and SIMONCELLI, E. P.: 'Mechanisms of visual motion detection', *Nature Neuroscience*, 2000, **3**(1), pp. 64–68

247 SCHUNCK, B. G.: 'Image flow segmentation and estimation by constraint line clustering', *IEEE Trans. on Pattern Analysis and Machine Intelligence*, 1989, **11**(10)

248 SCHUNCK, B. G.: 'Image flow: fundamentals and future research', IEEE Conference on *Computer Vision and Pattern Recognition (CVPR85)*, San Francisco, June 1985, pp. 560–571

249 SCHWARTZ, J. T., and SHARIR, M.: 'A survey of motion planning and related geometric algorithms', *Artificial Intelligence Journal*, 1988, **37**, pp. 156–169

250 SCOTT, C., and NOWAK, R.: 'Maximum likelihood estimation', available from http://cnx.rice.edu/content/m11446/latest/, cited 28 April 2004, last updated 31 October 2003

251 SIM, D.-G., and PARK, R.-H.: 'Robust reweighted MAP motion estimation', *IEEE Transactions on Pattern Analysis and Machine Intelligence*, 1998, **20**(4), pp. 353–363

252 SIMONCELLI, E. P.: 'Course-to-fine estimation of visual motion', Presented at IEEE Signal Processing Society, 8th Workshop on *Image and Multidimensional Signal Processing*, Cannes, France, September 1993

253 SIMONCELLI, E. P.: 'Distributed representation and analysis of visual motion', Ph.D. Thesis, Dept. Electrical Engineering and Computer Science, MIT, 1993

254 SIMONCELLI, E. P., ADELSON, E. H., and HEEGER, D. J.: 'Probability distributions of optical flow', *IEEE Proc. CVPR*, 1991, pp. 310–315

255 SMITH, P. W., and NANDHAKUMA, N.: 'Accurate structure and motion computation in the presence of range image distortions due to sequential acquisition', Proc. IEEE Conf. *Computer Vision and Pattern Recognition*, Seattle, Washington, USA, 20–23 June, 1994, pp. 925–928

256 SMITH, S. M.: 'ASSET-2: real-time motion segmentation and object tracking', Technical Report TR95SMS2b, Oxford Centre for Functional Magnetic Resonance Imaging of the Brain, Oxford University, 1995

257 SMITH, S. W.: 'The scientist and engineer's guide to digital signal processing' (California Technical Publishing, California, USA, 1999, 2nd edn). See also www.dspguide.com

258 SNYDER, M. A.: 'On the mathematical foundations of smoothness constraints for the determination of optical flow and for surface reconstruction', *IEEE Trans. Pattern Analysis and Machine Intelligence*, 1991, **13**(11), pp. 1105–1114

259 SOMOGYI, J., and ZAVOTI, J.: 'Robust estimation with iteratively reweighted least-squares method', *Acta Geod. Geoph. Mont. Hung.*, 1993, **28**(3–4), pp. 413–420

260 SONG, B. C., and RA, J. B.: 'A fast multi-resolution block matching algorithm for motion estimation', *Signal Processing: Image Communication*, 2000, **15**, pp. 799–810

261 SPETSAKIS, M. E.: 'Optical flow estimation using discontinuity conforming filters', *Computer Vision and Image Understanding*, 1997, **8**(3), pp. 276–289
262 SPIES, H., and BARRON, J.: 'Evaluating the range flow motion constraint', Proceedings 16th International Conference on *Pattern Recognition*, Quebec City, August 2002, **3**, pp. 517–520
263 SPIES, H., JAHNE, B., and BARRON, J. L.: 'Dense range flow from depth and intensity data', International Conference on *Pattern Recognition*, Barcelona, Spain, 3–8 September, 2000, pp. 131–134
264 SPOERRI, A.: 'The early detection of motion boundaries', Master Thesis, Department of Brain and Cognitive Sciences, MIT Artificial Intelligence Laboratory, 1991
265 STEIN, M.: 'Crossing the abyss: asynchronous signals in a synchronous world', *EDN Australia*, Reed Electronics Group, September, 2003, pp. 30–35
266 STEIN, M.: 'Crossing the abyss: asynchronous signals in a synchronous world', *EDN Australia*, Reed Electronics Group, September, 2003, pp. 30–35
267 STEWARD, C. V.: 'Robust parameter estimation in computer vision', *SIAM Review*, 1999, **41**(3), pp. 513–537
268 STEWART, C. V.: 'Bias in robust estimation caused by discontinuities and multiple structures', *IEEE Trans. Pattern Analysis and Machine Intelligence*, 1997, **19**(8), pp. 818–833
269 STILLER, C.: 'Object-based estimation of dense motion fields', *IEEE Transactions on Image Processing*, 1997, **6**(2), pp. 234–250
270 STILLER, C.: 'Towards intelligent automotive vision systems', Chapter 5 in 'Intelligent vehicle technologies: theory and applications', VLACIC, L., PARENT, M., and HARASHIMA, F. (Eds) (SAE International, Warrendale, USA, 2001)
271 STILLER, C., and KONRAD, J.: 'Estimating motion in image sequences: a tutorial on modeling and computation of 2D motion', *IEEE Signal Processing Magazine*, 1999, pp. 70–98
272 STÖFFLER, N. O., and FARBER, G.: 'An image processing board with an MPEG processor and additional confidence calculation for fast and robust optic flow generation in real environments', Proc. Int. Conf. on *Advanced Robotics* (ICAR'97), Monterey, California, USA, 1997, 7–9 July, pp. 845–850
273 STÖFFLER, N., BURKERT T., and FÄRBER, G.: 'Real-time obstacle avoidance using an MPEG-processor-based optic flow sensor', Proc. 15th Int. Conf. on *Pattern Recognition*, 2000, **4**, pp. 161–166
274 STOKES, J.: 'RAM guide', last edited July 2000, cited 3 August 2004. http://arstechnica.com/paedia/r/ram_guide/ram_guide.part1-5.html
275 SUDHIR, G., SUBHASHIS, B., BISWAS, K. K., and BAHL, R.: 'A cooperative integration of stereopsis and optic flow computation', *J. of Opt. Soc. of Am. Series A*, 1995, **12**, pp. 2564–2572
276 SUN, S., PARK, H.-W., HAYNOR, D. R., and KIM, Y.: 'Fast template matching using correlation based adaptive predictive search', *International Journal of Imaging Systems Technology*, 2003, **13**, pp. 169–178

277 TAI, P.-L., HUANG, S.-Y., LIU, C.-T., and WANG, J.-S.: 'Computationally-aware scheme for software based block motion estimation', *IEEE Trans. Circuits and Systems for Video Technology*, 2003, **13**(9), pp. 901–913
278 TEKALP, D. M.: 'Digital video processing' (Prentice Hall, New Jersey, USA, 1995)
279 TERZOPOULOS, D.: 'Multilevel computational processes for visual surface reconstruction', *Computer Vision, Graphics and Image Processing*, 1983, **24**, pp. 52–96
280 TREVES, P., and KONRAD, J.: 'Motion estimation and compensation under varying illumination', ICIP'94, November 13–16, Austin, Texas, USA, 1994
281 THORPE, C., HERBERT, M. H., KANADE, T., and SHAFER, S. A.: 'Vision and navigation for the Carnegie-Mellon Navlab', *IEEE Transactions on Pattern Analysis and Machine Intelligence*, 1988, **10**(3), pp. 362–373
282 TIAN, T. Y., and SHAH, M.: 'Motion estimation and segmentation', *Machine Vision and Applications*, 1996, **9**, pp. 32–42
283 Timing Analyzer Guide, Xilinx Foundation 2.1i Documentation, 1999. Available from http://toolbox.xilinx.com/docsan/2_1i/
284 TISTARELLI, M., and SANDINI, G.: 'On the advantages of polar and log-polar mapping for direct estimation of time-to-impact from optical flow', *IEEE Trans. Pattern Analysis and Machine Intelligence*, 1993, **15**(4), pp. 401–410
285 TORR, P. H. S.: 'Motion segmentation and outlier detection', Ph.D. Thesis, Dept. Engineering Science, University of Oxford, 1995
286 TUNLEY, H., and YOUNG, D.: 'First order optical flow from log-polar sample images', *Lecture Notes in Computer Science* ECCV 94, 1994, **800**, pp. 132–137
287 TURAGA, D., snd ALKANHAL, M.: 'Search algorithms for block-matching in motion estimation', http://www.ece.cmu.edu/~ee899/topics.htm, cited 18 January 2004, last updated 1998
288 ULLMAN, S.: 'Analysis of visual motion by biological and computer systems', *IEEE Computer*, August, 1981, pp. 57–69
289 URAS, S., GIROSI, F., VERRI, A., and TORRE, V.: 'A computational approach to motion perception', *Biol. Cybern*, 1988, **60**, pp. 79–97
290 VHDL Reference Guide, Xilinx Foundation 2.1i Documentation, 1999. Available from http://toolbox.xilinx.com/docsan/2_1i/
291 VIEVILLE, T., and FAUGERAS, O. D.: 'The first order expansion of motion equations in the uncalibrated case', *Computer Vision and Image Understanding*, 1996, **64**(1), pp. 128–146
292 VLACIC, L., ENGWIRDA, A., HITCHINGS, M., and O'SULLIVAN, Z.: 'Intelligent autonomous systems: Griffith University's creation', in 'Intelligent autonomous systems, IAS-5', KAKAZU, Y., WADA, M., and SATO, T. (Eds) (IOS Press, Netherlands, 1998) pp. 53–60
293 WANDELL, B. A.: 'Foundations of vision' (Sinauer Associated Inc, Massachusetts, USA, 1995)
294 WANG, J. Y. A., and ADELSON, E. H.: 'Representing moving images with layers', *IEEE Trans. Image Processing Special Issue: Image Sequence Compression*, 1994, **3**(5), pp. 625–638

295 WANG, W., and DUNCAN, J. H.: 'Recovering the three-dimensional motion and structure of multiple moving objects from binocular image flows', *Computer Vision and Image Understanding*, 1996, **63**(3), pp. 430–446
296 WARREN, W.: 'Multiple robot path coordination using artificial potential fields', Proc. IEEE Int. Conf. *Robotics and Automation*, pp. 500–505, 1990
297 WATANABE, T. (Ed.): 'High level motion processing: computational, neurobiological and psychophysical perspectives' (MIT Press, Cambridge, Massachusetts, USA, 1998)
298 WATSON, A. B., and AHUMADA, A. J.: 'Model of human visual motion sensing', *Journal of the Optical Society of America A*, 1985, **2**(2), pp. 322–341
299 WEBER, J., and MALIK, J.: 'Robust computation of optical flow in a multi-scale differential framework', *International Journal of Computer Vision*, 1994, **2**, pp. 5–19
300 WECHSLER, H.: 'Computational vision' (Academic Press Inc, Boston, USA, 1990)
301 WEISS, Y., and ADELSON, E. H.: 'Perceptually organized EM: a framework for motion segmentation that combines information about form and motion', MIT Media Lab Perceptual Computing Section TR #315. Available for download from http://www-bcs.mit.edu/people/yweiss/weiss.html, 1994
302 WEISS, Y., and ADELSON, E. H.: 'Slow and smooth: a Bayesian theory for the combination of local motion signals in human vision', MIT Artificial Intelligence Lab, AI Memo 1624, February 1998
303 WILCOX, R. R.: 'Introduction to robust estimation and hypothesis testing' (Academic Press, San Diego, California, USA, 1997)
304 Wiles, C. S.: 'Closing the loop on multiple motions', Ph.D. Thesis, Dept. Engineering Science, University of Oxford, 1995
305 WILLERSIN, D., and Enkelmann W.: 'Robust obstacle detection and tracking by motion analysis', IEEE Conf. on *Intelligent Transport Systems*, 1997, pp. 717–722
306 Xilinx FPGA Tools Course Notes, Frankfurt, Germany, 1999
307 XIONG, Y., and SHAFER, S. A.: 'Moment and hypergeometric filters for high precision computation of focus, stereo and optical flow', *International Journal of Computer Vision*, 1997, **22**(1), pp. 25–59
308 XU, S.: 'Motion and optical flow in computer vision', Ph.D. Thesis, Dissertation No. 428, Department of Electrical Engineering, Linkoping University, Sweden, 1996
309 YACOOB, Y., and DAVIS, L. S.: 'Temporal multiscale models for flow and acceleration', *International Journal of Computer Vision*, 1999, **32**(2), pp. 147–163
310 YAMADA, K., and SOGA, M.: 'A compact integrated visual motion sensor for ITS applications', *IEEE Trans. Intelligent Transportation Systems*, 2003, **4**(1), pp. 35–42
311 YAMAMOTO, M., BOULANGER, P., BERALDIN, J., and RIOUX, M.: 'Direct estimation of range flow on deformable shape from a video range

camera', *IEEE Trans. Pattern Analysis and Machine Intelligence*, 1993, **15**(1), pp. 82–89

312 YE, M., and HARALICK, R. M.: 'Two-stage robust optical flow estimation', Proc. IEEE Conf. *Computer Vision and Pattern Recognition*, 2000, **2**, pp. 623–628

313 ZHANG, J., and HANAUER, G. G.: 'The application of mean field theory to image motion estimation', *IEEE Trans. Image Processing*, 1995, **4**(1), pp. 19–33

314 ZHENGYOU, Z.: 'Parameter estimation techniques: a tutorial with application to conic fitting', cited 20 February 2004, last updated 8 February 1996, http://www-sop.inria.fr/robotvis/personnel/zzhang/Publis/Tutorial-Estim/node23.html

315 ZHU, L., FUJIMOTO, H., SANO, A., and YAMAKAWA, S.: 'Adaptive visual tracking of moving objects with neural PID controller', *Proc. SPIE*, 2000, **4197**, pp. 340–350

Index

ψ – conversion factor, 107
α – line process energy, 56, 89
ζ – pixel pitch, 105, 109, 262
λ – regularisation parameter, 31, 72, 88, 118, 156
$\rho(\ )$ – estimator function, 25, 50, 77
$\psi(\ )$ – influence function, 51, 54
$\mathbf{\Omega} = (\Omega_X, \Omega_Y, \Omega_Z)$ – rotational velocity, 107
ζ_{eff} – effective pixel pitch, 112
$\omega_x, \omega_y, \omega_t$ – spatiotemporal frequency, 85, 86, 105

A
accumulator cells, 28
address unpacking, 238, 245
Analog Devices Sharc DSP, 236
annealing M estimator, 142
Anscombe's method, 143
aperture problem, 69, 73, 79, 82, 126
apparent motion/velocity, 64, 67, 69, 70, 72, 81, 107
    calculation in simulation, 157
    relationship to optical flow, 107
    scale space, 111
    vertical motion, 123
    versus 3D motion, 119
    worst case, 110
arithmetic mean, 21
average
    as central tendency, 21
    expectation, 21, 22
    Gaussian distribution, 21
    mean, *see* mean
    median, 16, 21, 42
    mode, 21, 26
    robust average, *see* robust estimation
    sample mean, 22

B
back annotation, 177
background subtraction, 69
    accumulated difference images, 69
barrel distortion, 274
Bayes theorem, 16, 92
bias, 42, 147
    asymptotically unbiased, 42
Black, 89, 96, 113, 142
blobs, 123, 134, 156, 167
block based motion estimation, *see* token based motion estimation
block memory, 186
BlockRAM, 241, 250, 251, 257, 266
boundary conditions, 32, 37
    Dirichlet boundary conditions, 37
    Neumann boundary conditions, 37
bounded influence estimator, 56
breakdown point, 47
    of LAR estimator, 50
    of LMedS estimator, 61
    of LTS estimator, 61
    of M estimator, 53
brightness constancy assumption (BCA), 66, 69, 76, 93, 131
brightness offset field
    additive $C(x, y, t)$, 80, 81
    multiplicative $M(x, y, t)$, 80, 81
buffers, 156, 237, 245
    Calibration_Buffer, 238, 249, 251, 256
    Camera_Buffer, 222, 251
    first-in-first-out (FIFO), 222, 238, 242, 245, 250, 251, 253

over/under runs, 238, 247, 248, 250
PC_Output_Buffer, 253
Process_Side, 251
Processing_Input_Buffer, 252
Processing_Output_Buffer, 252
RAM_Side, 251
Range_Buffer, 252
tristate, *see* tristate buffers
use of BlockRAM, 241, 250
bus arbitration, 238, 247
centralised parallel arbitration, 247
round robin polling, 247

C

$C(x, y, t)$ – additive brightness offset field, *see* brightness offset field
camera calibration, 238, 244, 247
calibration process, 247
storage of calibration data, 242
zero order calibration, 247
Camera_Interface, 251, 256
central limit theorem, 19
chequerboard update, 38
clique potential function, 92
clock gating, 207
combinatorial logic block (CLB), 278
conjugate gradients method, 36
constraint equation
data constraint, 23, 32, 39
developed constraint equation, 125
linear constraint, 23
optical flow constraint equation (OFCE), *see* optical flow constraint equation (OFCE)
propagation of smoothness, 33
smoothness constraint, 31, 33, 56, 89, 92
constraint line, 28
constraint line clustering, 88
convergence, 34
accelerating, 36
choice of step size, 34
effect of temporal integration, 155
Jacobi and Gauss Siedel iteration, 38
of LTSV estimator, 136, 148
steepest versus direct descent, 36
conversion factor ψ, 107
convex function, 34
convex hull peeling (CHP), 144
cores, 175, 264, 265
counters, 208
Cramer–Rao bound, 48
cumulative distribution function (CDF), 17

D

data constraint, *see* constraint equation
data fusion, 64, 97, 100, 119
data space, 27
data term E_d, 72
data paths, 247, 253
camera data path, 221, 256
Camera_Interface, 251, 256
dynamic_scale_space block, 256
Filter block, 256, 257, 260
motion estimation data path, *see* motion estimation data path
PC data path, 255
Range data path, 255
dependent variable, 23
depth derivatives Z_x, Z_y, Z_t, 97
derivative
backward difference, 74
Barron's approximation, 75, 126
estimation, 73
evaluation of approximations, 126
first order backward difference (FOBD), 74
first order central difference (FOCD), 37, 75, 126, 260, 267
first order forward difference (FOFD), 74, 126
forward difference, 73
Horn and Schunck approximation, 76, 126
Prewitt mask, 76
Simoncelli's approximation, 75, 126
Sobel mask, 76
deterministic process, *see* process
dimensionality
dimensionality
linear regression, 23
location estimation, 39
direct descent, 34
Dirichlet boundary conditions, 37
displaced block difference (DBD), 78, 83
displaced frame difference (DFD), 76, 93
pel recursive strategy, 78
relationship with OFCE, 77
distance, *see* norm
distributed memory, 186
distribution, *see* probability density function (PDF)
cumulative distribution function (CDF), 17
Gaussian distribution, 19
joint distribution function, 18

marginal distribution function, 19
normal distribution, 19
D_{max} – maximum object range, 109, 263
D_{min} – minimum object range, 109, 159
downhill simplex method (DSM), 144
dynamic scale space, 103, 113, 123, 155, 163, 262
   implementation, 153, 256
   issues, 114
   variations, 278

E

E_d – data term, 72
effective pixel pitch ζ_{eff}, 112, 125
encoding
   binary, 183, 217
   enumerated types, 183
   one-hot, 183, 217
energy function, 24, 31, 92
ensemble, 20, 24
enumerated types, 183
equation
   constraint equation, *see* constraint equation
   optical flow constraint equation (OFCE), *see* optical flow constraint equation (OFCE)
   update equation, *see* update equation
error, 12
   residual, 23
error function, 24
E_s – smoothness term, 72
estimation
   bias, 42
   consistent estimator, 42
   data constraint, *see* constraint equation
   energy function, 24
   error function, 24
   estimation problems, 11, 23
   estimator function $\rho(\ )$, 25, 50, 140
   Gaussian distribution, 26
   Hough transform, 27
   linear regression, *see* linear regression
   location estimation, *see* location estimation
   maximum *a posteriori* (MAP), 91
   minimum mean squared error estimator (MMSE), 26
   MLE, *see* maximum likelihood estimate (MLE)
   multiple linear regression, 29
   regularisation, *see* regularisation
   residual, *see* residual

   robust estimation, *see* robust estimation
   smoothness constraint, *see* constraint equation
   trimming, 138
   variance, 42
   winsorisation, 138
estimator
   annealing M estimator, 142
   GM estimator, 56
   least absolute residuals estimators (LAR), *see* least absolute residuals estimator (LAR)
   least median of squares (LMedS), *see* least median of squares (LMedS)
   least squares (LS), *see* least squares (LS)
   least trimmed squares (LTS), *see* least trimmed squares (LTS)
   least trimmed squares variant (LTSV), *see* least trimmed squares variant (LTSV)
   L estimator, 139
   Mallow's estimator, 56
   M estimator, *see* M estimator
   properties of estimators, 42
   Schweppe estimators, 56
   total least squares (TLS), 25
estimator function $\rho(\ )$, 25, 50, 77, 140
Euclidian distance, 39
event, 12
   compliment of, 14
   independent, 16
Exner, 63
expectation, 21, 22, 42
experiment, 12
explanatory variables, 23

F

f – focal length, 167, 262
F – frame interval, 105, 107, 262
feature based motion estimation, *see* token based motion estimation
field of view (FOV), 98, 119, 120, 122, 167, 234, 235, 256
field programmable gate array (FPGA)
   block memory, 186
   distributed memory, 186
field programmable gate array (FPGA), 173, 231
   back annotation, 177
   buffer, *see* buffers
   combinatorial logic block (CLB), 278
   constraints, *see* user constraints
   cores, 175, 264, 265

definition, 174
design entry, 174
floor planner, 175
intellectual properties (IPs), 175, 264, 265
low level editor, 175
mapping process, 176
place and route, 177
prototype development platform, 236
schematic entry, 174
SignalMaster development platform, 236
slice, 278
state machine, *see* state machine
static timing analysis, 177
synthesis, 175
system-on-a-chip, 232
time in FPGA design, 177
Xilinx Virtex, 236
field programmable gate array (FPGA)BlockRAM, *see* BlockRAM
Filter block, 256
finite difference method (FDM), 32
first-in-first-out (FIFO), *see* buffers
first order backward difference (FOBD), *see* derivative
first order central difference (FOCD), *see* derivative
first order forward difference (FOFD), *see* derivative
Fleet and Jepson, 86
floor planner, 175
focal length (f), 106, 167, 262
focal length f, 109
frame interval F, 105–107, 262
frequency domain motion estimation, 65, 85
   Fleet and Jepson, 86
   phase, 86
   temporal aliasing, 104
Fuga 15D camera, 232, 270
   barrel distortion, 274
   calibration, 233
function
   clique potential function, 92
   convex function, 34
   cumulative distribution function (CDF), 17
   energy function, 24, 31, 92
   error function, 24
   estimator function $\rho(\ )$, 25, 50, 77, 140
   influence function $\psi(\ )$, *see* influence function $\psi(\ )$
   joint distribution function, 18
   likelihood function $l(\ )$, 22
   marginal distribution function, 19
   partition function $Z(\ )$, 92
   probability density function (PDF), 17, 19, 22
   spatiotemporal intensity function $I(x, y, t)$, 69, 80, 85 105
functional simulation, 175, 198
fundamental formula, 107

G

Gaussian distribution, 19, 22
   central limit theorem, 19
   in estimation, 26
Gauss–Siedel iteration, 38
generalised M estimator, 56
Gibbs distribution, 91
Gibbs random field, 91
Gibson, 64
global constraint, 30, 72, 89
global minimum, 33
GM estimator, 56
gradient based motion estimation, 65, 69
   application to range data, 119
   global constraints (regularisation), 71
   Horn and Schunck, *see* Horn and Schunck
   local constraints (regression), 71
   local versus global methods, 73
   Lucas and Kanade, *see* Lucas and Kanade
   normal flow, 69
   optical flow, 65, 69, 107
   optical flow constraint equation (OFCE), *see* optical flow constraint equation (OFCE)
   temporal aliasing, *see* temporal aliasing
graduated non-convexity (GNC), 57, 90, 92, 141, 147
Gupta, 89

H

hardware description languages (HDLs), 175
   logical simulation, 176, 186
   synthesis, 175
   verilog, 175
   VHDL, *see* VHDL
hardware motion estimation, 98
   vision chip, 98
Helmholtz, 64
Horn and Schunck, 72, 78, 123
Hough transform, 27, 88
   accumulator cells, 28
   constraint line, 28
   data space, 27
   parameter space, 28
   solving, 28

I

$I(x, y, t)$ – spatiotemporal intensity function, 69, 80, 85 105
ICSL CAMR Cooperative Autonomous Vehicles, 109, 270
ill-posed problems, 30, 71, 72
illumination, 66, 131
illumination flattening, 68
image formation models, 79
image pyramid, 111
image resolution (x_r, y_r), 109
image stabilisation, 122
implicit synchronisation, 238, 253
independent, identically distributed (iid), 19, 26
independent events, 16
independent variables, 23
influence function
    monotone, 51
    redescending, 51
influence function $\psi(\ )$, 51, 54, 140
inlier, 134
input variables, 23
intellectual properties (IPs), 175, 264, 265
iterated conditional modes (ICM), 94
iteration, 34
    effect of temporal integration, 96, 155
    Gauss–Siedel, 38
    how number of iterations affects an estimate, 147
    Jacobi, 38, 141
    update equation, 73, 78
iteratively reweighted least squares (IRLS), 57, 140

J

Jacobi iteration, 38, 141
joint distribution function, 18

L

$l(\ )$ – likelihood function, 22
L_1 norm, 40
L_2 norm, 39
laser range sensor, 119, 234
    IBEO LD Automotive, 235
    simulation of, 234
least absolute residuals estimator (LAR), 41, 50, 53, 77
    breakdown point of, 50
    robustness to outliers, 51
    scale equivariance, 50
least median of squares (LMedS), 59, 90, 134
    scale equivariance, 61
least squares (LS), 25, 26, 40, 77
    optimality, 43
    robustness, 44
    scale equivariance, 49
least trimmed squares (LTS), 61, 90, 134
    complexity of 1D location estimate, 153
    Monte Carlo study, 147
least trimmed squares variant (LTSV), 134
    algorithm, 136
    comparison to other estimators, 137
    computational complexity, 153
    improvements, 280
    linear regression, 137
    Monte Carlo study, 146
    relation with M estimators, 140
L estimator, 139
leverage point, 43, 45
likelihood, 22
    function $l(\ )$, 22
    log likelihood, 22
line process, 56, 89, 92, 94
line process energy α, 56, 89
linear regression using LTSV estimator, 137
linear regression, 23, 28
    dependent variable, 23
    dimensionality, 23
    explanatory variables, 23
    for motion estimation, 71
    independent variables, 23
    input variables, 23
    line-of-best-fit, 23, 24
    multiple linear regression, 29
    output variable, 23
    residual, 24
    response variable, 23
    solving, 26
    versus location estimation, 40
line-of-best-fit, 23, 24
local constraint, 30, 71, 90
local minimum, 33
location constraint, *see* user constraints
location estimation, 39
    dimensionality, 39
    multivariate, 39
    norms, 39
    residual, 39
    solving, 40
    versus linear regression, 40
logical simulation, 176, 186
long-range process, *see* process

Lorentzian, 51
Lucas and Kanade, 71, 78, 123

M

M(x, y, t) – multiplicative brightness offset field, *see* brightness offset field
Mahalanobis distance, 40
Mallow's estimator, 56
mapping process, 176
marginal distribution function, 19
Markov random field (MRF), 91
    clique potential function, 92
    energy function, 92
    Gibbs random field, 91
    partition function – Z( ), 92
    similarity with M estimator, 92
MATLAB, 157
maximum *a posteriori* (MAP) estimator, 91
maximum likelihood estimate (MLE), 26, 27, 48, 50
maximum object range D_{max}, 109, 263
maximum trimmed likelihood (MTL), 146
maximum velocity V_{max}, 109, 110, 159
mean, 20, 26, 41
    arithmetic mean, 21
    Gaussian distribution, 19, 21
    mean squared value, 22
    sample mean, 21, 22
mean field annealing, 58, 92
mean shift method (MSM), 145
measurement, 12, 16
median, 16, 21, 42
median absolute deviation from the median (MAD), 49
membrane model, 31
    boundary conditions, 37
    for motion estimation, 72, 89
    relationship with M estimators, 56
    similarity with FDM, 32
    smoothing, 31
    solving, 33
    weak membrane model, 56
memory management, 238, 240
M estimator, 50, 89, 92
    breakdown point, *see* breakdown point
    computational complexity, 153
    influence function $\psi()$, 51, 54, 140
    iteratively reweighted least sqaures, 57
    Lorentzian, 51, 59
    monotone, 51
    Monte Carlo study, 147
    properties, 51
    redescending, 51
    relationship with membrane model, 56
    scale equivariance, 51
    sensitivity to leverage points, 53
    sensitivity to outliers, 54
    solving, 57
    truncated quadratic, 51, 59
metastability, 204
minimum
    global, 33
    local 33
minimum mean squared error (MMSE), 26
minimum object range D_{min}, 109, 159
minimum volume ellipsoid (MVE), 143
Minpran, 62
mixture models, 94
mode, 21, 26
model
    image formation model, 79
    membrane, *see* membrane model
    mixture models, 94
    parametric motion model, 65, 79
    perspective rigid body motion model, 107
    pinhole camera model, 106
    reflectance model, 66
    rigid ground plan motion model, 108
    weak membrane, *see* membrane model
motion estimation, 65
    aperture problem, 69, 73, 79, 82, 126
    apparent motion/velocity, *see* apparent motion/velocity
    applications, 63
    background subtraction, 69
    brightness constancy assumption (BCA), 66, 69, 76, 93, 131
    constraint equation, *see* constraint equation
    constraint line clustering, 88
    data structure, 70
    developed algorithm, 156
    developed constraint equation, 125
    discontinuity preserving, 87
    displaced frame difference (DFD), 76
    dynamic scale space, *see* dynamic scale space
    effect of shadows, 67, 80, 132, 265, 281
    frequency domain, 65, 85
    frequency domain characteristics, 103
    from single blurred image, 98
    gradient based, 65, 69
    hardware, 98
    hierarchical motion estimation, 111
    Horn and Schunck, *see* Horn and Schunck
    Hough transform, 88

image stabilisation, 122
long-range process, *see* process
Lucas and kanade, *see* Lucas and kanade
Markov random field, 91
membrane model, 72
M estimator, 89
mixture models, 94
motion from defocus, 98
motion noise, 121
multiple motions, 87
occlusion, 67
on hills, 122
optical flow, 65, 69, 107
parametric model, 65, 79
pel recursive, 78
prediction error, 157
real-time, 103
reconstruction error, 157
robust estimation, 78, 89
scale space, *see* scale space
short range process, *see* process
structured light, 98
temporal aliasing, *see* temporal aliasing
temporal integration, *see* temporal integration
token based, 81, *see* token based motion estimation
transparent motion, 67
using laser speckle, 98
using range data, 97, 118
vision chip, 98
visual motion estimation, 65
visual stereo, 96
weak membrane model, 89
with space variant camera, 98
motion estimation data path, 258
   `Average_buffer`, 269
   derivative calculation, 260
   `Edge_Buffer`, 269
   `Edge_buffer_2`, 269
   `Optical_Flow` block, 262
   `Processing_Input_Buffer`, 260
   `Processing_Output_Buffer`, 270
   `Range_Buffer`, 267
   robust smoothing, 266
   `Robust_Column-wise_Average` block, 260, 266, 282
   `Robust_Smoothing` block, 269
   solving the constraint equation, 261
   temporal integration, 269
   `Velocity_buffer`, 265
   `Zero_Crossings` block, 267
motion models, 79

motion sensor
   address space, 243
   address unpacking, 238, 245
   bi-phase clocks, 238
   BlockRAM, *see* BlockRAM
   boot process, 238
   buffer, *see* buffers
   bus arbitration, *see* bus arbitration
   camera calibration, *see* camera calibration
   components, 232
   constraint equation, 125, 261
   data paths, *see* data paths
   dynamic scale space, *see* dynamic scale space
   experimental results, 270
   FUGA 15D camera, 232
   implicit synchronisation, 238, 253
   improving performance, 278, 281
   laser range sensor, *see* laser range sensor
   memory management, 238, 240
   practical issues, 270
   primary storage, 240
   processing, 237
   prototype development platform, 236
   RAM allocation, 241
   RAM timing, 248
   RAM utilisation, 249
   `RAMIC`, *see* `RAMIC`
   top level design, 237
MPEG, 100
multiple instruction stream, multiple data stream (MIMD), 254
multiple linear regression, 30
   ill-posed problems, 30
   region of support, 30
multivariate location estimation, 39

N
navigation
   exclusion zone, 168
   state-time space, 169
   using motion estimates, 164
Netravali–Robin algorithm, 78
Neumann boundary conditions, 37
noise, 12, 23
norm, 39
   Euclidian distance, 39
   L_1 norm, 40
   L_2 norm, 39
   Mahalanobis distance, 40
   sum of absolute differences (SAD), 40
   sum of squared differences (SSD), 39

normal distribution, *see* Gaussian distribution
normal flow, 69
Nyquist frequency, 104

O

observation, 12
occlusion, 67
optical flow, 65, 69, 107
optical flow constraint equation (OFCE), 69, 125
    Cornelius and Kanade variant, 80
    departure from BCA, 79
    integral form, 79
    Negahdaripour variant, 80
    parametric motion model, 79
    relationship with DFD, 77
    relationship with frequency domain, 86
    Schunck variant, 79
    variations, 79
optimisation
    conjugate gradients method, 36
    descent algorithms, 33
    direct descent, 34
    downhill simplex method (DSM), 144
    graduated non-convexity (GNC), *see* graduated non-convexity (GNC)
    iterated conditional modes (ICM), 94
    mean field annealing, 58, 92
    optimisation problem, 23
    simulated annealing, 58, 93
    steepest descent, 35, 78
    successive overrelaxation (SOR), 36
    update equation, *see* update equation
order statistics, 16
    median, 16
outcome, 12
outlier, 43, 51, 134
    effect of, 44
    leverage point, 43, 45
    multiple outliers, 45
    structured outliers, 46
output variable, 23

P

parallel processing
    concepts, 253
    data parallel, 254
    distributed processing, 254
    MIMD versus SIMD, 255
    multiple instruction stream, multiple data stream (MIMD), 254
    pipeline, 255
    processing element (PE), 254
    single instruction stream, single data stream (SIMD), 254
    speed up, 255
parameter space, 28
parametric motion model, 65, 79
partition function $Z(\ )$, 92
pel recursive motion estimation, 78
    Netravali-Robbin algorithm, 78
    Walker-Rao algorithm, 78
perspective rigid body motion model, 107
pinhole camera model, 106
    focal length, 167
pipeline, 208, 217, 232, 253, 255–257, *see* state machine, *see* data paths
pixel pitch ζ, 105, 109, 111, 262
place and route, 177
polyray, 157
Prewitt mask, 76
probability
    AND rule, 15, 16
    a posteriori, posterior, 13, 15, 16
    a priori, prior, 13, 16
    compliment, 14
    computation of probability, 13
    conditional probability, 13, 16
    OR rule, 14
    total probability, 14
probability density function (PDF), 17–19, 22
    joint distribution function, 18
    marginal distribution function, 19
probability distribution function, *see* probability density function (PDF)
process, 19
    deterministic process, 19
    long-range, 87
    random process, 19
    short-range, 87
    stationary process, 20
    stochastic process, 19
    strictly stationary, 21
    wide sense stationary, 21
processing element (PE), 254
propogation delay, 211

R

radar, 119
RAMIC, 237, 244
    camera calibration, *see* camera calibration
    contention for RAM, 247

RAMIC_Operations, 244, 245
RAMIC_Poll, 244, 245, 247
SDRAM_Interface, 244, 245
random process, *see* process
random variable, 16, 91
range flow (U, V, W), 97, 119
RANSAC, 62
real time, 118
realisation, 20, 24
real-time, 103
   latency, 103
   temporal aliasing, 103, 104, 109
reflectance
   Lambertian reflection, 66
   simple model, 66
   specular reflection, 66
regression, *see* linear regression
regularisation, 30
   for motion estimation, 72
   ill-posed problems, 30
   membrane model, 31
   regularisation parameter λ, 31, 72 88, 118, 156
   smoothness constraint, *see* constraint equations
   solving, 33
regularisation parameter λ
relative efficiency, 48
residual, 24, 29, 39, 42, 43, 49
response variable, 23
rigid ground plane motion model, 108
robust average, 132
robust estimation, 43
   annealing M estimator, 142
   Anscombe's method, 143
   bounded influence estimator, 56
   breakdown point, *see* breakdown point
   convex hull peeling (CHP), 144
   GM estimator, 56
   influence function $\psi()$, 51, 54, 140
   inliers, *see* inliers
   LAR, *see* least absolute residuals estimator (LAR)
   least median of squares (LMedS), *see* least median of squares (LMedS)
   least trimmed squares (LTS), *see* least trimmed squares (LTS)
   least trimmed squares variant (LTSV), *see* least trimmed squares variant (LTSV)
   L estimator, 139
   leverage point, *see* leverage point
   Mallow's estimator, 56
   maximum trimmed likelihood (MTL), 146
   mean shift method (MSM), 145
   M estimator, *see* M estimator
   minimum volume ellipsoid (MVE), 143
   Minpran, 62
   motion estimation, *see* motion estimation
   outliers, *see* outlier
   properties of robust estimators, 47
   RANSAC, 62
   regression versus location estimation, 48
   relative efficiency, 48
   scale, 49
   scale equivariance, *see* scale
   Schweppe estimators, 56
rotational velocity $\mathbf{\Omega} = (\Omega_X, \Omega_Y, \Omega_Z)$, 107

S
S – safety margin, 109
safety margin, 111, 122, 124, 168
safety margin S, 109
sample, 12
sample mean, 21, 22
sample space, 12
   continuous, 12
   discrete, 12
   non-uniform, 12
   uniform, 12
scale
   for robust estimators, 49
   for scale space, 112
   in scale space, 233
scale equivariance, 48, 50
   LS estimator, 49
   of LMedS estimator, 61
   of LTS estimator, 61
   of M esimators, 51
scale space, 111
   course to fine, 112
   dynamic scale space, *see* dynamic scale space
   Gaussian scale space, 111
   image pyramid, 111
   parameters, 234
   quad tree, 112
   space variant mesh, 112
scaling factor SU_x, 126, 159, 262, 264, 265
schematic entry, 174, 178, 227
   functional simulation, 175
   nets, 227
Schweppe estimators, 56
semi-custom digital logic, *see* field programmable gate array (FPGA)

shadows
　　effect on motion estimates, 67, 80, 132, 265, 281
Sharc DSP, 236
short range process, *see* process
signal to noise ratio (SNR), 147
SignalMaster development platform, 236
Simoncelli, 75, 86, 113, 126
simulated annealing, 58, 93
simulation
　　functional simulation, 175, 198
　　logical simulation, 176, 186
　　of motion estimation algorithm, 157
　　timing simulation, 177
single instruction stream, single data stream (SIMD), 254
slice, 278
Smith, 82
smoothness constraint, *see* constraint equations
smoothness term E_s, 72
Sobel mask, 76
space variant cameras, 98
spatial aliasing, 104
spatiotemporal frequency (ω_x, ω_y, ω_t), 86, 85
spatiotemporal intensity function I(x, y, t), 69, 80, 85, 105
standard deviation, 21
state machines, 175, 177, 178, 212
　　actions, 214
　　clocks, 215
　　encoding, 212
　　entry, 214
　　fail safe, 217
　　loops, 193, 219
　　Moore and Mealy machines, 212
　　operation sequencing, 219
　　optimisation, 212
　　pipelines, 217
　　ports, 215
　　reset transition, 215
　　signals, 215
　　state variable, 217
　　states, 214
　　transition conditions, 214
　　transitions, 214
　　variables, 215
　　versus sequential VHDL, 212
state variable, 217
static timing analysis, 177
stationary
　　process, 20
　　strictly stationary, 21
　　wide sense stationary, 21
`std_ulogic` type, 188
steepest descent, 35, 78
stochastic process, *see* process
structured light, 98
successive overrelaxation (SOR), 36
sum of absolute differences (SAD), 40
sum of squared differences (SSD), 39
SU_x – scaling factor, 126, 159, 262, 264, 265
synchroniser, 204
synthesis, 175, 178

T

TC – temporal integration coefficient, 159
temporal aliasing, 65, 103, 109
　　in gradient and frequency domain motion estimation, 104
temporal integration, 95, 132, 160, 280
　　coefficient TC, 155, 159, 162, 270
　　implementation, 154, 269
terahertz imaging, 119
timing constraint, *see* user constraints
token based motion estimation, 66, 81
　　block based, 82
　　cross correlation, 84
　　feature based, 81
　　full search, 84
　　matching pixel count, 84
　　partial search, 84
　　Smith, 82
　　Torr, 82
　　Wiles, 82
Torr, 82
total least squares (TLS), 25
transparent motion, 67
trial, 12
trimmed mean, 138
tristate buffers, 174
　　as multiplexers, 206
　　bus control, 224, 229, 246
　　correct use, 206
　　inference, 189, 190, 199, 201
　　timing, 202
truncated quadratic, 51

U

U, V, W – range flow, 119
ultrasonic sensor, 119
update equation, 35, 37, 38
　　boundary conditions, 37
　　chequerboard update, 38
　　for Horn and Schunck optical flow, 72

for pel recursive motion estimation, 78
Gauss–Siedel iteration, 38
Jacobi iteration, 38
user constraints, 177
    clock-to-pad path, 202
    clock-to-setup path, 202, 214, 218
    location constraints, 177
    slow exception constraint, 217
    timing constraints, 177, 178, 201
    timing path, 202

V

variable
    dependent variable, 23
    explanatory, 23
    in state machines, 215
    in VHDL, 195
    independent, 23
    input variable, 23
    output variable, 23
    random variable, 16, 91
    response variable, 23
variance, 20–22
    estimator variance, 42
    Gaussian distribution, 19
VHDL, 173, 178
    aggregate assignment, 185
    architecture, 179, 190
    arithmetic operators, 187
    array base type, 183
    arrays, 183
    bit string literal, 185
    bit type, 180
    `boolean` type, 182
    case statement, 197
    character literals, 183
    combinatorial process, 194
    component instantiation, 191
    concatentation operator, 185
    concurrent signal assignment, 190, 217
    concurrent statements, 190
    conditional signal assignment, 190
    constrained array, 183
    D-flip-flop inference, 200
    D-latch inference, 200
    downto directive, 184
    drivers, 181, 188, 195
    element assignment, 185
    entity, 179, 190
    enumerated types, 183
    enumerated types encoding, 183
    enumeration literals, 183

    enumeration literals – overloaded, 183
    event attribute, 195, 200
    IEEE library, 187
    if statement, 196
    inference, 199
    integer type, 180, 186
    libraries, 187
    logical operators, 185
    loops, 193
    natural type, 182
    numeric type, 186
    `numeric_std` package, 187
    packages, 188
    port direction, 180
    ports, 179, 215
    positive type, 186
    process, 190, 192, 193
    range directive, 186
    relational operators, 185, 187
    representing high impedance, 188
    resolved types, 188
    selected signal assignment, 191
    sensitivity list, 194, 213
    sequential process, 194
    sequential signal assignment, 195
    sequential statements, 190, 193, 212
    signal, 180, 181, 195, 215
    signed type, 187
    slice, 185
    state machines, *see* state machines
    `std_logic` type, 180, 187, 188, 201
    `std_logic_1164` package, 187, 190
    `std_logic_arith` package, 188, 190
    `std_logic_vector` type, 180
    subprograms, 193
    synthesis, 178
    timing constraints, *see* user constraints
    to directive, 184
    tristate buffers, *see* tristate buffers
    type conversion, 189
    types, 180
    unconstrained array, 183
    unresolved types, 188
    unsigned type, 188
    variable, 195
    variables, 215
vision chip, 98
visual stereo, 96, 119
VLSI, 100
V_{max} – maximum velocity, 109, 110, 159

W

Walker–Rao algorithm, 78
weak membrane model, 56
    for motion estimation, 89
    line process, 56, 89, 92, 94
Wiles, 82
winsorised mean, 138

X

Xilinx Virtex FPGA, 236
x_r, y_r – image resolution, 109

Z

$Z(\ )$ – partition function, 92
Z_x, Z_y, Z_t – depth derivatives, 97

UNIVERSITY OF STRATHCLYDE

30125 00747947 9

Books are to be returned on or before
the last date below.

DUE
12 NOV 2010